SCHAUM'S OUTLINE OF

THEORY AND PROBLEMS

OF

ACOUSTICS

•

BY

WILLIAM W. SETO

Associate Professor of Mechanical Engineering
San Jose State College

•

SCHAUM'S OUTLINE SERIES

McGRAW-HILL BOOK COMPANY

New York, St. Louis, San Francisco, Düsseldorf, Johannesburg, Kuala Lumpur, London, Mexico, Montreal, New Delhi, Panama, Rio de Janeiro, Singapore, Sydney, and Toronto

07-056328-4

3 4 5 6 7 8 9 10 SH SH 7 5 4 3 2 1 0 6

Preface

This book is designed primarily to supplement standard texts in physical or applied acoustic at the senior undergraduate level, based on the belief that numerous solved problems constitute one of the best means for clarifying and fixing in mind basic principles. Moreover, the statements of theory and principle are sufficiently complete that, with proper handling of lecture-problem time, the book could be used as a text. It should be of considerable value to the physics and engineering students who are interested in the science of sound and its applications. The practicing engineers could also make frequent references to the book for its numerical solutions of many realistic problems in the area of sound and vibration.

Throughout the book emphasis is placed on fundamentals, with discussions and problems extending into many phases and applications of acoustics. The subject matter is divided into chapters covering duly-recognized areas of theory and study. Each chapter begins with pertinent definitions, principles and theorems which are fully explained and reinforced by solved problems. Then a graded set of problems are solved followed by supplementary problems. The solved problems amplify the theory, present methods of analysis, provide practical examples, illustrate the numerical details, and bring into sharp focus those fine points which enable the students to apply the basic principles correctly and with confidence. Numerous proofs of theorems and derivations of basic results are included among the solved problems. The supplementary problems with answers serve as a complete review of the material of each chapter.

The essential requirements to use this book are knowledge of the fundamental principles of mechanics, electricity, strength of materials, and undergraduate mathematics including calculus and partial differential equations.

Topics covered are vibrations and waves, plane and spherical acoustic waves, transmission of sound, loudspeaker and microphone, sound and hearing, architectural acoustics, underwater acoustics and ultrasonics. To make the book more flexible, considerably more material has been included here than can be covered in most semester courses.

I wish to thank Mr. Daniel Schaum for his utmost patience and kind assistance.

W. W. SETO

San Jose State College
December, 1970

CONTENTS

CONTENTS

Chapter 1

Vibrations and Waves

NOMENCLATURE

a = speed of wave propagation, m/sec; acceleration, m/sec^2
A = area, m^2
A_0 = amplitude of wave, m
A, B = constants
c = damping coefficient, nt-sec/m
C, D = constants
d = diameter, m
f = frequency, cyc/sec
f_b = beat frequency, cyc/sec
h = length, m
I_0 = Bessel hyperbolic function of the first kind of order zero
J_0 = Bessel function of the first kind of order zero
k = spring constant, nt/m
K_0 = Bessel hyperbolic function of the second kind of order zero
m = mass, kg
p_n = natural frequencies, cyc/sec
P = period, sec
P_b = beat period, sec
r = frequency ratio; radial distance, m
S = tension, nt
SHM = simple harmonic motion
t' = thickness, m
W = work done, joules/cyc
Y = Young's modulus of elasticity, nt/m^2
ω = circular frequency, rad/sec
ω_d = damped circular frequency, rad/sec
ω_n = natural circular frequency, rad/sec
λ = wavelength, m
ζ = damping factor
θ, ϕ = angles, rad
ρ = density, kg/m^3
ρ_a = mass/area, kg/m^2
ρ_L = mass/length, kg/m
μ = Poisson's ratio
σ = stress, nt/m^2
ϵ = strain

INTRODUCTION

Acoustics is the physics of sound. Although the fundamental theory of acoustics treats of vibrations and wave propagation, we can consider the subject as a multidisciplinary science.

Physicists, for example, are investigating the properties of matter by using concepts of wave propagation in material media. The acoustical engineer is interested in the fidelity of reproduction of sound, the conversion of mechanical and electrical energy into acoustical energy, and the design of acoustical transducers. The architect is more interested in the absorption and isolation of sound in buildings, and in controlled reverberation and echo prevention in auditoriums and music halls. The musician likes to know how to obtain rhythmic combinations of tones through vibrations of strings, air columns, and membranes.

On the other hand, physiologists and psychologists are actively studying the characteristics and actions of the human hearing mechanism and vocal cords, hearing phenomena and reactions of people to sounds and music, and the psychoacoustic criteria for comfort of noise level and pleasant listening conditions. Linguists are interested in the subjective perception of complex noises and in the production of synthetic speech.

Ultrasonics, a topic in acoustics dealing with sound waves of frequencies above 15,000 cycles per second, has found increasing application in oceanography, medicine and industry.

Moreover, because of the general awareness and resentment of the increasing high level of noise produced by airplanes, automobiles, heavy industry, and household appliances, and its adverse effects such as ear damage and physical and psychological irritation, greater demand is made for better understanding of sound, its causes, effects and control.

WAVES

Waves are caused by an influence or disturbance initiated at some point and transmitted or propagated to another point in a predictable manner governed by the physical properties of the elastic medium through which the disturbance is transmitted.

As a vibrating body moves forward from its static equilibrium position, it pushes the air before it and compresses it. At the same time, a rarefaction occurs immediately behind the body, and air rushes in to fill this empty space left behind. In this way the compression of air is transferred to distant parts and air is set into a motion known as *sound waves*. The result is *sound*. To the human ear, sound is the auditory sensation produced by the disturbance of air. Because fluids and solids possess inertia and elasticity, they all transmit sound waves.

Sound waves are *longitudinal waves*, i.e. the particles move in the direction of the wave motion. Propagation of sound waves involves the transfer of energy through space. The energy carried by sound waves is partly kinetic and partly potential; the former is due to the motion of the particles of the medium, the latter is due to the elastic displacement of the same particles. While sound waves spread out in all directions from the source, they may be reflected and refracted, scattered and diffracted, interfered and absorbed. A medium is required for the propagation of sound waves, the speed of which depends on the density and temperature of the medium. (See Problems 1.1-1.7.)

SIMPLE HARMONIC MOTION

For a particle in rectilinear motion, if its acceleration a is always proportional to its distance x from a fixed point on the path and is directed toward the fixed point, then the particle is said to have *simple harmonic motion* (SHM), which is the simplest form of *periodic motion*. In differential equation form, simple harmonic motion is represented by

$$a = -\omega^2 x \quad \text{or} \quad d^2x/dt^2 + \omega^2 x = 0$$

with solution
$$x(t) = A \sin \omega t + B \cos \omega t$$

or
$$x(t) = \sqrt{A^2+B^2} \sin(\omega t + \theta), \qquad x(t) = \sqrt{A^2+B^2} \cos(\omega t - \phi)$$

where A, B are arbitrary constants, ω is the circular frequency in rad/sec, and θ, ϕ are phase angles in radians.

Simple harmonic motion can be either a *sine* or *cosine function* of time, and can be conveniently represented by rotating vectors as shown in Fig. 1-1. The vector r of constant magnitude is rotating counterclockwise at constant angular velocity ω; its projections on the x and y axes are respectively cosine and sine functions of time. (See Problem 1.8.)

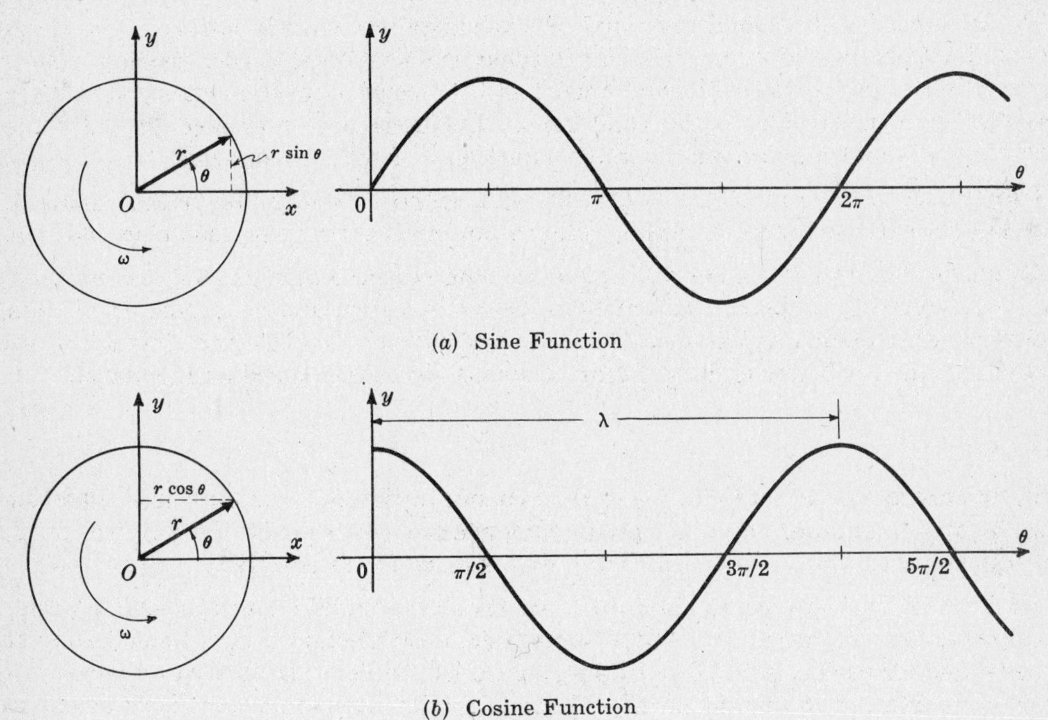

(*a*) Sine Function

(*b*) Cosine Function

Fig. 1-1

A *harmonic wave* is one whose profile or shape (displacement configuration) is sinusoidal, i.e. a sine or cosine curve. A harmonic wave moving in the positive x direction with velocity c is given by

$$u(x,t) = \begin{cases} A_0 \sin m(x-ct) \\ A_0 \cos m(x-ct) \end{cases}$$

whereas a harmonic wave moving in the negative x direction with velocity c is given by

$$u(x,t) = \begin{cases} A_0 \sin m(x+ct) \\ A_0 \cos m(x+ct) \end{cases}$$

where A_0 is the *amplitude* of the wave. These are known as *harmonic progressive waves.*

A spherical wave diverging from the origin of the coordinate with a velocity c is represented by

$$u(r,t) = (A_0/r)\, f(ct-r)$$

Similarly, a *spherical harmonic progressive wave* is designated by

$$u(r,t) = (A_0/r)e^{i(\omega t - kr)}$$

where $i = \sqrt{-1}$ and $k = 1/\lambda$ is the wave number, i.e. the number of cycles of the wave per unit length. The wave profile repeats itself after a distance $\lambda = 2\pi/m$ which is called the *wavelength.*

VIBRATIONS

Systems possessing mass and elasticity are capable of relative motion. If the motion of such systems repeats itself after a given interval of time, such periodic motion is known as *vibration.* To analyze vibration, the system is first idealized and simplified in terms of *mass m, spring k,* and *dashpot c,* which represents the body, the elasticity, and the friction of the system respectively. The *equation of motion* then expresses displacement of the system as a function of time. The *period P* is the time in seconds required for a periodic motion to repeat itself, and the frequency f is the number of cycles per unit time.

Free vibration, or *transient,* is the periodic motion observed as the system is displaced from its static equilibrium position. The forces acting are the spring force, the friction force, and the weight of the mass. Due to friction the vibration will decrease with time and is given by

$$x_c(t) = e^{-\zeta\omega_n t}(A \sin \omega_d t + B \cos \omega_d t)$$

where ζ = damping factor,

ω_n = natural circular frequency in rad/sec,

ω_d = natural damped circular frequency in rad/sec,

A, B = arbitrary constants. (See Problems 1.9-1.10.)

When external forces, usually of the type $F(t) = F_0 \sin \omega t$ or $F_0 \cos \omega t$, are acting on the system during its vibratory motion, the resultant motion is called *forced vibration.* At forced vibration, the system will tend to vibrate at its own natural frequency as well as to follow the frequency of the excitation force. In the presence of damping, that portion of motion not sustained by the sinusoidal excitation force will gradually die out. As a result, the system will vibrate at the frequency of the excitation force regardless of the initial conditions or the natural frequency of the system. The resultant motion is called *steady state vibration* or *response* of the system, and is represented by

$$x_p(t) = \frac{F_0}{\sqrt{(k - m\omega^2)^2 + (c\omega)^2}} \cos(\omega t - \theta)$$

where F_0 = magnitude of the excitation force,

k = spring constant,

m = mass of the system,

c = damping coefficient,

ω = frequency of the excitation force in rad/sec,

$\theta = \tan^{-1} \dfrac{c\omega}{k - m\omega^2}$ = phase angle. (See Problem 1.11.)

Resonance occurs when the frequency of the excitation force is equal to the natural frequency of the system. When this happens, the amplitude of vibration will increase without bound and is governed only by the amount of damping present in the system.

ENERGY OF VIBRATION

During free vibration with damping, energy is being continuously absorbed by the damper and dissipated as heat. The system is therefore continuously losing energy, and as a result the amplitude of vibration will diminish. For free vibration without damping, the total energy is constant and is either equal to the maximum kinetic or potential energy; the system continues to vibrate.

During forced vibration with damping, energy is being continuously supplied from external sources to maintain steady state vibration. (See Problems 1.12-1.15.)

VIBRATION OF STRINGS

The string is a unique vibrator with continuous media characteristics and is also the simplest example of a medium of wave transmission. It has its mass uniformly spread along its length and is the simplest case of a system with an infinite number of frequencies.

The general differential equation of motion is given by

$$\frac{\partial^2 y}{\partial t^2} = a^2 \frac{\partial^2 y}{\partial x^2}$$

where y = deflection of the string,
x = coordinate along the longitudinal axis of the string,
$a = \sqrt{S/\rho_L}$ = speed of wave propagation,
S = tension,
ρ_L = mass per unit length of the string.

The general solution can be expressed as either standing waves or progressive waves as given in the following two equations:

$$y(x,t) = \sum_{i=1,2,\ldots}^{\infty} \left(A_i \sin\frac{p_i}{a}x + B_i \cos\frac{p_i}{a}x\right)(C_i \sin p_i t + D_i \cos p_i t)$$

where A_i, B_i are arbitrary constants to be evaluated by boundary conditions, C_i, D_i are arbitrary constants to be evaluated by initial conditions, and p_i are the natural frequencies of the system;

$$y(x,t) = f_1(x-at) + f_2(x+at)$$

where f_1 and f_2 are arbitrary functions. The first part $f_1(x-at)$ represents a wave of arbitrary shape traveling in the positive x direction with velocity a, whereas $f_2(x+at)$ represents a similar wave traveling in the negative x direction with velocity a. (See Problems 1.16-1.20.)

LONGITUDINAL VIBRATION OF BARS

A bar is a material body greatly elongated in one direction, made of homogeneous, isotropic material, and free of transverse constraints throughout. If a sudden blow is made in the direction of its axis, the elongation characteristics of any right section of the bar will vary periodically with time but with different amplitudes. This is *longitudinal vibration* of bars.

The general differential equation of motion is given by

$$\frac{\partial^2 u}{\partial t^2} = a^2 \frac{\partial^2 u}{\partial x^2}$$

where u = displacement of any cross section,

x = coordinate along the longitudinal axis,

$a = \sqrt{Y/\rho}$ = speed of wave propagation,

Y = Young's modulus of elasticity,

ρ = density.

The general solution is the same as that for the vibration of strings. (See Problems 1.21-1.25.)

VIBRATION OF MEMBRANES

A membrane is a material body of finite extent and uniform thickness, held under homogeneous tension in a rigid frame. It is completely flexible and its thickness is very small compared to its two other dimensions. When excited, free vibration without damping is assumed to take place perpendicular to the plane surface of the membrane.

A vibrating membrane is the most easily visualized physical example of wave motion in effectively *two-dimensional* space. Compared to its *one-dimensional* counterpart, the flexible string, the membrane has much more freedom of motion.

The general differential equation of motion is given by

$$\frac{\partial^2 y}{\partial x^2} + \frac{\partial^2 y}{\partial z^2} = \frac{1}{a^2}\frac{\partial^2 y}{\partial t^2}$$

where y = vertical deflection of the membrane,

$a = \sqrt{S/\rho_a}$ = speed of wave propagation,

S = tension,

ρ_a = mass per unit area of the membrane,

x, z = coordinates in the plane of the membrane.

The general solution can be expressed either as series solution or traveling-waves solution as follows:

$$y(x,z,t) = \sum_{i=1,2,\ldots}^{\infty} \left(A_i \sin\sqrt{(p_i/a)^2 - k_i^2}\,x + B_i \cos\sqrt{(p_i/a)^2 - k_i^2}\,x\right) \times (C_i \sin k_i z + D_i \cos k_i z)(E_i \sin p_i t + F_i \cos p_i t)$$

where A_i, B_i, C_i and D_i are arbitrary constants to be evaluated by boundary conditions, E_i and F_i are arbitrary constants to be evaluated by initial conditions, and p_i are the natural frequencies of the membrane;

$$y(x,z,t) = f_1(mx + nz - at) + f_2(mx + nz + at) \qquad \text{where } m^2 + n^2 = 1$$

This form of solution represents waves of the same arbitrary profile traveling in opposite directions along x and z axes with velocity a. (See Problems 1.26-1.31.)

VIBRATION OF CIRCULAR PLATES

The vibration of plates is the two-dimensional analog of the transverse vibration of beams. In contrast to a membrane, the thickness of a plate is not small compared to other dimensions. Moreover, stresses and strains resulting from the stiffness and bending of the plate will complicate greatly the almost limitless freedom of motion of the plate.

The general differential equation of motion is given by

$$\left[\frac{\partial^2 y}{\partial r^2} + \frac{1}{r}\frac{\partial y}{\partial r}\right]^2 + \frac{12\rho(1-\mu^2)}{Yt'^2}\frac{\partial^2 y}{\partial t^2} = 0$$

where y = deflection of plate,
r = radial distance from center of plate,
ρ = density of plate,
Y = Young's modulus of elasticity,
t' = thickness,
μ = Poisson's ratio.

The general solution for free vibration of a circular plate is

$$y(r,t) = [AJ_0(kr) + BI_0(kr)]e^{i\omega t}$$

where A and B are arbitrary constants, J_0 is the Bessel function of the first kind of order zero, and I_0 is the Bessel hyperbolic function. (See Problems 1.32-1.33.)

Solved Problems

WAVES

1.1. Prove each wave addition:

(a) $A\cos\omega t + B\sin\omega t = C\sin(\omega t+\theta)$

(b) $A\cos\omega t + B\sin\omega t = C\cos(\omega t-\phi)$

where $C = \sqrt{A^2+B^2}$, $\tan\theta = A/B$, and $\tan\phi = B/A$.

(a) $$C\sin(\omega t+\theta) = C(\sin\omega t\cos\theta + \cos\omega t\sin\theta) = (C\cos\theta)\sin\omega t + (C\sin\theta)\cos\omega t$$

Let $(C\cos\theta) = B$, $(C\sin\theta) = A$. Then $A^2+B^2 = C^2$ or $C = \sqrt{A^2+B^2}$, and $\tan\theta = A/B$. Thus

$$A\cos\omega t + B\sin\omega t = C\sin(\omega t+\theta) \quad \text{if } C=\sqrt{A^2+B^2} \text{ and } \tan\theta = A/B$$

(b) $$C\cos(\omega t-\phi) = C(\cos\omega t\cos\phi + \sin\omega t\sin\phi) = (C\cos\phi)\cos\omega t + (C\sin\phi)\sin\omega t$$

Let $(C\cos\phi) = A$, $(C\sin\phi) = B$. Then $A^2+B^2 = C^2$ or $C = \sqrt{A^2+B^2}$, and $\tan\phi = B/A$. Thus

$$A\cos\omega t + B\sin\omega t = C\cos(\omega t-\phi) \quad \text{if } C=\sqrt{A^2+B^2} \text{ and } \tan\phi = B/A$$

The above wave additions can also be found by considering the rotating vectors shown in Fig. 1-2.

Vectors A, B and C are rotating about point O with constant angular velocity ω. $AA_1 = OA\cos\omega t$, $BB_1 = OB\sin\omega t$, and $CC_1 = OC\sin(\omega t+\phi)$ are the projections on the y axis of vectors A, B and C respectively.

(a) Since vector C is the resultant of vectors A and B, we have

$$CC_1 = CC_2 + C_2C_1 = AA_1 + BB_1$$

or

$$OC\sin(\omega t+\phi) = OA\cos\omega t + OB\sin\omega t$$

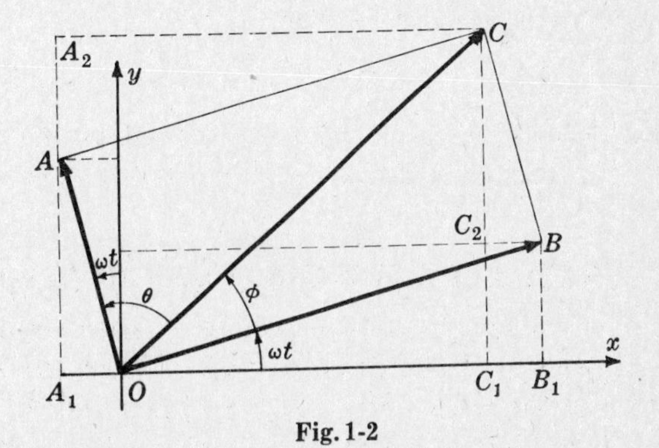

Fig. 1-2

Calling $OA = A$, $OB = B$, $OC = C$, then $C = \sqrt{A^2 + B^2}$, $\tan\phi = A/B$ and the required result follows.

(b) Similarly,

$$A_1A + AA_2 = A_1A + BB_1$$

or $C\cos(\omega t - \theta) = A\cos\omega t + B\sin\omega t$ where $C = \sqrt{A^2+B^2}$ and $\tan\theta = B/A$.

1.2. Two harmonic wave motions $x_1 = \sin(\omega t + 60°)$ and $x_2 = 2\sin\omega t$ are propagated in the same direction. Find the resultant wave motion.

The resultant wave motion is given by

$$\begin{aligned} x &= x_1 + x_2 = \sin(\omega t + 60°) + 2\sin\omega t \\ &= \sin\omega t\cos 60° + \cos\omega t\sin 60° + 2\sin\omega t \\ &= 2.5\sin\omega t + 0.866\cos\omega t = \sqrt{2.5^2 + 0.866^2}\sin(\omega t + \theta) \\ &= 2.66\sin(\omega t + 19°) \end{aligned}$$

since $\theta = \tan^{-1}(0.866/2.5) = 19°$.

The resultant wave motion can also be found by considering the rotating vectors shown in Fig. 1-3. All the three vectors A, B, C are rotating with constant angular velocity ω. The projections of vectors A and B on the x axis represent the two wave motions x_1 and x_2 respectively. The resultant wave motion is represented by the projection of the vector C on the x axis.

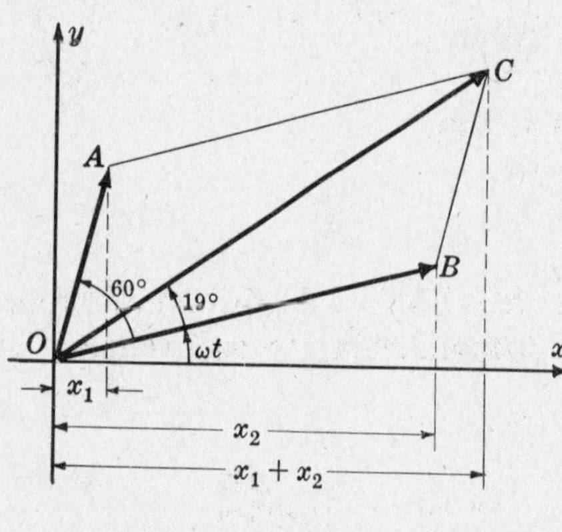

Fig. 1-3

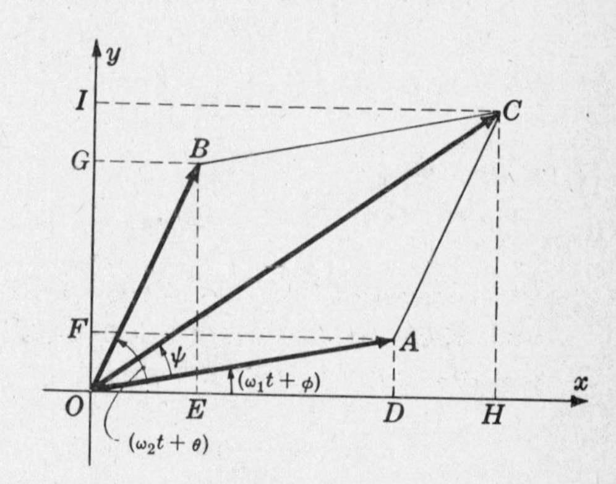

Fig. 1-4

1.3. Given two sine or cosine waves of different frequencies and amplitudes, determine their sum.

The addition of two or more sine or cosine waves is most conveniently done by rotating vectors as shown in Fig. 1-4. A and B are vectors of different lengths rotating about O with constant angular velocities ω_1 and ω_2 and initial phase angles ϕ and θ. The projections of vectors A and B on the x axis are respectively

$$OD = A\cos(\omega_1 t + \phi), \qquad OE = B\cos(\omega_2 t + \theta) \tag{1}$$

where A and B are the magnitudes of the vectors. The corresponding projections on the y axis are

$$OF = A\sin(\omega_1 t + \phi), \qquad OG = B\sin(\omega_2 t + \theta) \tag{2}$$

Similarly, the projections of vector C on the x and y axes are

$$OH = OD + DH = OD + OE, \qquad OI = OF + FI = OF + OG \tag{3}$$

From equations (1) and (2),

$$C\cos(\omega_1 t + \phi + \psi) = A\cos(\omega_1 t + \phi) + B\cos(\omega_2 t + \theta) \tag{4}$$

$$C\sin(\omega_1 t + \phi + \psi) = A\sin(\omega_1 t + \phi) + B\sin(\omega_2 t + \theta) \tag{5}$$

where the magnitude of vector C is $C = \sqrt{A^2 + B^2 + 2AB\cos[(\omega_1 - \omega_2)t + (\phi - \theta)]}$ which varies sinusoidally with time at a frequency equal to the difference between the given frequencies. The phase angle of the vector C is $\psi = \tan^{-1}\dfrac{CH}{CI} = \tan^{-1}\dfrac{A\sin(\omega_1 t + \phi) + B\sin(\omega_2 t + \theta)}{A\cos(\omega_1 t + \phi) + B\cos(\omega_2 t + \theta)}$.

Thus equation (4) represents the addition of two cosine waves whereas equation (5) represents the addition of two sine waves.

Fig. 1-5 shows the addition of two sine waves of different frequencies and amplitudes. The resultant wave is periodic but not harmonic.

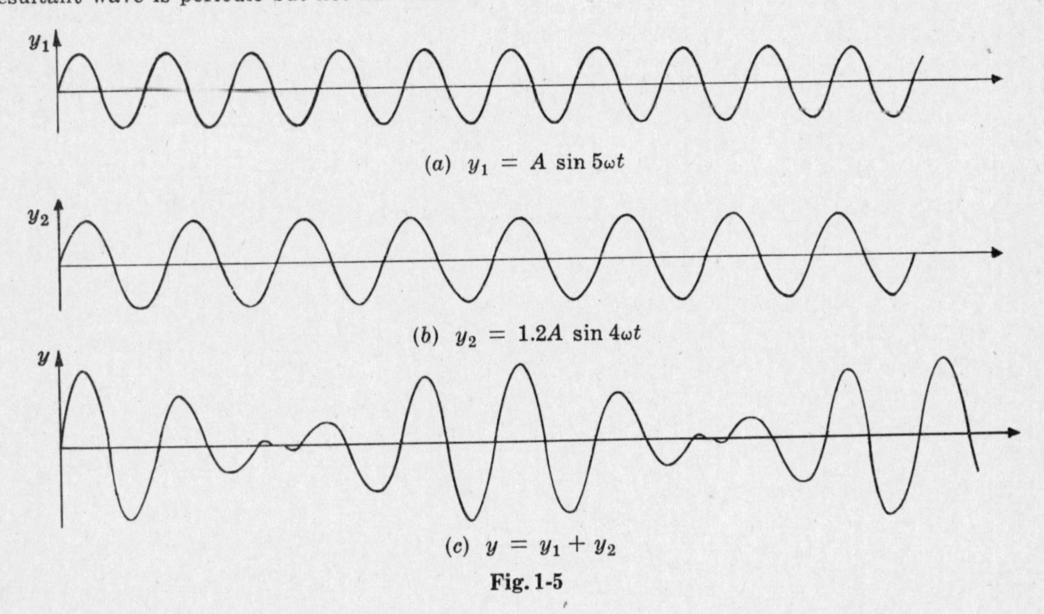

Fig. 1-5

1.4. Two wave motions $A = \cos(\omega t + 30°)$ and $B = 1.5\sin(\omega t + 30°)$ are propagated simultaneously from source O in directions perpendicular to each other. Determine the resultant wave motion.

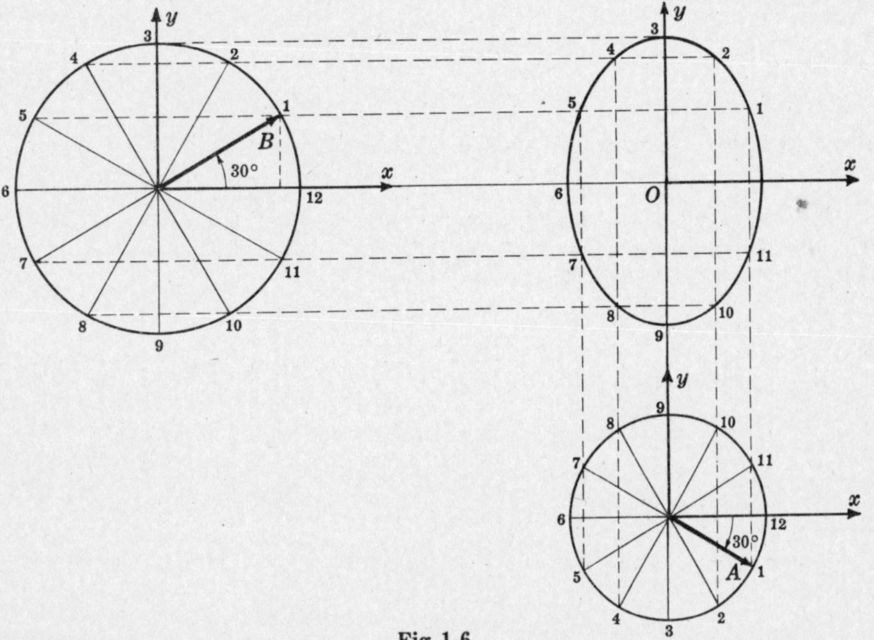

Fig. 1-6

The shape of the resultant wave motion can be found graphically by means of rotating vectors in the xy plane as shown in Fig. 1-6 above. The lengths of the vectors represent the amplitudes while their projections on the x and y axes represent the original shapes of the waves. The circumferences of both circles are marked for equal time intervals of the circular motion of the vectors. Then all these points are projected across the xy plane to form the locus of points, which is an ellipse.

1.5. Given two wave motions $A \cos 2\omega t$ and $A \sin 3\omega t$ in directions at right angle to each other, find the resultant motion.

Let $x = A \cos 2\omega t$, $y = A \sin 3\omega t$ as shown in Fig. 1-7. The resultant motion on the xy plane can be found graphically by means of rotating vectors. The lengths of the vectors represent the amplitudes of the wave motions while their projections on the x and y axes represent the original shapes of the waves.

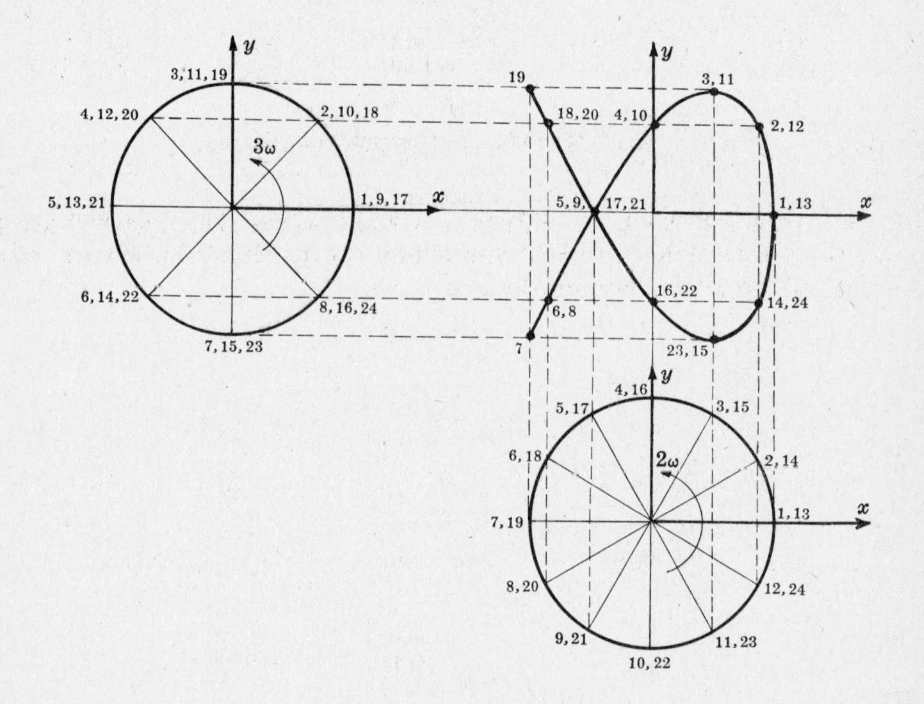

Fig. 1-7

The circumferences of both circles are marked for equal time intervals in the ratio of 3 : 2 which is the ratio of the circular speeds of the vectors. All these points, 1 to 24 on both circumferences, are projected across the xy plane to form the locus of points which is known as the Lissajou figure. Lissajou figures are useful when setting up a series of motions whose frequencies are harmonics of the fundamental.

1.6. Two harmonic motions of the same amplitude but of slightly different frequencies are imposed on a vibrating body. Analyze the motion of the body.

Let $x_1(t) = A_0 \cos \omega t$, $x_2(t) = A_0 \cos (\omega + \Delta\omega)t$ be the two harmonic motions. The motion of the body, then, is the superposition of the two given motions:

$$x(t) = x_1(t) + x_2(t) = A_0 \cos \omega t + A_0 \cos (\omega + \Delta\omega)t = A_0[\cos \omega t + \cos (\omega + \Delta\omega)t]$$

From trigonometry, $\cos x + \cos y = 2 \cos \frac{1}{2}(x + y) \cos \frac{1}{2}(x - y)$. Thus

$$x(t) = A_0[2 \cos \tfrac{1}{2}(\omega t + \omega t + \Delta\omega t) \cos (\Delta\omega/2)t] = [2A_0 \cos (\Delta\omega/2)t] \cos (\omega + \Delta\omega/2)t$$

The amplitude of $x(t)$ is seen to fluctuate between zero and $2A_0$ according to the $2A_0 \cos(\Delta\omega/2)t$ term, while the general motion of x is a cosine function of angular frequency $(\omega + \Delta\omega/2)$. This special pattern of motion is known as the *beating phenomenon.* Whenever the amplitude reaches a maximum, there is said to be a *beat.* The *beat frequency* as determined by two consecutive maximum amplitudes is equal to

$$f_b = \frac{\Delta\omega + \omega}{2\pi} - \frac{\omega}{2\pi} = \frac{\Delta\omega}{2\pi} \text{ cyc/sec}$$

and the *beat period* $P_b = 1/f_b = 2\pi/\Delta\omega$ sec. Sound waves of slightly different frequencies will also give rise to beats as described here.

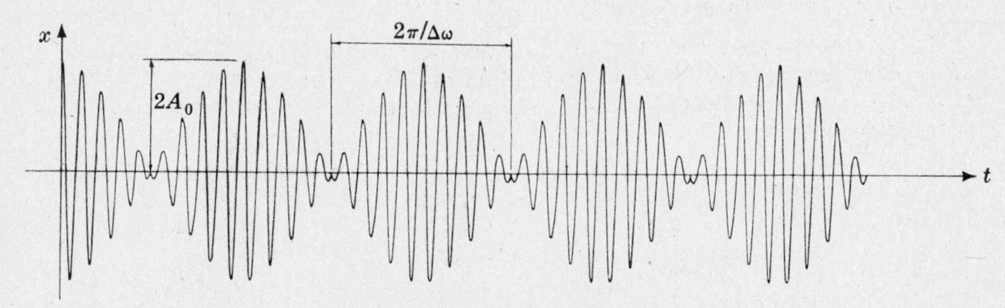

Fig. 1-8. The beating phenomenon

1.7. In each of Fig. 1-9(*a*)-(*j*), two identical triangle waves shown dashed are propagated in the same direction. In each case, study the resultant wave with respect to the indicated phase angle between the two waves.

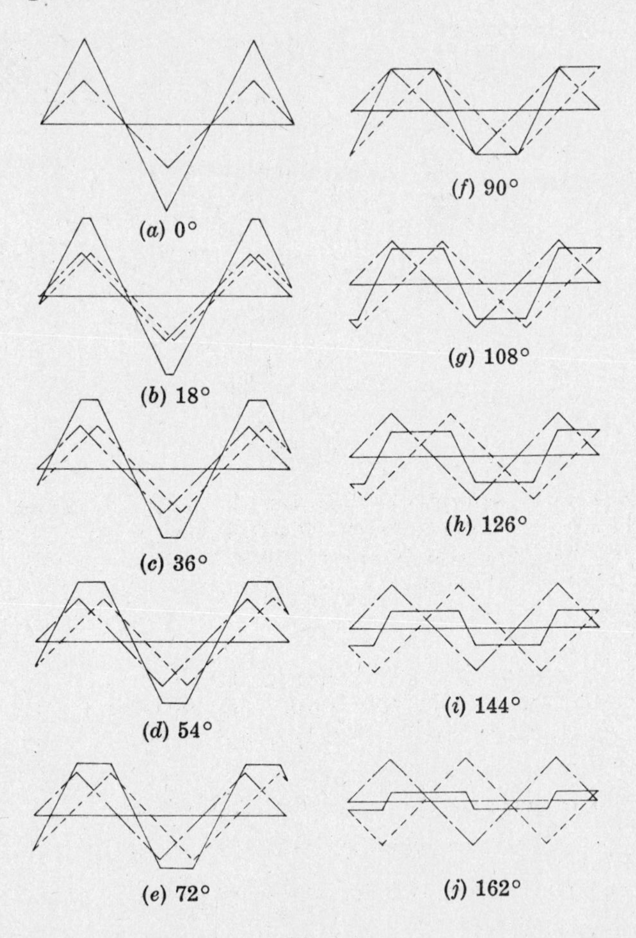

Fig. 1-9

The resultant wave (solid line) is obtained by adding the two waves graphically. We begin in Fig. 1-9(*a*) with zero phase angle between the two waves, i.e. the two waves are completely in phase with each other. The resultant amplitude is equal to twice the amplitude of the given waves.

Fig. 1-9(*b*) shows the addition of two identical waves with 18° phase difference between them. Similarly, Fig. 1-9(*c*) to Fig. 1-9(*j*) are the resultants of the additions of two identical waves with progressively greater values of phase angle between the two identical waves.

When the two identical waves are completely out of phase, i.e. the phase angle between the two waves is 180°, the resultant wave is zero. In other words, the two waves cancel each other.

SIMPLE HARMONIC MOTIONS

1.8. A simple harmonic motion is given as $x(t) = 10\sin(10t - 30°)$ where x is measured in meters, t in seconds, and the phase angle in degrees. Find (*a*) the frequency and period of the motion, (*b*) the maximum displacement, velocity and acceleration, (*c*) the displacement, velocity and acceleration at $t = 0$ and $t = 1$ seconds.

(*a*)
$$x(t) = 10\sin(10t - 30°) = A_0 \sin(\omega t - \theta)$$

Then $\omega = 10$ rad/sec, $f = \omega/2\pi = 1.6$ cyc/sec, and $p = 1/f = 0.63$ sec.

(*b*) Displacement is $x(t) = 10\sin(10t - 30°)$. Thus the maximum displacement is 10 m.

Velocity is $dx/dt = \omega A_0 \cos(\omega t - \theta)$. Thus the maximum velocity is $10(10) = 100$ m/sec.

Acceleration is $d^2x/dt^2 = -\omega^2 A_0 \sin(\omega t - \theta)$, and so the maximum acceleration is $10^2(10) = -1000$ m/sec^2.

(*c*) At $t = 0$:

$$x(0) = 10\sin(-30°) = 10(-0.5) = -5 \text{ m}$$
$$\dot{x}(0) = \omega A_0 \cos(-30°) = 10(10)(0.866) = 86.6 \text{ m/sec}$$
$$\ddot{x}(0) = -\omega^2 A_0 \sin(-30°) = -(10)^2(10)(-0.5) = 500 \text{ m/sec}^2$$

At $t = 1$:

$$x(1) = 10\sin(10 - 30°) \doteq 10\sin(570° - 30°) \doteq 10\sin 180° = 0$$
$$\dot{x}(1) = 10(10)\cos 180° = -100 \text{ m/sec}$$
$$\ddot{x}(1) = -(10)^2(10)\sin 180° \doteq 0$$

FREE VIBRATION

1.9. Determine the differential equation of motion and natural frequency of vibration of the simple single-degree-of-freedom spring-mass system shown in Fig. 1-10.

Apply Newton's law of motion, $\Sigma F = ma$. For vertical motion, the forces acting on the mass are the spring force $k(\delta_{st} + x)$ and the weight mg of the mass. Therefore the differential equation of motion is

$$m\ddot{x} = -k(\delta_{st} + x) + mg$$

where δ_{st} is the static deflection due to the weight of the mass acting on the spring. Then $mg = \delta_{st}k$, and the equation of motion becomes

$$m\ddot{x} + kx = 0$$

which is the differential equation for SHM. The general forms of solution for this equation are

$$x(t) = A\sin\sqrt{k/m}\,t + B\cos\sqrt{k/m}\,t$$
$$x(t) = C\sin(\sqrt{k/m}\,t + \phi)$$
$$x(t) = D\cos(\sqrt{k/m}\,t - \theta)$$

where A, B, C, D, ϕ and θ are arbitrary constants depending on initial conditions $x(0)$ and $\dot{x}(0)$. Two constants must appear in each of the general solutions because this is a second order differential equation.

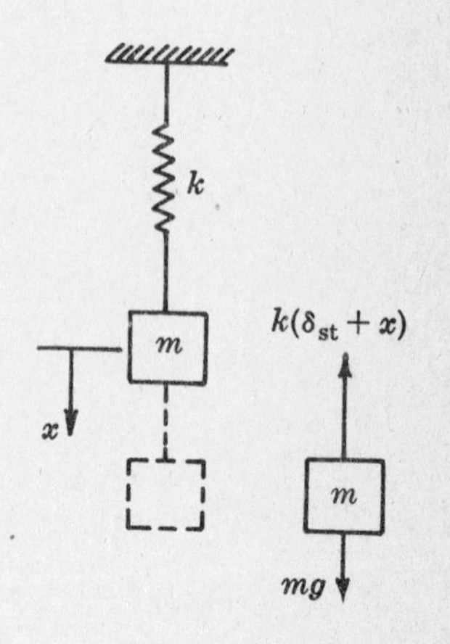

Fig. 1-10

For an initial displacement $x(0) = x_0$ and zero initial velocity $\dot{x}(0) = 0$, we have $A = 0$, $B = x_0$ and hence

$$x(t) = x_0 \cos \sqrt{k/m}\, t$$

Physically, this solution represents an undamped free vibration, one cycle of which occurs when $\sqrt{k/m}\, t$ varies through 360 degrees. Therefore the period P and the natural frequency f_n are

$$P = \frac{2\pi}{\sqrt{k/m}} \text{ sec} \quad \text{and} \quad f_n = 1/P = \frac{\sqrt{k/m}}{2\pi} \text{ cyc/sec}$$

where $\omega_n = \sqrt{k/m}$ rad/sec is the circular natural frequency of the system.

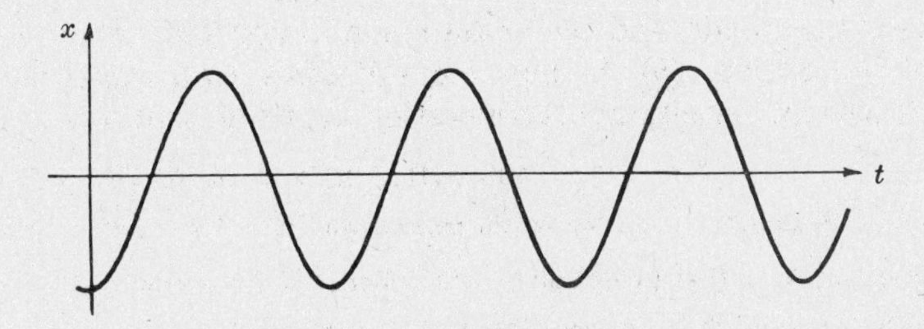

Fig. 1-11. Free vibration without damping

1.10. A generalized single-degree-of-freedom spring-mass system with damping is shown in Fig. 1-12. Investigate its general motion.

Employing Newton's law of motion $\Sigma F = ma$,

$$m\ddot{x} = -c\dot{x} - kx \quad \text{or} \quad m\ddot{x} + c\dot{x} + kx = 0$$

where k is the spring constant, m the mass, and c the damping coefficient.

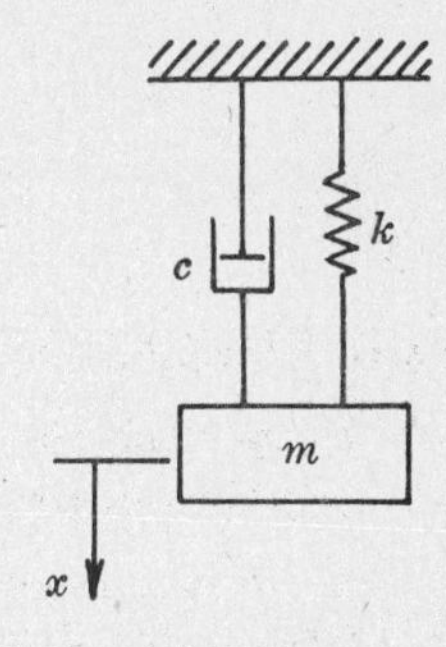

Fig. 1-12

We cannot assume solutions of the sine or cosine functions because of the term $c\dot{x}$. We assume $x = e^{rt}$; then $\dot{x} = re^{rt}$, $\ddot{x} = r^2e^{rt}$. Substituting these values into the differential equation of motion, we obtain

$$mr^2e^{rt} + cre^{rt} + ke^{rt} = 0 \quad \text{or} \quad mr^2 + cr + k = 0$$

The two values of r satisfying the above equation are

$$r_1, r_2 = \frac{-c \pm \sqrt{c^2 - 4mk}}{2m} = (-\zeta \pm \sqrt{\zeta^2 - 1})\omega_n$$

where $\omega_n = \sqrt{k/m}$, and $\zeta = c/2m\omega_n$ is called the damping factor. Thus the solution to the equation of motion is

$$x(t) = Ae^{r_1 t} + Be^{r_2 t}$$

where A and B are arbitrary constants determined by the two initial conditions imposed on the system.

Since the values of r depend on the magnitude of ζ, we have the following three cases of free vibration with damping:

Case 1: If ζ is greater than unity, the values of r are real and distinct; the amplitude of x is decreasing but will never change sign. Therefore oscillatory motion is not possible for the system regardless of initial conditions. This is *overdamped,* where

$$x(t) = Ae^{-r_1 t} + Be^{-r_2 t}$$

Case 2: If ζ is equal to unity, the values of r are real and negative, and are equal to $-\omega_n$. The motion of the system is again not oscillatory, and its amplitude will eventually reduce to zero. This is *critically-damped,* where

$$x(t) = (C + Dt)e^{-\omega_n t}$$

Case 3: If ζ is less than unity, the values of r are complex conjugates:

$$r_1 = \omega_n(-\zeta + i\sqrt{1-\zeta^2}), \qquad r_2 = \omega_n(-\zeta - i\sqrt{1-\zeta^2})$$

And if we define $\omega_d = \sqrt{1-\zeta^2}\,\omega_n$ as the damped natural frequency in rad/sec, we have

$$r_1 = -\zeta\omega_n + i\omega_d, \qquad r_2 = -\zeta\omega_n - i\omega_d$$

and
$$x(t) = e^{-\zeta\omega_n t}(Ee^{i\omega_d t} + Fe^{-i\omega_d t})$$

Expanding,
$$x(t) = e^{-\zeta\omega_n t}[(E+F)\cos\omega_d t + i(E-F)\sin\omega_d t]$$

Letting $E+F=G$ and $i(E-F)=H$, we finally obtain

$$x(t) = e^{-\zeta\omega_n t}(G\cos\omega_d t + H\sin\omega_d t)$$

As shown before, we may combine a cosine and sine function of the same frequency into a single sine or cosine function as

$$x(t) = Ie^{-\zeta\omega_n t}\sin(\omega_d t + \theta)$$

$$x(t) = Ie^{-\zeta\omega_n t}\cos(\omega_d t - \phi)$$

where $I = \sqrt{G^2+H^2}$, $\theta = \tan^{-1}(G/H)$, $\phi = \tan^{-1}(H/G)$.

The motion is oscillatory with angular frequency ω_d. The amplitude of motion will decrease exponentially with time because of the term $e^{-\zeta\omega_n t}$, which is known as the *decaying factor*. This is underdamped vibration. Refer to Fig. 1-13.

Hence it may be concluded that the motion of a dynamic system with damping and having free vibration depends on the amount of damping present in the system. The resulting motion will be periodic only if the amount of damping present is less than critical, and the system oscillates with angular frequency slightly less than the free natural frequency of the system.

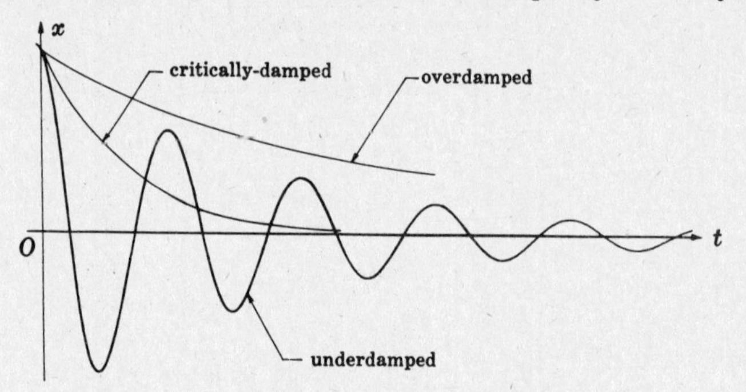

Fig. 1-13. Free vibration with damping

FORCED VIBRATION

1.11. Investigate the general motion of a simple spring-mass system with damping excited by a sinusoidal force $F_0 \cos \omega t$ as shown in Fig. 1-14.

Employing Newton's law of motion,

$$m\ddot{x} = \text{sum of forces in the } x \text{ direction}$$
$$= -k(x+\delta_{st}) + mg - c\dot{x} + F_0\cos\omega t$$

But $k\delta_{st} = mg$, the weight of the mass; hence the equation of motion takes its most general form

$$m\ddot{x} + c\dot{x} + kx = F_0\cos\omega t$$

The general solution for this second order differential equation with constant coefficients is

$$x = x_c + x_p$$

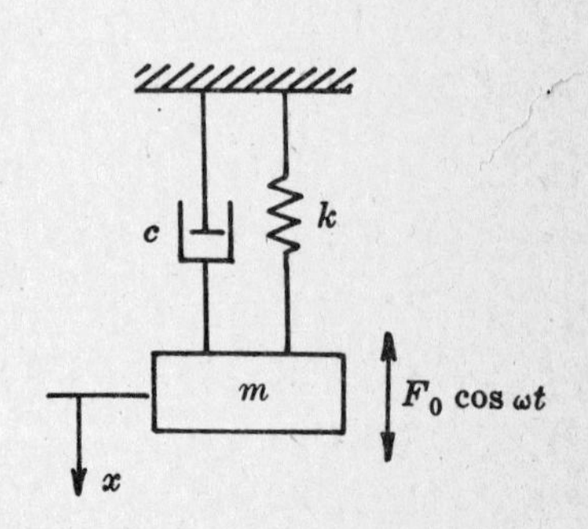

Fig. 1-14

where x_c is called the complementary solution, or the solution of the homogeneous equation, $m\ddot{x} + c\dot{x} + kx = 0$. x_p is the particular solution for the given equation.

The complementary solution, known as free vibration, has been solved previously in Problem 1-10. The particular solution, obtained from the nonhomogeneous part $F_0 \cos \omega t$ of the differential equation of motion, is

$$x_p(t) = A \sin \omega t + B \cos \omega t$$

and so

$$\dot{x}_p(t) = \omega A \cos \omega t - \omega B \sin \omega t$$

$$\ddot{x}_p(t) = -\omega^2 A \sin \omega t - \omega^2 B \cos \omega t$$

Substituting these expressions into the equation of motion, we obtain

$$(kA - mA\omega^2 - c\omega B) \sin \omega t + (kB - mB\omega^2 + c\omega A) \cos \omega t = F_0 \cos \omega t$$

Equating the coefficients,

$$(k - m\omega^2)A - c\omega B = 0, \qquad c\omega A + (k - m\omega^2)B = F_0$$

from which

$$A = \frac{F_0 \omega c}{(k - m\omega^2)^2 + (c\omega)^2}, \qquad B = \frac{F_0(k - m\omega^2)}{(k - m\omega^2)^2 + (c\omega)^2}$$

Then

$$x_p(t) = \frac{F_0 \omega c}{(k - m\omega^2)^2 + (c\omega)^2} \sin \omega t + \frac{F_0(k - m\omega^2)}{(k - m\omega^2)^2 + (c\omega)^2} \cos \omega t$$

We may combine these two sinusoidal functions of the same frequency either by rotating vectors or by trigonometric identities to obtain

$$x_p(t) = \frac{F_0}{\sqrt{(k - m\omega^2)^2 + (c\omega)^2}} \cos(\omega t - \phi)$$

or

$$x_p(t) = \frac{F_0/k}{\sqrt{(1 - r^2)^2 + (2\zeta r)^2}} \cos(\omega t - \phi)$$

where $r = \omega/\omega_n$, $\omega_n = \sqrt{k/m}$, and $\phi = \tan^{-1}\dfrac{c\omega}{k - m\omega^2} = \tan^{-1}\dfrac{2\zeta r}{1 - r^2}$.

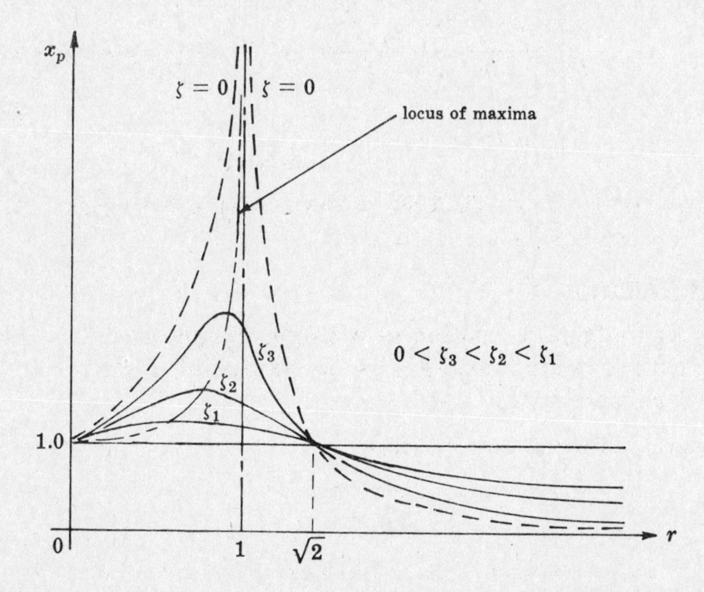

Fig. 1-15

Hence it may be concluded that the particular solution $x_p(t)$, which is known as the *steady state response* or *forced vibration*, is of the same frequency as that of the excitation force regardless of initial conditions. The amplitude of forced vibration depends on the amplitude and frequency of the excitation force, and the parameters of the systems. At *resonance*, i.e. when the forcing frequency is equal to the natural frequency, or $\omega/\omega_n = 1$, the amplitude of forced vibration is limited only by the damping factor ζ and hence the amount of damping present. Therefore resonance should be avoided at all times. Finally, the steady state response of the system is not in phase with the excitation force; its variation by the phase angle ϕ is due to the presence of damping in the system. Without damping, the steady state response is either in phase or 180° out of phase with the excitation force. See Fig. 1-15 to Fig. 1-19.

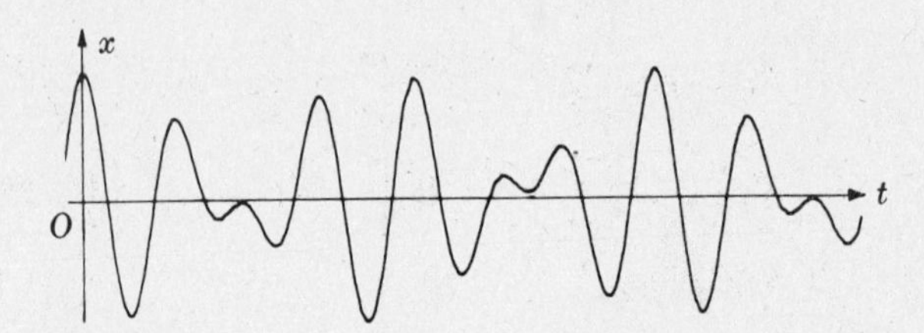

Fig. 1-16. Forced vibration without damping ($2f_n = 3f$)

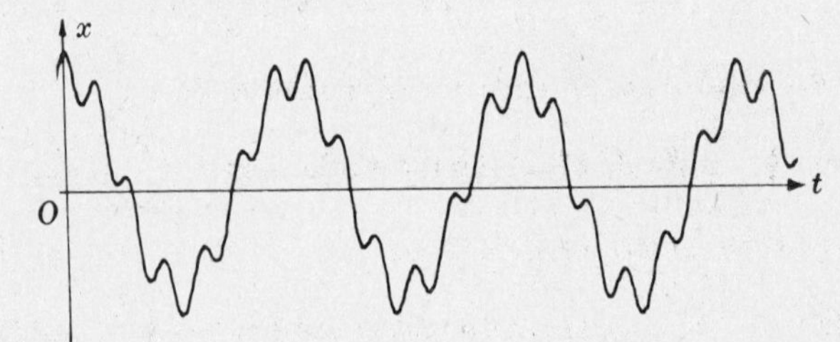

Fig. 1-17. Forced vibration without damping ($f_n = 6.28f$)

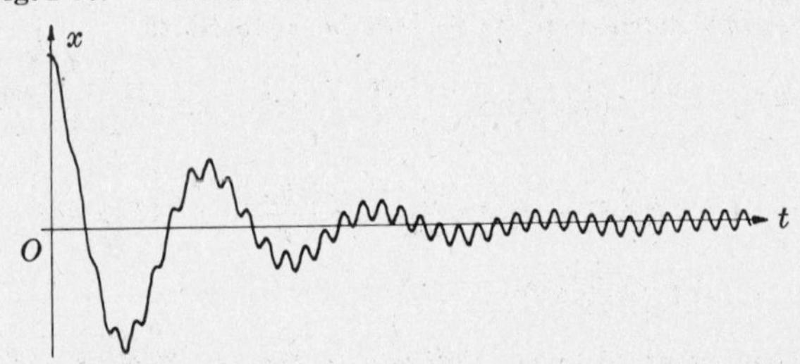

Fig. 1-18. Forced vibration with damping ($f_n = 6.28f$)

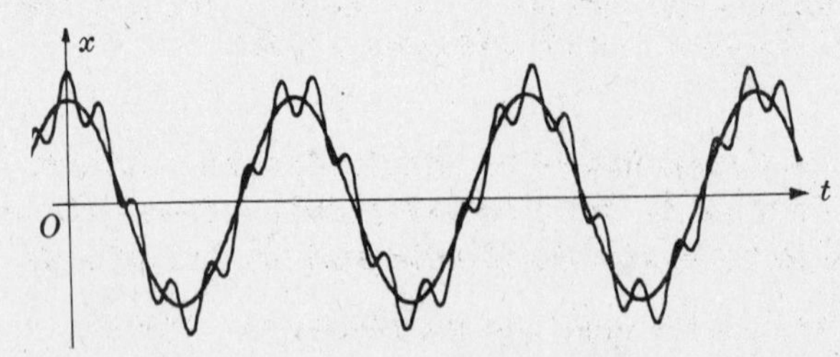

Fig. 1-19. Free and forced vibration ($f_n = 6.28f$)

ENERGY OF VIBRATION

1.12. Determine the power requirements for vibration testing and analysis.

In vibration testing, we have forced vibration. The work done is the product of the excitation and displacement, while power required is the rate of doing work. Let $F = F_0 \cos \omega t$ and $x = A \cos(\omega t - \phi)$; then the work done is

$$W = \int F\, dx = \int F_0 \cos \omega t[-A \sin(\omega t - \phi)\, d(\omega t)]$$

and work done per cycle of motion is

$$W = -F_0 A \int_0^{2\pi} \cos \omega t \sin (\omega t - \phi)\, d(\omega t)$$

as the angle ωt goes through a cycle of 2π. Since $\sin (\omega t - \phi) = \sin \omega t \cos \phi - \cos \omega t \sin \phi$, the work done per cycle of motion becomes

$$\begin{aligned} W &= F_0 A \sin \phi \int_0^{2\pi} \cos^2 \omega t\, d(\omega t) - F_0 A \cos \phi \int_0^{2\pi} \cos \omega t \sin \omega t\, d(\omega t) \\ &= F_0 A \sin \phi \left[\frac{\omega t}{2} + \frac{\sin 2\omega t}{4}\right]_0^{2\pi} - F_0 A \cos \phi \left[\frac{\sin^2 \omega t}{2}\right]_0^{2\pi} \\ &= F_0 A \sin \phi \left[\frac{\omega t}{2} + \frac{\sin 2\omega t}{4}\right]_0^{2\pi} - F_0 A \cos \phi \left[\frac{1}{4} - \frac{\cos 2\omega t}{2}\right]_0^{2\pi} \\ &= \pi A F_0 \sin \phi \end{aligned}$$

If $F = F_0 \sin \omega t$ and $x = A \sin (\omega t - \phi)$, then the work done is

$$W = \int F\, dx = \int F \frac{dx}{dt}\, dt = \int F \dot{x}\, dt$$

The expression for work done in one cycle of motion is then

$$W = \int_0^{2\pi/\omega} F_0 \sin \omega t [\omega A \cos (\omega t - \phi)\, dt] = \int_0^{2\pi/\omega} F_0 A \omega \sin \omega t \cos (\omega t - \phi)\, dt$$

Since $\cos (\omega t - \phi) = \cos \omega t \cos \phi + \sin \omega t \sin \phi$,

$$W = \int_0^{2\pi/\omega} \omega A F_0 \sin \omega t \cos \omega t \cos \phi\, dt + \int_0^{2\pi/\omega} \omega A F_0 \sin \phi \sin^2 \omega t\, dt$$

As shown earlier, the above expression can be reduced to

$$\begin{aligned} W &= \left[A F_0 \omega \cos \phi \frac{\sin^2 \omega t}{2} + A F_0 \omega \sin \phi \left(\frac{t}{2} - \frac{\sin 2\omega t}{4\omega}\right)\right]_0^{2\pi/\omega} \\ &= \left[A F_0 \omega \cos \phi \left(\frac{1}{4\omega} - \frac{\cos 2\omega t}{4\omega}\right) + A F_0 \omega \sin \phi \left(\frac{t}{2} - \frac{\sin 2\omega t}{4\omega}\right)\right]_0^{2\pi/\omega} \\ &= \pi A F_0 \sin \phi \end{aligned}$$

Thus the power required is proportional to the amplitude F_0 of the excitation force as well as to the amplitude A of the displacement. When there is no damping in the system, the work done by the driving force is zero because $\phi = 0°$ or $180°$. At resonance, energy is needed to build up the amplitude of vibration; and for this case, $\phi = 90°$.

1.13. The steady state response of a simple dynamic system to a sinusoidal excitation $10 \sin 0.1\pi t$ newtons is $0.1 \sin (0.1\pi t - 30°)$ meters. Determine the work done by the excitation force in (*a*) one minute and (*b*) one second.

(*a*) From Problem 1.12, the work done per cycle by the excitation force is given by

$$W = \int_0^{2\pi} F\, dx = \int_0^{P} F \dot{x}\, dt = \pi A F_0 \sin \phi$$

where $F_0 = 10$ newtons is the amplitude of the excitation force, $A = 0.1$ m is the amplitude of the steady state response, and $\phi = 30°$ is the phase angle. Hence work done by the excitation force is $W = 3.14(0.1)(10)(0.5) = 1.57$ joules/cyc. The angular frequency is 0.1π rad/sec and the period $P = 1/f = 20$ sec. In one minute, the excitation force will complete three cycles. Therefore work done by the excitation force in one minute is 4.71 joules.

(*b*) Work done per cycle is $W = \int_0^{20} F \dot{x}\, dt$. Then work done in one second is

$$W = \int_0^1 (10 \sin 0.1\pi t)(0.01\pi) \cos (0.1\pi t - 30°)\, dt = 0.05 \text{ joule}$$

1.14. Prove that the mean kinetic and potential energies of nondissipative vibrating systems are equal.

For free vibration without damping, the motion can be assumed harmonic and is given by

$$x(t) \quad = \quad A \sin \omega_n t$$

Kinetic energy $\text{KE} = \frac{1}{2}m\dot{x}^2 = \frac{1}{2}m(\omega_n^2 A^2 \cos^2 \omega_n t) = \frac{1}{2}kA^2 \cos^2 \omega_n t$, where $\omega_n^2 = k/m$.

Potential energy $\text{PE} = \frac{1}{2}kx^2 = \frac{1}{2}kA^2 \sin^2 \omega_n t$.

$$(\text{KE})_{\text{mean}} \quad = \quad \frac{1}{P}\int_0^P (\tfrac{1}{2}kA^2 \cos^2 \omega_n t)\, dt \quad = \quad \tfrac{1}{4}kA^2$$

$$(\text{PE})_{\text{mean}} \quad = \quad \frac{1}{P}\int_0^P (\tfrac{1}{2}kA^2 \sin^2 \omega_n t)\, dt \quad = \quad \tfrac{1}{4}kA^2$$

1.15. A uniform string fixed at both ends is displaced a distance h at the center and released from rest as shown in Fig. 1-20. Find the energy of transverse vibration of the string.

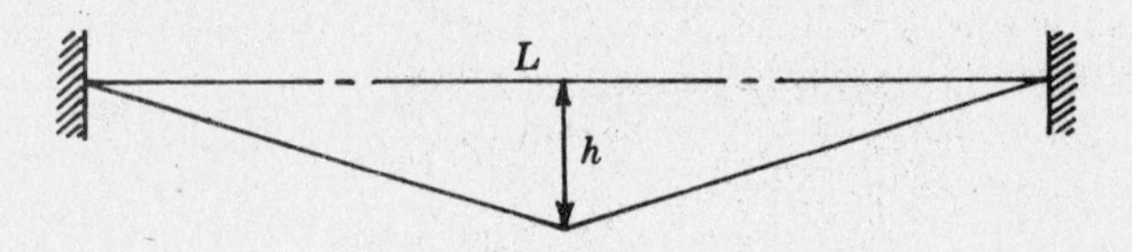

Fig. 1-20

The free transverse vibration of a uniform string can be expressed as

$$y(x,t) \quad = \sum_{i=1,2,\ldots}^{\infty} A_i \sin\frac{i\pi x}{L} \cos\left(\frac{i\pi a}{L}t + \theta_i\right)$$

where A_i is the amplitude of motion and θ_i is the phase angle. (See Problem 1.17.) Then

$$\text{KE} \quad = \quad \tfrac{1}{2}\rho_L \int_0^L \dot{y}^2\, dx \quad = \quad \frac{\rho_L a^2 \pi^2}{4L} \sum_{i=1,2,\ldots}^{\infty} i^2 A^2 \sin^2\left(\frac{i\pi a}{L}t + \theta_i\right)$$

$$\text{PE} \quad = \quad \tfrac{1}{2}S \int_0^L \left[\frac{\partial^2 y}{\partial x^2}\right]^2 dx \quad = \quad \frac{S\pi^2}{4L} \sum_{i=1,2,\ldots}^{\infty} i^2 A_i^2 \cos^2\left(\frac{i\pi a}{L}t + \theta_i\right)$$

or

$$\text{KE} + \text{PE} \quad = \quad \frac{\pi^2 a^2 \rho_L}{4L} \sum_{i=1,2,\ldots}^{\infty} i^2 A_i^2$$

where S is the tension in the string, ρ_L is the mass per unit length of the string, and $a = \sqrt{S/\rho_L}$ is the speed of wave propagation.

From the initial conditions $\dot{y}(x,0) = 0$ and $y(x,0) = \begin{cases} 2hx/L, & 0 \leq x \leq L/2 \\ 2h(1-x/L), & L/2 \leq x \leq L \end{cases}$ we obtain $A_i^2 = 64h^2/i^2\pi^2$. The expression for the energy of transverse vibration of the string becomes

$$\text{KE} + \text{PE} \quad = \quad 16\rho_L a^2 h^2/\pi^2 i^2 L, \qquad i \quad = \quad 1, 3, \ldots$$

Let the total energy associated with the fundamental mode of vibration be E_1, i.e.

$$E_1 \quad = \quad 16\rho_L a^2 h^2/L\pi^2$$

Then the energies associated with the first harmonic, second harmonic, third harmonic, ... are respectively

$$E_3 = E_1/9, \quad E_5 = E_1/25, \quad E_7 = E_1/49, \quad \ldots$$

Thus the main part of the energy of vibration is associated with the normal modes of low order. The quality of a tone is governed by the proportion of energy in each of the modes of vibration. Though the fundamental frequency may be the same, the energy distribution in the harmonics characterizes each musical instrument.

VIBRATION OF STRINGS

1.16. Investigate the transverse vibration of a stretched string of length L in a plane, assuming the tension S in the string remains constant.

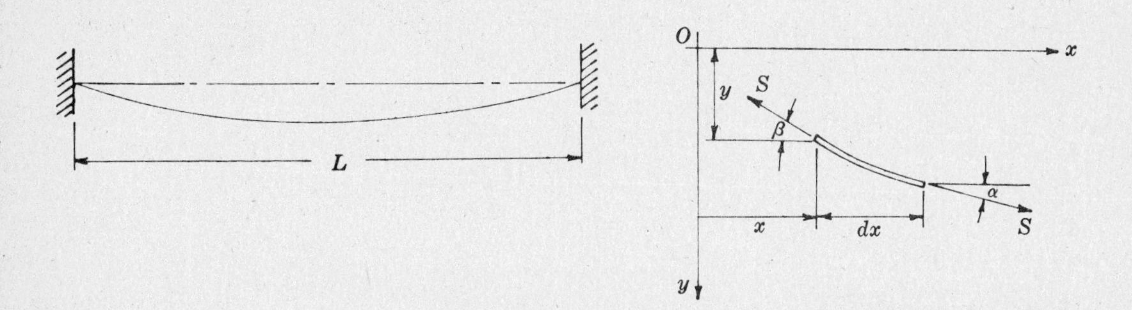

Fig. 1-21

In general, it can be assumed that the flexible string offers no resistance to bending nor to shear, and its tension is constant for small displacements.

The differential equation of motion for an infinitesimal element of the string as shown in Fig. 1-21 can be written as

$$\sum F = m\ddot{y}$$

or
$$(\rho_L \, \Delta x)\frac{\partial^2 y}{\partial t^2} = -S \sin\beta + S \sin\alpha$$

where ρ_L is the mass per unit length of the string and S is the tension in the string. Partial derivatives are used because there are two independent variables, x and t.

But $\left[\frac{\partial y}{\partial x}\right]_{x=x} = \tan\beta$, $\left[\frac{\partial y}{\partial x}\right]_{x=x+\Delta x} = \tan\alpha$. And for small displacements, $\sin\alpha \doteq \tan\alpha$ and $\sin\beta \doteq \tan\beta$. Hence

$$(\rho_L \, \Delta x)\frac{\partial^2 y}{\partial t^2} = -S\left[\frac{\partial y}{\partial x}\right]_{x=x} + S\left[\frac{\partial y}{\partial x}\right]_{x=x+\Delta x}$$

or
$$\frac{\partial^2 y}{\partial t^2} = \frac{(S/\rho_L)[(\partial y/\partial x)_{x+\Delta x} - (\partial y/\partial x)_x]}{\Delta x}$$

$$\frac{\partial^2 y}{\partial t^2} = \frac{S}{\rho_L}\frac{\partial^2 y}{\partial x^2}$$

which is generally known as the one-dimensional wave equation, and is usually written in the form

$$\frac{\partial^2 y}{\partial t^2} = a^2 \frac{\partial^2 y}{\partial x^2}$$

replacing $\sqrt{S/\rho_L}$ by the constant a.

The solution of this wave equation can be found by the "variables separable" method. Since y is a function of x and t, it can be represented as

$$y(x,t) = X(x) \cdot T(t)$$

Then
$$\frac{\partial^2 y}{\partial x^2} = T\frac{d^2 X}{dx^2}, \quad \frac{\partial^2 y}{\partial t^2} = X\frac{d^2 T}{dt^2}$$

and the wave equation becomes
$$X\frac{d^2 T}{dt^2} = a^2 T\frac{d^2 X}{dx^2}$$

Separating the variables,
$$\frac{d^2T/dt^2}{T} = a^2\frac{d^2X/dx^2}{X}$$

As X and T are independent of each other, the above expression must equal a certain constant. Let this constant be $-p^2$. This then leads to two ordinary differential equations,

$$\frac{d^2T}{dt^2} + p^2T = 0 \qquad \text{and} \qquad \frac{d^2X}{dx^2} + \frac{p^2}{a^2}X = 0$$

and the solution is

$$y(x, t) = \left(A \sin\frac{px}{a} + B \cos\frac{px}{a}\right)(C \sin pt + D \cos pt)$$

With both ends of the string fixed, the boundary conditions are

$$y(0, t) = 0 \tag{1}$$

$$y(L, t) = 0 \tag{2}$$

From condition (1),

$$0 = B(C \sin pt + D \cos pt) \qquad \text{or} \qquad B = 0$$

and from condition (2),

$$0 = (A \sin pL/a)(C \sin pt + D \cos pt)$$

Because A cannot equal zero all the time, $\sin pL/a$ must equal zero. Therefore the frequency equation is

$$\sin pL/a = 0$$

and the natural frequencies of the string are given by

$$p_i = i\pi a/L \qquad \text{where } i = 1, 2, 3, \ldots$$

It is clear that there are an infinite number of natural frequencies; this is in agreement with the fact that all continuous systems are composed of an infinite number of mass particles.

For this particular configuration of the vibrating string, i.e. with both ends fixed, the normal function $X(x)$ is therefore given by

$$X_i(x) = \sin i\pi x/L$$

and

$$y(x, t) = (A \sin px/a)(C \sin pt + D \cos pt)$$

In general, the expression for the vibrating string is given by

$$y(x, t) = \sum_{i=1,2,\ldots}^{\infty} \left(\sin\frac{i\pi x}{L}\right)(C_i \sin p_i t + D_i \cos p_i t)$$

in which the principle of superposition is used to represent the many natural modes of vibration of the string. C_i and D_i are arbitrary constants to be evaluated by the initial conditions of the system.

1.17. A uniform string of length L and high initial tension is statically displaced h units from the center and released as shown in Fig. 1-22. Find its subsequent displacements.

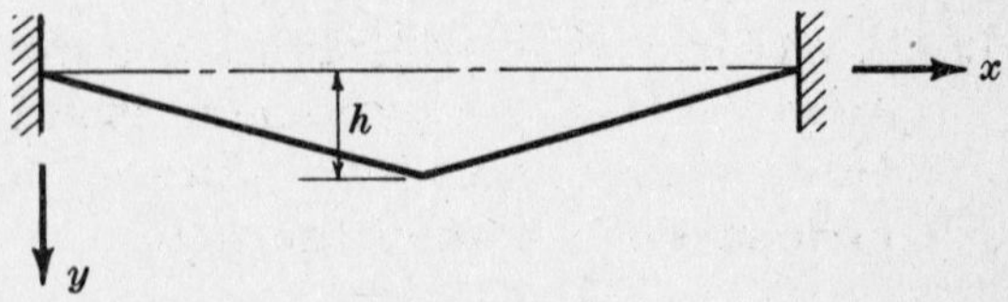

Fig. 1-22

The general expression for the free vibration of a string fixed at both ends is

$$y(x, t) = \sum_{i=1,2,\ldots}^{\infty} \left(\sin\frac{i\pi x}{L}\right)(A_i \sin p_i t + B_i \cos p_i t)$$

The initial conditions are

$$\dot{y}(x, 0) = 0, \qquad y(x, 0) = \begin{cases} 2hx/L, & 0 \leq x \leq L/2 \\ 2h(1 - x/L), & L/2 \leq x \leq L \end{cases}$$

which are equal to

$$y(x, 0) = \sum_{i=1,2,\ldots}^{\infty} B_i \sin\frac{i\pi x}{L}, \qquad \dot{y}(x, 0) = \sum_{i=1,2,\ldots}^{\infty} A_i p_i \sin\frac{i\pi x}{L}$$

Hence $A_i = 0$, and

$$\sum_{i=1,2,\ldots}^{\infty} B_i \sin\frac{i\pi x}{L} = \begin{cases} 2hx/L, & 0 \le x \le L/2 \\ 2h(1-x/L), & L/2 \le x \le L \end{cases}$$

Multiplying both sides of the above equation by $\sin i\pi x/L$ and integrating between the limits $x = 0$ and $x = L$, we obtain

$$\int_0^L B_i \sin\frac{i\pi x}{L}\sin\frac{i\pi x}{L}\,dx = \int_0^{L/2} \frac{2hx}{L}\sin\frac{i\pi x}{L}\,dx + \int_{L/2}^{L} 2h\left(1-\frac{x}{L}\right)\sin\frac{i\pi x}{L}\,dx$$

or

$$LB_i/2 = \frac{2h}{L}\left[\int_0^{L/2} x\sin\frac{i\pi x}{L}\,dx + \int_{L/2}^{L}(L-x)\sin\frac{i\pi x}{L}\,dx\right]$$

and thus

$$B_i = (-1)^{(i-1)/2}\frac{8h}{i^2\pi^2} \qquad \text{where } i = 1, 3, \ldots$$

The natural frequencies are given by

$$p_i = \frac{i\pi a}{L} \quad \text{or} \quad p_1 = \frac{\pi a}{L}, \quad p_3 = \frac{3\pi a}{L}, \quad p_5 = \frac{5\pi a}{L}, \quad \ldots$$

Therefore the expression for the displacement of the string is

$$y(x,t) = \sum_{i=1,3,\ldots}^{\infty} (-1)^{(i-1)/2}\left[\frac{8h}{i^2\pi^2}\right]\sin\frac{i\pi x}{L}\cos\frac{i\pi a}{L}t$$

where $a = \sqrt{S/\rho_L}$ is the speed of wave propagation, S is the tension in the string, and ρ_L is the density per unit length of the string.

1.18. A flexible string of length 0.99 m and mass 0.001 kg is stretched to a tension S newtons. If the string vibrates in three segments at a frequency of 500 cyc/sec, find the unknown tension S.

If the string vibrates in 3 segments, the wavelength is $\lambda = 2L/3 = 2(0.99)/3 = 0.66$ m and the speed of transverse wave propagation $a = \lambda f = 0.66(500) = 330$ m/sec.

Now $a^2 = S/\rho_L$ where ρ_L is the mass of the string per unit length. Hence

$$S = a^2\rho_L = (330)^2(0.001/0.99) = 110 \text{ newtons}$$

1.19. A uniform string of length L and fixed at both ends is released at zero initial velocity from the displaced position as shown in Fig. 1-23(a) below. By means of the wave-travel method, sketch the shape of the string at time intervals of $L/8a$ for one half cycle of the motion of the string.

As shown in the following figures, solid lines represent the actual shape of the string, and dotted lines the traveling waves in opposite directions. At any time under consideration, the shape of the string is the resultant configuration of the traveling waves.

The shape of the traveling wave is determined by the initial displacement of the string. Here, as shown in Fig. 1-23(b), it is the shape of a triangle of height $h/2$. The initial configuration of the string is made up of two identical traveling waves on top of each other but traveling in opposite directions.

At the end of the first time interval $L/8a$ (where a is the velocity of the traveling waves), the traveling waves have moved a distance of $L/8$, one to the right and the other to the left. The configuration of the string at this moment is the resultant of the two traveling waves and is shown in Fig. 1-23(c).

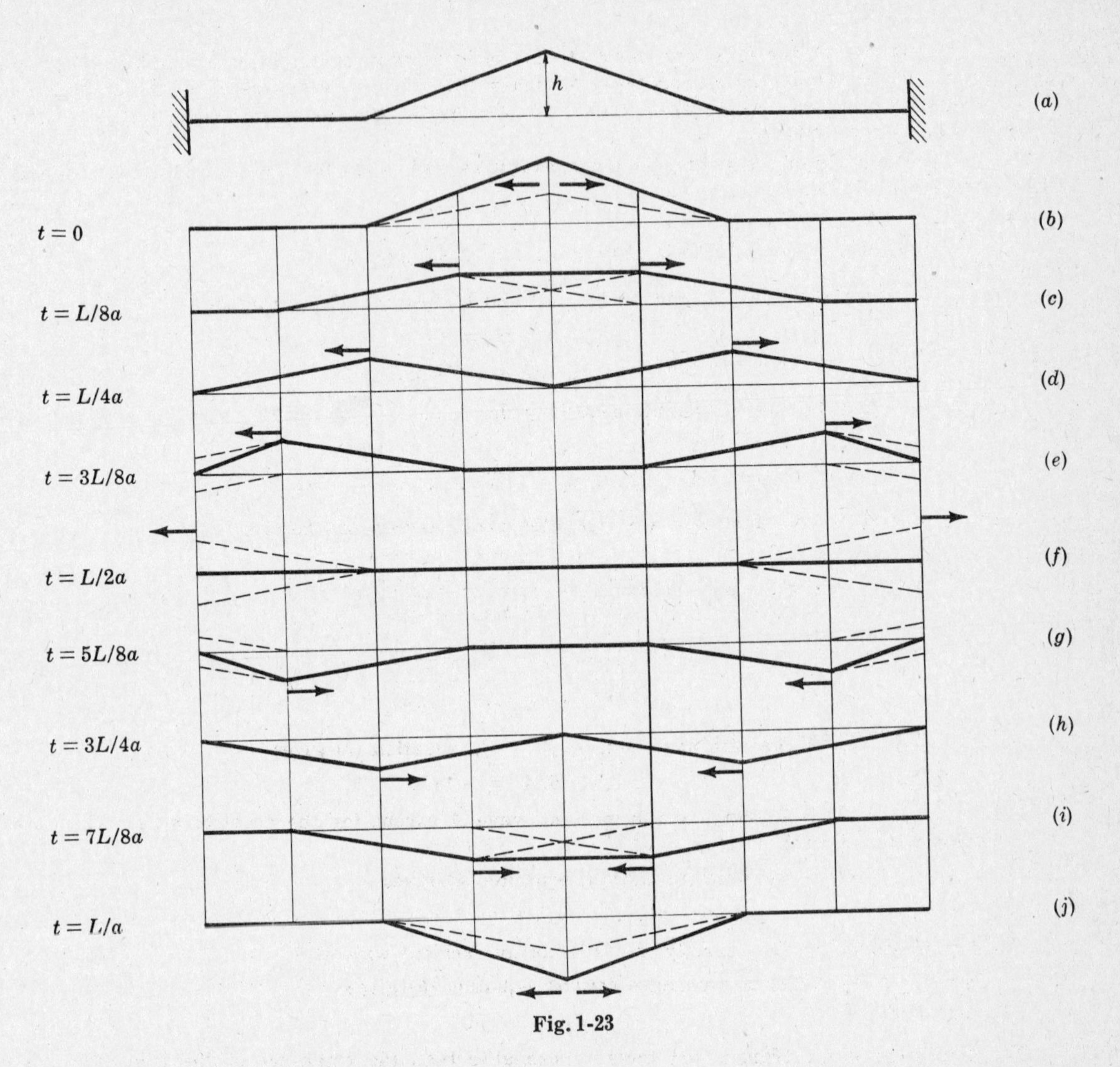

Fig. 1-23

When the waves reach the fixed ends of the string as shown in Fig. 1-23(e), they reflect and change sign. Then the waves just keep moving as shown in the rest of the figures. This procedure goes on for the rest of the cycle. At the end of the cycle, i.e. when $t = 2L/a$, the cycle repeats itself. In the absence of damping, this procedure will continue indefinitely and the amplitudes as well as the shapes of the waves will remain the same.

The traveling wave representation of the transverse vibration of a string, however, becomes very involved if the initial velocity is not equal to zero.

1.20. Investigate the wave motion and energy transmission of the transverse vibration of a *compound string* as shown in Fig. 1-24.

To account for the change of phase and mass density of the string, we use the complex exponential to represent the harmonic progressive waves of the string:

$$y_1(x,t) = Ae^{i\omega(t-x/a_1)} + Be^{i\omega(t+x/a_1)} \quad (1)$$

$$y_2(x,t) = Ce^{i\omega(t-x/a_2)} \quad (2)$$

where $a_1 = \sqrt{S/(\rho_L)_1}$, $a_2 = \sqrt{S/(\rho_L)_2}$; S is the tension in the string and ρ_L is the mass per unit length of the string. In the right hand

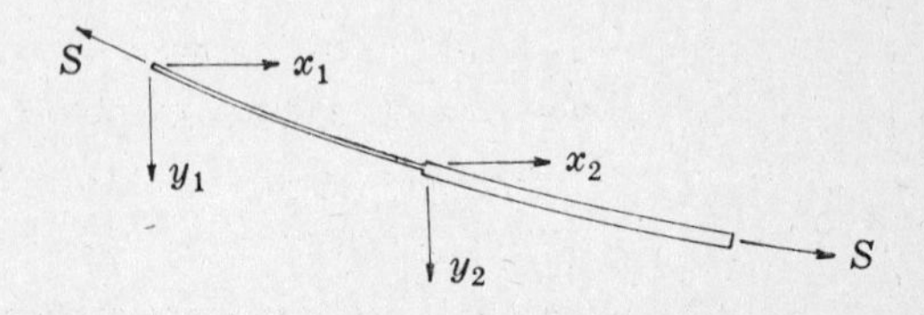

Fig. 1-24

side of equation (1), the first term refers to the incident wave traveling in the positive x direction with velocity a_1 while the second term refers to the reflected wave traveling in the negative x direction with velocity a_1. $y_2(x, t)$ represents the transmitted wave traveling in the positive x direction with velocity a_2.

At the junction of the string, the displacement as well as the force given by the two expressions y_1 and y_2 should be the same, i.e.

$$(y_1)_{x=0} = (y_2)_{x=0} \tag{3}$$

$$S(\partial y_1/\partial x)_{x=0} = S(\partial y_2/\partial x)_{x=0} \tag{4}$$

Substituting equations (1) and (2) into (3) and (4) respectively, we obtain

$$A + B = C \tag{5}$$

$$(A - B)/a_1 = C/a_2 \tag{6}$$

Solving equations (5) and (6) simultaneously yields

$$\frac{B}{A} = \frac{a_1 - a_2}{a_1 + a_2} \quad \text{and} \quad \frac{C}{A} = \frac{2a_1}{a_1 + a_2}$$

Putting $a_1 = \sqrt{S/(\rho_L)_1}$ and $a_2 = \sqrt{S/(\rho_L)_2}$, the above expressions become

$$\frac{B}{A} = \frac{\sqrt{(\rho_L)_1} - \sqrt{(\rho_L)_2}}{\sqrt{(\rho_L)_1} + \sqrt{(\rho_L)_2}} \tag{7}$$

$$\frac{C}{A} = \frac{2\sqrt{(\rho_L)_1}}{\sqrt{(\rho_L)_1} + \sqrt{(\rho_L)_2}} \tag{8}$$

If $(\rho_L)_2$ is very large (for fixed end, $(\rho_L)_2 = \infty$), equation (7) gives

$$B/A = -1$$

The reflected wave B is equal to the incident wave A except for the negative sign. This means reflection with reversal.

If $(\rho_L)_1 = (\rho_L)_2$ (for uniform string), equation (8) gives

$$C/A = 1$$

The transmitted wave C is exactly the same as the incident wave A.

If $(\rho_L)_2 > (\rho_L)_1$ (for non-uniform string), equation (8) gives

$$C < A$$

The amplitude of the transmitted wave C is smaller than the amplitude of the incident wave A.

If $(\rho_L)_2$ is very small (for free end, $(\rho_L)_2 = 0$), equation (7) gives

$$B = A$$

The reflected wave is exactly the same as the incident wave A.

The energy per unit length of the string for each of the three different waves is given by

$$\text{incident energy} = \tfrac{1}{2}(\rho_L)_1 A^2\omega^2$$

$$\text{reflected energy} = \tfrac{1}{2}(\rho_L)_1 B^2\omega^2$$

$$\text{transmitted energy} = \tfrac{1}{2}(\rho_L)_2 C^2\omega^2$$

From the principle of conservation of energy, the rate of energy approaching the junction must equal the rate of energy leaving the junction. Thus

$$\tfrac{1}{2}(\rho_L)_1 A^2\omega^2 a_1 = \tfrac{1}{2}(\rho_L)_1 B^2\omega^2 a_1 + \tfrac{1}{2}(\rho_L)_2 C^2\omega^2 a_2$$

or

$$Z_1 A^2 = Z_1 B^2 + Z_2 C^2 \tag{9}$$

where $Z = (\rho_L)a$ is called the *mechanical impedance.*

From equations (6) and (9), we obtain

$$\frac{\text{reflected energy}}{\text{incident energy}} = \frac{(Z_1 - Z_2)^2}{(Z_1 + Z_2)^2}, \quad \frac{\text{transmitted energy}}{\text{incident energy}} = \frac{4Z_1Z_2}{(Z_1 + Z_2)^2}$$

In order to obtain maximum transmission of energy, the two impedances must match each other. In other words, when $Z_1 = Z_2$ there is no reflected energy, and transmitted energy is equal to incident energy.

LONGITUDINAL VIBRATION OF BARS

1.21. Derive the differential equation of motion for the longitudinal vibration of uniform bars and investigate its general solution.

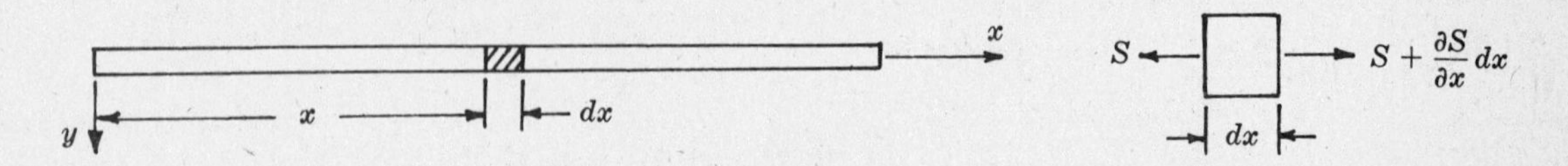

Fig. 1-25

Let u be the displacement of any cross section dx of the bar as shown in Fig. 1-25. Then the strain ϵ_x at any point x is

$$\epsilon_x = \frac{\partial u}{\partial x}$$

For an elastic bar, the stress is $\sigma_x = Y\epsilon_x$, where Y is the modulus of elasticity. Thus the tensile force at x is

$$S = \int_A \sigma_x \, dA = YA\frac{\partial u}{\partial x}$$

and the inertia force is $\rho A\,dx\,\frac{\partial^2 u}{\partial t^2}$, where ρ is the density of the bar and A is the area of cross section of the bar. Balancing the two forces, we have

$$S + \frac{\partial S}{\partial x}dx = S + \rho A\frac{\partial^2 u}{\partial t^2}dx \qquad \text{or} \qquad \frac{\partial^2 u}{\partial t^2} = a^2\frac{\partial^2 u}{\partial x^2} \tag{1}$$

where $a = \sqrt{Y/\rho}$ is the speed of wave propagation.

For the solution of this partial differential equation of motion for the longitudinal vibration of bars, let us look for a solution in the form of $u(x,t) = X(x)\,T(t)$. Substituting this expression into equation (1) yields

$$a^2\frac{d^2X/dx^2}{X} = \frac{d^2T/dt^2}{T} \tag{2}$$

Since the left-hand side of equation (2) is a function of x alone, and the right-hand side of equation (2) a function of t alone, each side must be equal to a constant. Let this constant be $-p^2$. This leads to two ordinary differential equations

$$d^2T/dt^2 + p^2T = 0 \qquad \text{and} \qquad d^2X/dx^2 + (p/a)^2X = 0$$

the solutions of which are

$$T(t) = A\cos pt + B\sin pt, \qquad X(x) = C\cos(p/a)x + D\sin(p/a)x$$

where A, B, C and D are arbitrary constants.

As $X(x)$ is a function of x alone and determines the shape of the normal mode of vibration under consideration, it is called a *normal function.* Thus the general solution is

$$u(x,t) = \sum_{i=1,2,\ldots}^{\infty} (A_i\cos p_it + B_i\sin p_it)\left(C_i\cos\frac{p_i}{a}x + D_i\sin\frac{p_i}{a}x\right) \tag{3}$$

where A_i and B_i are arbitrary constants to be evaluated by the boundary conditions, C_i and D_i are arbitrary constants to be evaluated by the initial conditions, and p_i are the natural frequencies of the system.

1.22. Determine the free longitudinal vibration of a uniform bar of length L fixed at both ends.

For longitudinal vibration of bars, the general solution is given by equation (3) of Problem 1.21.

The displacements of this bar at its ends are equal to zero, i.e. the boundary conditions are $u(0,t) = u(L,t) = 0$. Substituting these boundary conditions into the general solution, we have

$$u(0,t) = (A_i \cos p_i t + B_i \sin p_i t)C_i = 0 \quad \text{or} \quad C_i = 0$$

$$u(L,t) = (A_i \cos p_i t + B_i \sin p_i t)D_i \sin(p_i L/a) = 0$$

or $\sin(p_i L/a) = 0$ and $p_i = i\pi a/L$, $i = 1, 2, \ldots$.

The free vibration is

$$u(x,t) = \sum_{i=1,2,\ldots}^{\infty} \sin\frac{i\pi x}{L}(A_i' \cos p_i t + B_i' \sin p_i t)$$

where A_i' and B_i' are arbitrary constants to be evaluated by the initial conditions and p_i are the natural frequencies of vibration of the bar.

1.23. Determine the free longitudinal vibration of a uniform bar of length L free at both ends.

For free longitudinal vibration of bars, the general solution is given by equation (*3*) of Problem 1.21.

The forces at the ends of this bar during vibration are equal to zero, i.e. the boundary conditions are $\partial u/\partial x = 0$ at $x = 0$ and at $x = L$. Substituting these boundary conditions into the general solution, we get

$$\frac{\partial u(0,t)}{\partial x} = \frac{D_i p_i}{a}(A_i \cos p_i t + B_i \sin p_i t) = 0 \quad \text{or} \quad D_i = 0$$

$$\frac{\partial u(L,t)}{\partial x} = -\frac{p_i C_i}{a}\sin\frac{p_i L}{a}(A_i \cos p_i t + B_i \sin p_i t) = 0$$

or $\sin(p_i L/a) = 0$, and $p_i = i\pi a/L$, $i = 1, 2, \ldots$.

The free vibration is

$$u(x,t) = \sum_{i=1,2,\ldots}^{\infty} \cos\frac{i\pi x}{L}(A_i' \cos p_i t + B_i' \sin p_i t)$$

where A_i' and B_i' are arbitrary constants to be evaluated by the initial conditions and p_i are the natural frequencies.

1.24. Obtain an expression for the free longitudinal vibration of a uniform bar of length L, one end of which is fixed and the other end free.

For free longitudinal vibration of bars, the general solution is given by equation (*3*) of Problem 1.21.

The tensile force at the free end of this bar is equal to zero while the displacement at the fixed end of the bar is also equal to zero, i.e. the boundary conditions are $(u)_{x=0} = 0$, $(\partial u/\partial x)_{x=L} = 0$. Substituting these boundary conditions into the general solution, we obtain

$$u(0,t) = C_i(A_i \cos p_i t + B_i \sin p_i t) = 0 \quad \text{or} \quad C_i = 0$$

$$\frac{\partial u(L,t)}{\partial x} = \frac{p_i D_i}{a}\cos\frac{p_i L}{a}(A_i \cos p_i t + B_i \sin p_i t) = 0$$

or $\cos(p_i L/a) = 0$ as D_i cannot be equal to zero. Hence $p_i = i\pi a/2L$ where $i = 1, 3, \ldots$.

The free vibration is

$$u(x,t) = \sum_{i=1,3,\ldots}^{\infty} \sin\frac{i\pi x}{2L}(A_i' \cos p_i t + B_i' \sin p_i t)$$

where A_i' and B_i' are arbitrary constants to be evaluated by the initial conditions and p_i are the natural frequencies.

1.25. A bar of length L is fixed at one end and has a concentrated mass M attached at the other end as shown in Fig. 1-26. Derive the frequency equation for the free longitudinal vibration of this bar.

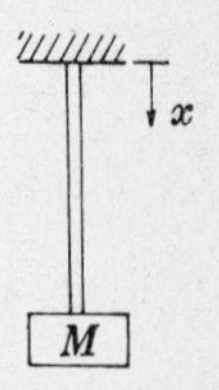

Fig. 1-26

For free longitudinal vibration of bars, the general solution is given by equation (3) of Problem 1.21.

There is no displacement at the fixed end of this bar, and a dynamic force in the bar at the free end is equal to the inertia force of the concentrated mass M, i.e. the boundary conditions are

$$(u)_{x=0} = 0, \qquad AY\left(\frac{\partial u}{\partial x}\right)_{x=L} = -M\left(\frac{\partial^2 u}{\partial t^2}\right)_{x=L}$$

where A is the cross-sectional area of the bar and Y is the Young's modulus of elasticity.

From the first of these two boundary conditions,

$$u(0,t) = C_i(A_i \cos p_i t + B_i \sin p_i t) = 0 \quad \text{or} \quad C_i = 0$$

and from the second boundary condition,

$$\frac{AYp_i}{a}\cos\frac{p_i L}{a} = Mp_i^2 \sin\frac{p_i L}{a} \quad \text{or} \quad \frac{p_i L}{a}\tan\frac{p_i L}{a} = \frac{A\rho L}{M} = M_{\text{bar}}/M$$

where $a = \sqrt{Y/\rho}$ and ρ is the density of the bar.

When $M_{\text{bar}}/M \to \infty$, i.e. when the mass M is small compared to the mass of the bar, the frequency equation becomes $\cos(p_i L/a) = 0$. The system becomes that of a bar fixed at one end and free at the other end. (See Problem 1.24.)

When M is large compared to the mass of the bar, it can be shown that $p_1 = \sqrt{AY/ML}$. This corresponds to the natural frequency of a simple spring-mass system of mass M and spring constant AY/L.

VIBRATION OF MEMBRANES

1.26. Derive the differential equation of motion for the transverse vibration of uniform membranes and investigate its general solution.

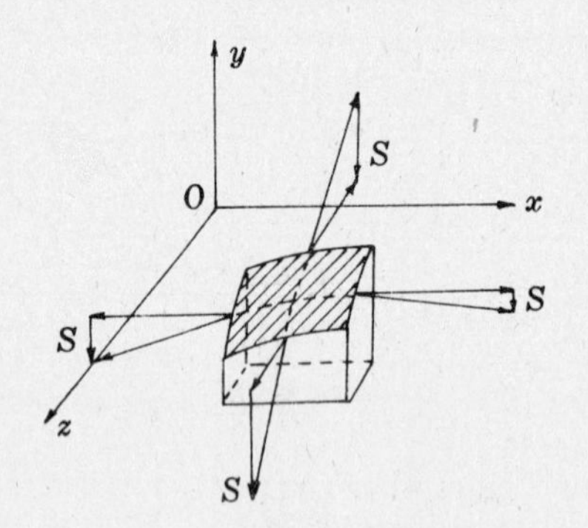

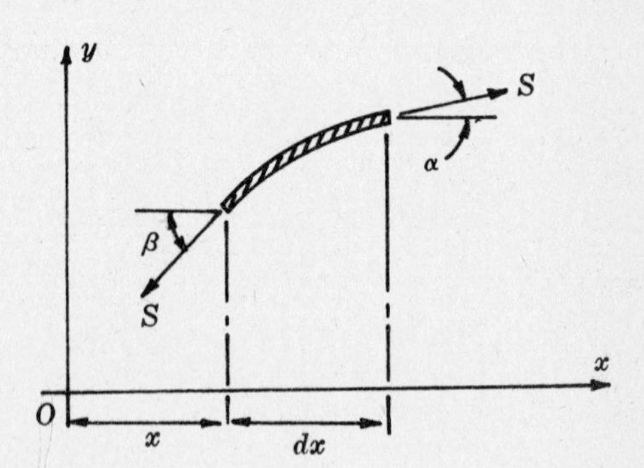

Fig. 1-27

Assume an ideal two-dimensional membrane with a completely flexible surface of extremely small uniform thickness which offers no resistance to bending or to shear. The tension is assumed to remain constant in magnitude and uniform everywhere in all directions, and is not affected by the small deflections taking place perpendicular to the membrane. In its rest or equilibrium position, the membrane is assumed to be a plane surface, i.e. in the xz plane.

Consider the differential element $dx\,dz$ of a membrane as shown in Fig. 1-27. The forces acting are those resulting from the uniform tension S per unit length of the edge of the element due to the deflection of the membrane from the equilibrium xz plane.

As in the case of the flexible string, the total restoring force is equal to the product of the mass times the acceleration, i.e. $\sum F = m\ddot{y}$.

The restoring force as shown in Fig. 1-27 is $(-S\sin\beta + S\sin\alpha)\,dz$. For small displacements, slopes are small, $\sin\beta \doteq \tan\beta = \left[\frac{\partial y}{\partial x}\right]_{x=x}$ and $\sin\alpha \doteq \tan\alpha = \left[\frac{\partial y}{\partial x}\right]_{x=x+dx}$, and the restoring force is

$$S\,dz\left[\left(\frac{\partial y}{\partial x} + \frac{\partial^2 y}{\partial x^2}dx\right) - \frac{\partial y}{\partial x}\right] \quad = \quad S\frac{\partial^2 y}{\partial x^2}dx\,dz$$

Similarly, the restoring force along edges dx is $S\frac{\partial^2 y}{\partial z^2}dx\,dz$ and the differential equation of motion is given by

$$S\left(\frac{\partial^2 y}{\partial x^2} + \frac{\partial^2 y}{\partial z^2}\right)dx\,dz \quad = \quad \rho_a\frac{\partial^2 y}{\partial t^2}dx\,dz \tag{1}$$

where ρ_a is the mass per unit area of the membrane. The two-dimensional wave equation is therefore

$$\frac{\partial^2 y}{\partial x^2} + \frac{\partial^2 y}{\partial z^2} \quad = \quad \frac{1}{a^2}\frac{\partial^2 y}{\partial t^2} \tag{2}$$

where $a = \sqrt{S/\rho_a}$ is the speed of wave propagation.

The solution of this two-dimensional wave equation can be obtained by the "variables separable" method. Since y is a function of x, z and t, it can be represented as

$$y(x,z,t) \quad = \quad X(x)\,Z(z)\,T(t) \tag{3}$$

Then

$$\frac{\partial^2 y}{\partial x^2} = ZT\frac{d^2X}{dx^2}, \qquad \frac{\partial^2 y}{\partial z^2} = XT\frac{d^2Z}{dz^2}, \qquad \frac{\partial^2 y}{\partial t^2} = XZ\frac{d^2T}{dt^2} \tag{4}$$

Substituting (4) into (3) gives

$$a^2ZT\frac{d^2X}{dx^2} + a^2XT\frac{d^2Z}{dz^2} \quad = \quad XZ\frac{d^2T}{dt^2} \tag{5}$$

Dividing (5) by XZT,

$$\frac{a^2}{X}\frac{d^2X}{dx^2} + \frac{a^2}{Z}\frac{d^2Z}{dz^2} \quad = \quad \frac{1}{T}\frac{d^2T}{dt^2} \tag{6}$$

Because X, Z, T are independent of one another, and because the right-hand side of equation (6) contains only t, both sides of (6) must be equal to a certain constant. Let this constant be $-p^2$. This then leads to the following two differential equations:

$$\frac{d^2T}{dt^2} + p^2T = 0 \quad \text{with solution} \quad T(t) = E\sin pt + F\cos pt \tag{7}$$

$$\frac{a^2}{X}\frac{d^2X}{dx^2} + \frac{a^2}{Z}\frac{d^2Z}{dz^2} = -p^2 \quad \text{or} \quad \frac{1}{X}\frac{d^2X}{dx^2} + \frac{p^2}{a^2} = -\frac{1}{Z}\frac{d^2Z}{dz^2} \tag{8}$$

Now each side of equation (8) involves only one variable, and so both sides must be equal to some constant. Let this constant be k^2. This leads to the following two ordinary differential equations in x and z,

$$\frac{d^2X}{dx^2} + \left(\frac{p^2}{a^2} - k^2\right)X = 0, \qquad \frac{d^2Z}{dz^2} + k^2Z = 0 \tag{9}$$

with solutions

$$X(x) \quad = \quad A\sin\sqrt{(p^2/a^2) - k^2}\,x + B\cos\sqrt{(p^2/a^2) - k^2}\,x \tag{10}$$

$$Z(z) \quad = \quad C\sin kz + D\cos kz \tag{11}$$

A solution of the two-dimensional wave equation is therefore given by

$$y(x,z,t) \quad = \quad (A\sin\sqrt{(p^2/a^2) - k^2}\,x + B\cos\sqrt{(p^2/a^2) - k^2}\,x)(C\sin kz + D\cos kz)(E\sin pt + F\cos pt)$$

The general solution is the sum of an arbitrary number of such solution, i.e.

$$y(x,z,t) \quad = \quad \sum_{i=1,2,\ldots}^{\infty}(A_i\sin\sqrt{(p_i^2/a^2) - k_i^2}\,x + B_i\cos\sqrt{(p_i^2/a^2) - k_i^2}\,x)(C_i\sin k_iz + D_i\cos k_iz)(E_i\sin p_it + F_i\cos p_it) \tag{12}$$

where A_i, B_i, C_i, D_i are arbitrary constants to be evaluated by the boundary conditions, E_i and F_i are arbitrary constants to be evaluated by the initial conditions, and p_i are the natural frequencies.

1.27. A uniform rectangular membrane is rigidly fixed at all its edges as shown in Fig. 1-28. Determine the general solution for the free transverse vibration of the membrane.

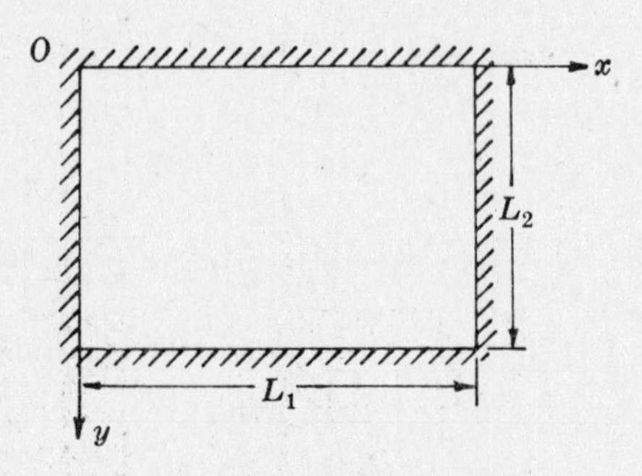

Fig. 1-28

The two-dimensional wave equation for the free transverse vibration of a uniform membrane is

$$\frac{S}{\rho_a}\left[\frac{\partial^2 y}{\partial x^2} + \frac{\partial^2 y}{\partial z^2}\right] = \frac{\partial^2 y}{\partial t^2}$$

with general solution given by

$$y(x, z, t) = (A \sin \sqrt{(p^2/a^2) - k^2}\, x + B \cos \sqrt{(p^2/a^2) - k^2}\, x)(C \sin kz + D \cos kz)(E \sin pt + F \cos pt)$$

where $a = \sqrt{S/\rho_a}$ is the speed of wave propagation.

The four boundary conditions are

$$(1)\ y(0, z, t) = 0, \quad (2)\ y(L_1, z, t) = 0, \quad (3)\ y(x, 0, t) = 0, \quad (4)\ y(x, L_2, t) = 0$$

i.e. there is no deflection at the edges.

From boundary condition (1) we obtain

$$y(0, z, t) = B(C \sin kz + D \cos kz)(E \sin pt + F \cos pt) = 0 \quad \text{or} \quad B = 0$$

From boundary condition (2),

$$y(L_1, z, t) = A \sin \sqrt{(p^2/a^2) - k^2}\, L_1(C \sin kz + D \cos kz)(E \sin pt + F \cos pt) = 0$$

or $\sin \sqrt{(p^2/a^2) - k^2}\, L_1 = 0$, i.e. $\sqrt{(p^2/a^2) - k^2} = m\pi/L_1 = \gamma$, $m = 0, 1, 2, \ldots$.

From boundary condition (3),

$$y(x, 0, t) = A \sin \gamma x(E \sin pt + F \cos pt)D = 0 \quad \text{or} \quad D = 0$$

From boundary condition (4),

$$y(x, L_2, t) = A \sin \gamma x(C \sin kL_2)(E \sin pt + F \cos pt) = 0 \quad \text{or} \quad \sin kL_2 = 0$$

i.e. $k = n\pi/L_2$, $n = 0, 1, 2, \ldots$. Thus $p^2 = a^2(m^2\pi^2/L_1^2 + k^2)$ or $p_{mn} = (a\pi/L_1L_2)\sqrt{L_1^2 n^2 + L_2^2 m^2}$, $m = 1, 2, \ldots$, $n = 1, 2, \ldots$, and the general solution becomes

$$y(x, z, t) = A \sin \gamma x(C \sin kz)(E \sin pt + F \cos pt)$$

Combine the constants into $ACE = M$ and $ACF = N$. Since there are many possible solutions, the most general solution will be the superposition of all possible solutions,

$$y(x, z, t) = \sum_{m=1,2,\ldots}^{\infty} \sum_{n=1,2,\ldots}^{\infty} \sin \gamma_m x \sin k_n z(M_{mn} \sin p_{mn}t + N_{mn} \cos p_{mn}t)$$

where $M_{mn} = A_m C_n E_{mn}$, $N_{mn} = A_m C_n F_{mn}$, and γ_m, k_n and p_{mn} are defined as above.

Fig. 1-29 below shows the modes of vibration of a rectangular membrane fixed at all its edges. Shaded and unshaded areas are in opposite phase.

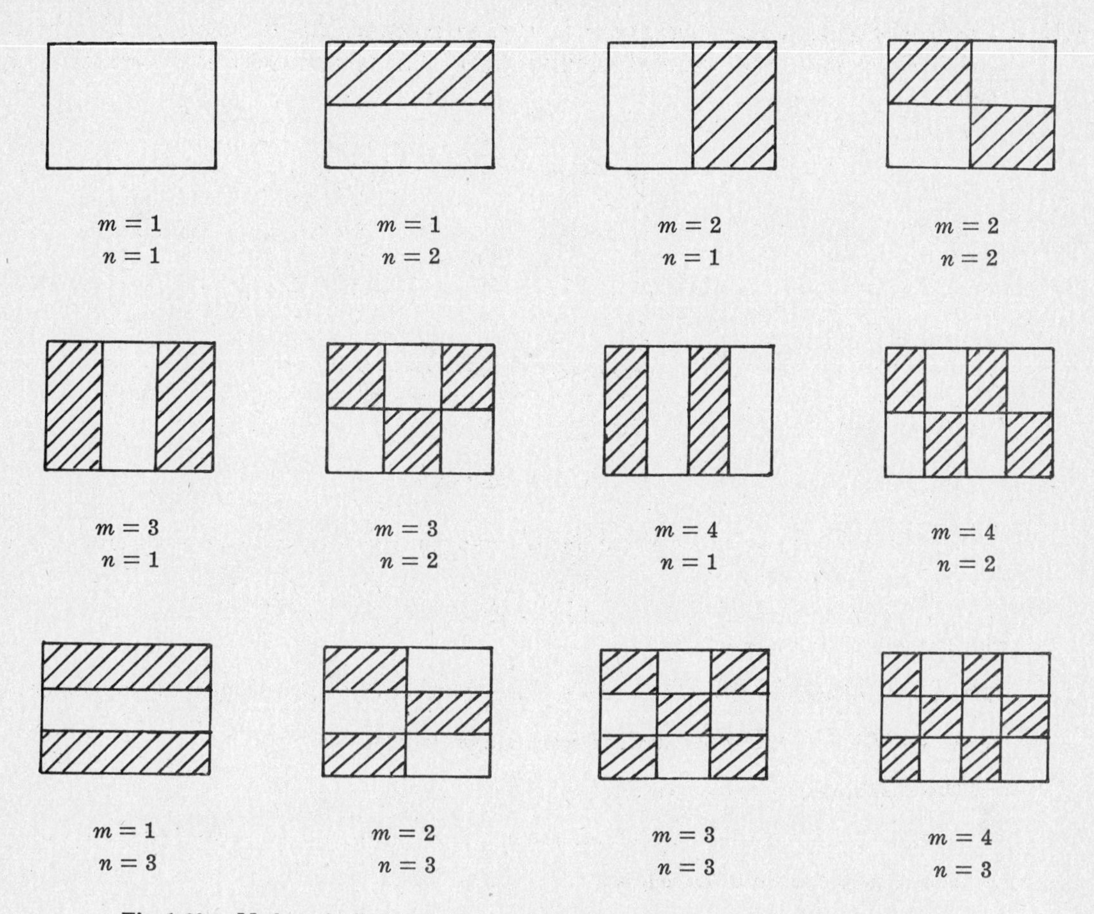

Fig. 1-29. Modes of vibration of a rectangular membrane fixed at all its edges.

1.28. A uniform circular membrane of radius d_0 is rigidly fixed at its circumference as shown in Fig. 1-30. Determine the general solution for the free transverse vibration of the membrane.

The general two-dimensional wave equation in cartesian coordinates for the free transverse vibration of uniform membranes is

$$a^2\left(\frac{\partial^2 y}{\partial x^2} + \frac{\partial^2 y}{\partial z^2}\right) = \frac{\partial^2 y}{\partial t^2} \tag{1}$$

where $a = \sqrt{S/\rho_a}$ is the speed of wave propagation, S is the tension, and ρ_a is the density per unit area of the membrane. For circular boundary, equation (1) can be transformed into polar coordinates as

$$\frac{\partial^2 y}{\partial r^2} + \frac{1}{r}\frac{\partial y}{\partial r} + \frac{1}{r^2}\frac{\partial^2 y}{\partial \theta^2} = \frac{1}{a^2}\frac{\partial^2 y}{\partial t^2} \tag{2}$$

by using the transformation equations

$$x = r\cos\theta, \qquad z = r\sin\theta$$

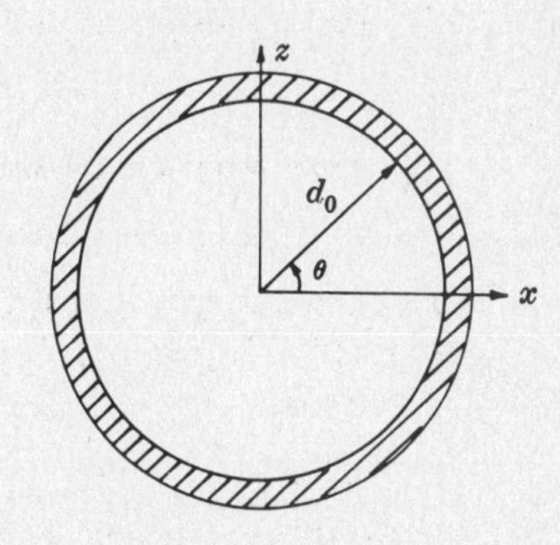

Fig. 1-30

Due to the symmetry of circular membrane with respect to its geometric center, $\partial y/\partial\theta = 0$ and equation (2) becomes

$$\frac{\partial^2 y}{\partial r^2} + \frac{1}{r}\frac{\partial y}{\partial r} = \frac{1}{a^2}\frac{\partial^2 y}{\partial t^2} \tag{3}$$

Since y is a function of r and t, the "variables separable" method leads to the following form of solution

$$y(r,t) = R(r)\,T(t)$$

and equation (3) becomes
$$\frac{a^2}{R}\left(\frac{d^2R}{dr^2} + \frac{1}{r}\frac{dR}{dr}\right) = \frac{1}{T}\frac{d^2T}{dt^2} \qquad (4)$$

Since each side of (4) contains only one independent variable, both sides must equal the same constant. Let this constant be $-p^2$. Then we have

$$\frac{d^2T}{dt^2} + p^2T = 0 \quad \text{with solution} \quad T(t) = C\sin pt + D\cos pt \qquad (5)$$

and
$$\frac{d^2R}{dr^2} + \frac{1}{r}\frac{dR}{dr} + \beta^2R = 0 \quad \text{where} \quad \beta^2 = p^2/a^2$$

or
$$r^2\frac{d^2R}{dr^2} + r\frac{dR}{dr} + r^2\beta^2R = 0 \qquad (6)$$

Using the transformation $\gamma = r\beta$, rewrite equation (6) as

$$\gamma^2\frac{d^2R}{d\gamma^2} + \gamma\frac{dR}{d\gamma} + \gamma^2R = 0 \qquad (7)$$

which is known as the Bessel differential equation of zero order. The solution is given by

$$R(\gamma) = AJ_0(r\beta) + BK_0(r\beta) \qquad (8)$$

where A and B are arbitrary constants, J_0 is the Bessel function of the first kind of order zero, and K_0 is the Bessel function of the second kind of order zero. Therefore the solution of (3) becomes

$$y(r,t) = [AJ_0(r\beta) + BK_0(r\beta)](C\sin pt + D\cos pt) \qquad (9)$$

where
$$J_0(\gamma) = \sum_{k=0,1,2,\ldots}^{\infty} \frac{(-1)^k}{(K!)^2}(r/2)^{2k}$$

$$K_0(\gamma) = J_0(\gamma)\ln\gamma - \sum_{k=0,1,2,\ldots}^{\infty} \frac{(-1)^k}{(K!)^2}\left(\frac{r}{2}\right)^{2k} f(k)$$

$$f(k) = \sum_{n=1,2,\ldots}^{\infty} (1/n)$$

The boundary condition implies that the displacement at the center of the membrane must be finite, i.e. $y(0,t) \neq 0$. Now $K_0(0) = \ln 0 = -\infty$, so B must be zero. Then

$$y(r,t) = (E\sin pt + F\cos pt)J_0(r\beta) \qquad (10)$$

where the new constants $E = AC$ and $F = AD$.

The other boundary condition is $y(d_0,t) = 0$ or $J_0(d_0\beta)(E\sin pt + F\cos pt) = 0$ from which $J_0(d_0\beta) = 0$, i.e. $d_0\beta_1 = 2.4$, $d_0\beta_2 = 5.5$, $d_0\beta_3 = 8.7, \ldots$, and since $\beta^2 = p^2/a^2$, $p_i = (a/\sqrt{d_0})\sqrt{d_0\beta_i}$, $i = 1, 2, \ldots$.

The complete solution for the free transverse vibration of a circular membrane fixed at its edges is therefore given by

$$y(r,t) = \sum_{i=1,2,\ldots}^{\infty} J_0(r\beta_i)(E_i\sin p_it + F_i\cos p_it) \qquad (11)$$

where p_i are the natural frequencies and J_0 is the Bessel function of the first kind of order zero. E_i and F_i are arbitrary constants to be evaluated by initial conditions.

Fig. 1-31 below shows the modes of vibration of a rigidly stretched uniform circular membrane. Shaded and unshaded areas are in opposite phase.

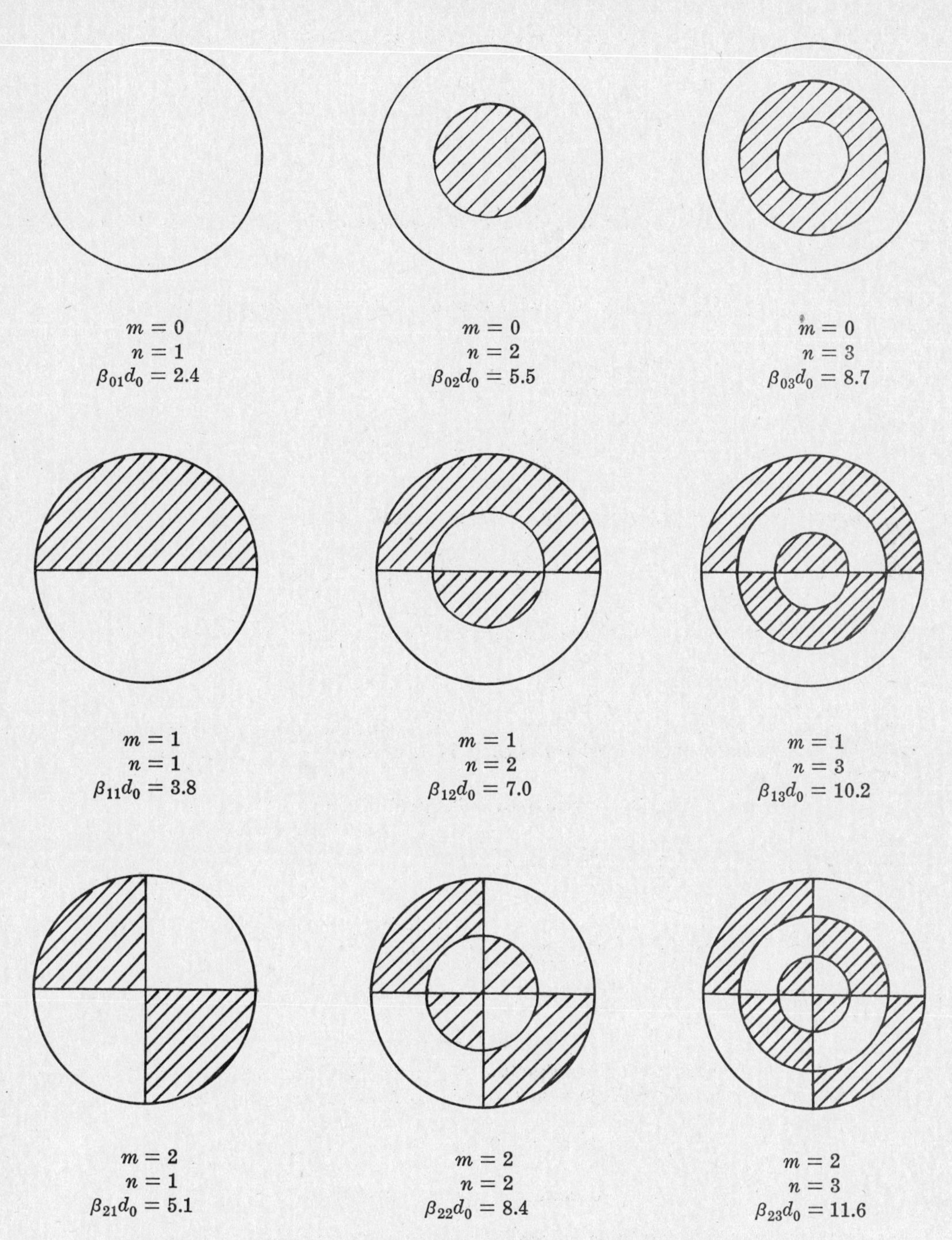

Fig. 1-31. Modes of vibration of a rigidly stretched circular membrane.

1.29. The displacement amplitude of a driven uniform circular membrane of a microphone is given as $y = (P_0/k^2S)[J_0(kr)/J_0(kr_0) - 1]$. Find the corresponding average displacement y_{av} of the surface of the membrane.

The average displacement may be defined as

$$y_{av} = \frac{1}{\pi r_0^2}\int_{S'} y(r)\, dS' \tag{1}$$

where $S' = \pi r_0^2$ is the area of the surface of the membrane. Then

$$y_{av} = \frac{1}{\pi r_0^2}\int_0^{r_0} (P_0/k^2S)\left[\frac{J_0(kr)}{J_0(kr_0)} - 1\right] 2\pi r\, dr \qquad (2)$$

where P_0 = amplitude of driven force, S = tension of membrane, r_0 = radius of membrane, and J_0 = Bessel function of the first kind of order zero.

Rewrite equation (2) as

$$y_{av} = \frac{2P_0}{r_0^2k^2SJ_0(kr_0)}\int_0^{r_0} J_0(kr)r\, dr - \frac{2P_0}{r_0^2k^2S}\int_0^{r_0} r\, dr \qquad (3)$$

Now $\int xJ_0(x)\, dx = xJ_1(x)$ or $\int (kr)J_0(kr)k\, dr = (kr)J_1(kr)$. Thus equation (3) becomes

$$y_{av} = \frac{2P_0}{r_0^2k^4SJ_0(kr_0)}\int_0^{r_0} (kr)J_0(kr)k\, dr - \frac{2P_0}{r_0^2k^2S}\int_0^{r_0} r\, dr \qquad (4)$$

Performing the indicated integrations, we obtain

$$y_{av} = \frac{2P_0}{r_0^2k^4SJ_0(kr_0)}\Big[(kr)J_1(kr)\Big]_0^{r_0} - \frac{2P_0}{r_0^2k^2S}\Big[(r^2/2)\Big]_0^{r_0}$$

or

$$y_{av} = \frac{2P_0}{r_0^2k^4SJ_0(kr_0)}(kr_0)J_1(kr_0) - P_0/k^2S \qquad (5)$$

From a table of Bessel functions, $J_2(kr_0) = 2J_1(kr_0)/kr_0 - J_0(kr_0)$; then equation (5) can be written as

$$y_{av} = \frac{2P_0kr_0J_1(kr_0) - P_0r_0^2k^2J_0(kr_0)}{r_0^2k^4SJ_0(kr_0)} = \frac{P_0}{k^2SJ_0(kr_0)}\left[\frac{2J_1(kr_0)}{kr_0} - J_0(kr_0)\right] = \frac{P_0}{k^2S}\frac{J_2(kr_0)}{J_0(kr_0)}$$

1.30. A uniform circular membrane of radius r_0 is tightly stretched along its circumference. A sinusoidal driving force $F_0 \sin \omega t$ is acting uniformly over one side of the membrane. If the coefficient of the damping force present is c, determine the resulting vibration.

The general differential equation for the free transverse vibration of a circular membrane in polar coordinates is given by

$$\frac{\partial^2 y}{\partial t^2} = a^2\left(\frac{\partial^2 y}{\partial r^2} + \frac{1}{r}\frac{\partial y}{\partial r}\right)$$

where $a = \sqrt{S/\rho_a}$ and r = radial distance from center of membrane.

With the presence of the damping force $c(\partial y/\partial t)$ and the driving force $F_0 \sin \omega t$, the equation of motion becomes

$$\frac{\partial^2 y}{\partial t^2} = \frac{S}{\rho_a}\left(\frac{\partial^2 y}{\partial r^2} + \frac{1}{r}\frac{\partial y}{\partial r}\right) - \frac{c}{\rho_a}\frac{\partial y}{\partial t} + \frac{F_0}{\rho_a}\sin \omega t \qquad (1)$$

Using complex exponential notation, we have

$$\frac{\partial^2 y}{\partial t^2} = \frac{S}{\rho_a}\left(\frac{\partial^2 y}{\partial r^2} + \frac{1}{r}\frac{\partial y}{\partial r}\right) - \frac{c}{\rho_a}\frac{\partial y}{\partial t} + \frac{F_0}{\rho_a}e^{i\omega t} \qquad (2)$$

Assume a steady state solution $y = Ye^{i\omega t}$ and substitute the assumed solution into (2):

$$\frac{S}{\rho_a}\left(\frac{d^2Y}{dr^2} + \frac{1}{r}\frac{dY}{dr}\right) + \left(\omega^2 - \frac{ic\omega}{\rho_a}\right)Y = -F_0/\rho_a \qquad (3)$$

which can be written as

$$\frac{d^2Y}{dr^2} + \frac{1}{r}\frac{dY}{dr} + k^2Y = -F_0/S \qquad (4)$$

where $k^2 = (\rho_a\omega^2 - ic\omega)/S$.

The complete solution of equation (4) is the sum of the complementary and particular solutions. The complementary solution is obtained by solving

$$\frac{d^2Y}{dr^2} + \frac{1}{r}\frac{dY}{dr} + k^2Y = 0 \qquad (5)$$

with solution

$$Y(r) = AJ_0(kr) + BK_0(kr) \qquad (6)$$

where A and B are arbitrary constants, J_0 is the Bessel function of the first kind of order zero, and K_0 is the Bessel function of the second kind of order zero. For a stretched circular membrane, $B = 0$. (See Problem 1.28.)

The particular solution is $Y(r) = -F_0/k^2S$. Thus the complete solution is

$$Y(r) \quad = \quad AJ_0(kr) - F_0/k^2S \tag{7}$$

Now the deflection at the boundary is zero, i.e. $Y = 0$ at $r = r_0$; then from equation (7), $Y(r_0) = AJ_0(kr_0) - F_0/k^2S = 0$ or $A = F_0/k^2SJ_0(kr_0)$. Hence (7) becomes

$$Y(r) \quad = \quad \frac{F_0J_0(kr)}{k^2SJ_0(kr_0)} - \frac{F_0}{k^2S}\, e^{i\omega t} \tag{8}$$

and the steady state vibration of the membrane is given by the imaginary part of equation (8),

$$y(r,t) \quad = \quad \frac{F_0}{k^2S}\left[\frac{J_0(kr)}{J_0(kr_0)} - 1\right]\sin\omega t$$

1.31. The diaphragm of a condenser microphone is made of a circular sheet of aluminum. If its radius is 0.01 m and its thickness is 0.00001 m, find the maximum allowable tension in nt/m to which this diaphragm may be stretched. What is the fundamental frequency when stretched to this maximum tension? Determine the displacement amplitude at the center of the diaphragm when it is acted upon by a sound wave of frequency 100 cyc/sec and pressure amplitude 2.0 nt/m². What is the average displacement amplitude?

The maximum allowable tension S_{max} is equal to the area times allowable stress, i.e. $S_{max} = \sigma A$. If allowable stress $\sigma = 10^8$ nt/m², then $S_{max} = 10^8(0.00001) = 1000$ nt/m.

The fundamental frequency of a uniform circular membrane is

$$f_1 \quad = \quad \frac{2.4}{2\pi R}\sqrt{S/\rho_a} \quad = \quad 7350 \text{ cyc/sec}$$

where $R = 0.01$ m is the radius, $S = S_{max} = 1000$ nt/m is the tension,

$$\rho_a \quad = \quad 2700(10)^{-5} \quad = \quad 0.027 \text{ kg/m}^2$$

is mass per unit area of the membrane, and $\rho = 2700$ kg/m³ is the density of aluminum.

The displacement amplitude at the center of the diaphragm is

$$y(0,t) \quad = \quad \frac{P_0}{S}\left[\frac{J_0(0) - J_0(kR)}{k^2J_0(kR)}\right]$$

where $k = \omega/a = \omega/\sqrt{S/\rho_a} = 100(2\pi)/\sqrt{1000/0.027} = 3.26$ or $k^2 = 10.06$, $J_0(0) = 1$, $J_0(kR) = J_0[(3.26)(0.01)] = J_0(0.0326) = 0.9997$ are the Bessel functions of the first kind and order zero. Hence

$$y(0,t) \quad = \quad \frac{2}{1000}\left[\frac{1 - 0.9997}{10.06(0.9997)}\right] \quad = \quad 6(10)^{-8} \text{ m}$$

The average displacement amplitude is given by

$$y_{av} \quad = \quad \frac{1}{\pi R^2}\int_0^R (P_0/k^2S)\,\frac{J_0(kr) - J_0(kR)}{J_0(kR)}\,2\pi r\,dr \quad = \quad \frac{P_0}{k^2S}\left[\frac{J_2(kR)}{J_0(kR)}\right]$$

where $J_2(kR) = J_2(0.0326) = 0.00015$ is the Bessel function of the first kind of order two and $S = S_{max} = 1000$ nt/m. Thus

$$y_{av} \quad = \quad \frac{2(0.00015)}{10.06(1000)} \quad = \quad 3(10)^{-8} \text{ m}$$

VIBRATION OF CIRCULAR PLATES

1.32. A thin uniform circular plate of radius R and thickness t_0 is rigidly clamped all around its circumference. Investigate the free transverse vibration of the plate.

The differential equation for the free transverse vibration of a thin uniform circular plate is given by

$$\left[\frac{\partial^2 y}{\partial r^2} + \frac{1}{r}\frac{\partial y}{\partial r}\right]^2 + \frac{12\rho(1-\mu^2)}{Yt_0^2}\frac{\partial^2 y}{\partial t^2} \quad = \quad 0 \quad \text{or} \quad \nabla_r^4 y + \frac{12(1-\mu^2)}{Yt_0^2}\frac{\partial^2 y}{\partial t^2} \quad = \quad 0 \tag{1}$$

where ρ = density of the plate, μ = Poisson's ratio, Y = Young's modulus, and t_0 = thickness of the plate.

Assume a periodic motion in the following form

$$y = \bar{Y}e^{i\omega t} \tag{2}$$

where $\bar{Y}$ is a complex function of r alone. Then equation (1) reduces to

$$\nabla_r^4 \bar{Y} = \frac{12\omega^2\rho(1-\mu^2)}{Yt_0^2}\bar{Y} \quad \text{or} \quad (\nabla_r^4 - k^4)\bar{Y} = 0 \tag{3}$$

where $k^4 = 12\omega^2\rho(1-\mu^2)/Yt_0^2$. Since $\nabla_r^4 - k^4 = (\nabla_r^2 + k^2)(\nabla_r^2 - k^2)$, the solution of (3) consists of the sum of the solutions of $\nabla_r^2 + k^2 = 0$, given by $\bar{Y} = AJ_0(kr)$, and the solution of $\nabla_r^2 - k^2 = 0$, given by $\bar{Y} = BJ_0(ikr) = BI_0(kr)$ where I_0 is the Bessel hyperbolic function. Thus

$$\bar{Y}(r) = AJ_0(kr) + BI_0(kr) \tag{4}$$

For a plate rigidly clamped at the edges, the boundary conditions are $\bar{Y}(R) = 0$ and $\partial\bar{Y}(R)/\partial r = 0$. Substituting these into equation (4) and its derivative, we have

$$AJ_0(kR) + BI_0(kR) = 0 \tag{5}$$

$$-AkJ_1(kR) + BkI_1(kR) = 0 \tag{6}$$

Divide equation (5) by (6) to obtain

$$\frac{J_0(kR)}{J_1(kR)} = -\frac{I_0(kR)}{I_1(kR)} \tag{7}$$

where $kR = n\pi$, $n = 1, 2, \ldots$. Then

$$\omega^2 = \frac{Yt_0^2k^4}{12\rho(1-\mu^2)} = \frac{Yt_0^2(n\pi/R)^4}{12\rho(1-\mu^2)} \tag{8}$$

and the free transverse vibration of the plate is

$$y(r,t) = [AJ_0(kr) + BI_0(kr)]e^{i\omega t}$$

1.33. The diaphragm of a telephone receiver is a circular steel plate of radius 0.015 m and uniform thickness 0.0001 m. If the diaphragm is rigidly clamped at its edges, find its fundamental frequency of transverse vibration.

From Problem 1.32, the fundamental frequency of a circular thin plate clamped at its edges is

$$f_1 = \frac{0.47t_0}{R^2}\sqrt{\frac{Y}{\rho(1-\mu^2)}} = 1100 \text{ cyc/sec}$$

where $t_0 = 0.0001$ m is the thickness of the plate, $R = 0.015$ m is the radius of the plate, $Y = 19.5(10)^{10}$ nt/m² is Young's modulus of steel, $\rho = 7700$ kg/m³ is the density of steel, and $\mu = 0.28$ is Poisson's ratio.

Supplementary Problems

WAVES

1.34. Show that $A\cos\omega t + A\cos(\omega t + 120°) + A\cos(\omega t + 240°) = 0$.

1.35. Given two harmonic motions $x_1 = 10\cos\omega t$ and $x_2 = \cos(\omega t + 60°)$, find X and ϕ in $X\cos(\omega t + \phi) = x_1 + x_2$. *Ans.* $X = 10.6$, $\phi = 39.5°$

1.36. Given two harmonic motions $x_1 = 20\sin 22t$ and $x_2 = 30\sin 23t$, find the beat frequency and beat period. *Ans.* $f_b = 0.16$ cyc/sec, $P_b = 6.28$ sec

1.37. If P_1 and P_2 are the periods of two harmonic waves x_1 and x_2 respectively, and $mP_1 = nP_2$, find the period of $x_1 + x_2$. *Ans.* $P = mP_1 = nP_2$

1.38. Given $u(x,t) = f(x-ct) + g(x+ct)$ and $u(0,t) = u(L,t) = 0$. If the waves are confined between $x = 0$ and $x = L$, what is the period of the functions f and g? *Ans.* $P = 2L/c$

VIBRATIONS

1.39. A simply-supported beam of length L is acted upon by a mass M_0 at midspan. If the mass of the beam is negligible compared to M_0, find the natural frequency of vibration of the beam.
Ans. $\omega_n = \sqrt{48YI/M_0L^3}$ rad/sec

1.40. A homogeneous square plate of side L and mass M_0 is suspended from the midpoint of one of the sides. Find its frequency of vibration. *Ans.* $\omega_n = \sqrt{6g/5L}$ rad/sec

1.41. A U-shape tube has a uniform bore of cross-sectional area A. If a column of liquid of length L and density ρ is set into motion, find the frequency of the resultant motion of the liquid column.
Ans. $\omega_n = \sqrt{2g/L\rho}$ rad/sec

1.42. An electric circuit contains a capacitor C, an inductor L, and a switch in series. The capacitor has initially a charge q_0 and the switch is open at time $t < 0$. If the switch is closed at $t = 0$, find the subsequent charge on the capacitor. *Ans.* $q(t) = q_0 \cos\sqrt{1/LC}\,t$

1.43. If a simple spring-mass system is subjected to an impulsive excitation F_i, find the response of the system. *Ans.* $x(t) = (F_i/\sqrt{km})\sin\sqrt{k/m}\,t$

VIBRATION OF STRINGS

1.44. Obtain an expression for the potential energy of a uniform vibrating string of length L, considering that the tension S is not constant. *Ans.* $PE = \frac{1}{2}\int_0^L S(\partial y/\partial x)^2\,dx$

1.45. A uniform string of length L is fixed at both ends, and a damping force proportional to the velocity of the string acts upon all points of the string. Find the free vibration of the string.

Ans. $y(x,t) = \sum_{i=1,2,\ldots}^{\infty} \sin\frac{i\pi x}{L}(e^{-ct/2\rho})(A_i \sin p_i t + B_i \cos p_i t)$ where $p_i = \sqrt{i^2\pi^2a^2/L^2 - c^2/4\rho^2}$

1.46. A taut uniform string of length L is fixed at both ends and is acted upon by a uniformly distributed sinusoidal excitation $F_0 \cos\omega t$. Determine the steady state vibration of the string.

Ans. $y(x,t) = (F_0/\rho\omega^2)\left(\cos\frac{\omega}{a}x + \tan\frac{\omega L}{2a}\sin\frac{\omega}{a}x - 1\right)\cos\omega t$

1.47. Find the motion in terms of traveling waves of a uniform string of length L fixed at both ends. The string is displaced a distance h at the center and released without initial velocity.

Ans. $y(x,t) = \frac{4h}{\pi^2}\left[\sin\frac{\pi a}{L}\left(\frac{x}{a} - t\right) + \sin\frac{\pi a}{L}\left(\frac{x}{a} + t\right) - \frac{1}{9}\sin\frac{3\pi a}{L}\left(\frac{x}{a} - t\right) - \frac{1}{9}\sin\frac{3\pi a}{L}\left(\frac{x}{a} + t\right) + \cdots\right]$

1.48. A uniform string fixed at both ends is struck at the center so as to obtain an initial velocity which varies linearly from zero at the ends to v_0 at the center. Find the resulting free vibration.

Ans. $y(x,t) = \frac{8v_0L}{a\pi^3}\sum_{i=1,2,\ldots}^{\infty}\frac{1}{i^3}\sin\frac{i\pi}{2}\sin\frac{i\pi x}{L}\sin\frac{i\pi a}{L}t$

LONGITUDINAL VIBRATION OF BARS

1.49. Show that the differential equation of motion for the free longitudinal vibration of a bar of variable cross section A is given by $\frac{\partial^2 u}{\partial x^2} + \frac{1}{A}\frac{\partial A}{\partial x}\frac{\partial u}{\partial x} = \frac{\rho}{Y}\frac{\partial^2 u}{\partial t^2}$.

1.50. A uniform bar of length L is moving in a horizontal plane with velocity v_0. If the bar hits a solid wall with one end and stops, what will be the free longitudinal vibration of the bar?

Ans. $u(x,t) = \frac{8v_0L}{\pi^2 a}\sum_{i=1,3,\ldots}^{\infty}\frac{1}{i^2}\sin\frac{i\pi x}{2L}\sin\frac{i\pi a}{2L}t$

1.51. A uniform bar of length L is fixed at one end and the free end is stretched uniformly to L_0 and released at $t = 0$. Find the resulting free longitudinal vibration of the bar.

Ans. $u(x,t) = \dfrac{8(L_0 - L)}{\pi^2} \displaystyle\sum_{i=1,3,\ldots}^{\infty} (-1)^{(i-1)/2} \frac{1}{i^2} \sin\frac{i\pi x}{2L} \cos\frac{i\pi a}{2L} t$

1.52. What is the effect of a constant longitudinal force on the natural frequency of a uniform bar undergoing longitudinal vibration? *Ans.* No effect

1.53. A uniform bar of length L is free at one end and is forced to follow a sinusoidal movement $A \sin \omega t$ at the other end. Find the steady state vibration of the bar.

Ans. $u(x,t) = A\left(\cos\dfrac{\omega}{a}x + \tan\dfrac{\omega L}{a} \sin\dfrac{\omega}{a}x\right) \sin \omega t$

VIBRATION OF MEMBRANES

1.54. A rectangular membrane of sides L and $2L$ is clamped at its edges. What are the lowest degenerate modes of free transverse vibration of the membrane? *Ans.* (2, 2) and (4, 1)

1.55. Show that the fundamental frequency of free transverse vibration of an equilateral triangle membrane tightly stretched at all its edges is $f_1 = 4.77\sqrt{S/A\rho_a}$ where A is the area of the membrane.

1.56. A circular membrane of radius 10 cm and density 1.0 kg/m² is stretched to a uniform tension of 10,000 nt/m. Compute the three lowest natural frequencies of transverse vibration of the membrane.
Ans. $f_1 = 380$, $f_2 = 870$, $f_3 = 1460$ cyc/sec

1.57. A uniform square membrane of sides L is fixed at two adjacent edges. It has an initial displacement $y(x,z,0) = y_0 \sin(2\pi x/L) \sin(3\pi z/L)$. Obtain an expression for the free transverse vibration of the membrane. *Ans.* $y(x,z,t) = y_0 \sin\dfrac{2\pi x}{L} \sin\dfrac{3\pi z}{L} \cos\dfrac{13\pi S}{\rho_a L} t$

1.58. A uniform rectangular membrane of sides L_1 and L_2 is firmly fixed at all its edges. The membrane is under the action of a constant force F_0 over its entire surface. If the force is suddenly removed, find the resulting free transverse vibration of the membrane.

Ans. $y(x,z,t) = \displaystyle\sum_{m=1,3,\ldots}^{\infty} \sum_{n=1,3,\ldots}^{\infty} \frac{16F_0}{mn\pi^2 p_{mn}^2} \sin\frac{m\pi x}{L_1} \sin\frac{n\pi z}{L_2} \cos p_{mn} t$

VIBRATION OF PLATES

1.59. The diaphragm of an electromagnetic sonar transducer is a circular steel plate of radius 0.09 m and thickness 0.004 m. Find its fundamental frequency of free transverse vibration.
Ans. $f_1 = 1230$ cyc/sec

1.60. Determine the average displacement amplitude of a uniform circular plate vibrating transversely in its fundamental mode. *Ans.* $y_{av} = 0.31 y_0$

1.61. A uniform circular steel plate of radius 12 inches and thickness 1.0 inch is clamped at the boundary. What is the lowest natural frequency? *Ans.* $f_1 = 700$ cyc/sec

1.62. A uniform rectangular steel plate of lengths 8×4 ft and thickness $\frac{1}{2}$ inch is simply-supported at all the edges. Determine its lowest natural frequency. *Ans.* 25 cyc/sec

Chapter 2

Plane Acoustic Waves

NOMENCLATURE

A = area, m^2
AL = acceleration level, db
B = bulk modulus, nt/m^2
c = speed of wave propagation, m/sec
e = end correction factor, m
E = energy density, $joules/m^3$
f = frequency, cyc/sec
I = acoustic intensity, $watts/m^2$
IL = intensity level, db
k = wave number
L = length, m
p = acoustic pressure, nt/m^2
P = period, sec
PWL = sound power level, db
r = specific acoustic resistance, rayls
s = condensation
SPL = sound pressure level, db
T = absolute temperature
u = instantaneous displacement, m
v = speed of observer, m/sec
V = volume, m^3
VL = velocity level, db
w = speed of medium, m/sec
W = power, watts
x = specific acoustic reactance, rayls
Y = Young's modulus of elasticity, nt/m^2
z = specific acoustic impedance, rayls
ω = circular frequency, rad/sec
ρ = density, kg/m^3
γ = ratio of the specific heat of air at constant pressure to that at constant volume
μ = Poisson's ratio
λ = wavelength, m
α = coefficient of expansion of air

INTRODUCTION

Sound waves are produced when air is disturbed, and travel through a three-dimensional space commonly as progressive longitudinal sinusoidal waves. Assuming no variation of pressure in the y or z direction, we can define *plane acoustic waves* as one-dimensional free progressive waves traveling in the x direction. The wavefronts are infinite planes perpendicular to the x axis, and they are parallel to one another at all time.

In fact, when a small body is oscillating in an extended elastic medium such as air, the sound waves produced will spread out in widening spheres instead of planes. The longitudinal wave motion of an infinite column of air enclosed in a smooth rigid tube of constant cross-sectional area closely approximates plane acoustic wave motion.

WAVE EQUATION

In the analysis of plane acoustic wave motion in a rigid tube, we make the following assumptions: (*a*) zero viscosity, (*b*) homogeneous and continuous fluid medium, (*c*) adiabatic process, and (*d*) isotropic and perfectly elastic medium. Any disturbance of the fluid medium will result in the motion of the fluid along the longitudinal axis of the tube, causing small variations in pressure and density fluctuating about the equilibrium state. These phenomena are described by the *one-dimensional wave equation*

$$\frac{\partial^2 u}{\partial t^2} = c^2 \frac{\partial^2 u}{\partial x^2}$$

where $c = \sqrt{B/\rho}$ is the speed of wave propagation, B the bulk modulus, ρ the density, and u the instantaneous displacement.

Since this partial differential equation of motion for plane acoustic waves has exactly the same form as those for free longitudinal vibration of bars and free transverse vibration of strings, practically everything deduced for waves in strings and bars is valid for plane acoustic waves.

The general solution for the one-dimensional wave equation can be written in *progressive waves* form

$$u(x,t) = f_1(x-ct) + f_2(x+ct)$$

which consists of two parts: the first part $f_1(x-ct)$ represents a wave of arbitrary shape traveling in the positive x direction with velocity c, and the second part $f_2(x+ct)$ represents a wave also of arbitrary shape traveling in the negative x direction with velocity c. In *complex exponential* form, the general solution can be written as

$$u(x,t) = Ae^{i(\omega t - kx)} + Be^{i(\omega t + kx)}$$

where $k = \omega/c$ is the wave number, $i = \sqrt{-1}$, and A and B are arbitrary constants (real or complex) to be evaluated by initial conditions. In *sinusoidal* sine and cosine series, the general solution is

$$u(x,t) = \sum_{i=1,2,\ldots}^{\infty} \left(A_i \sin\frac{p_i}{c}x + B_i \cos\frac{p_i}{c}x\right)(C_i \sin p_i t + D_i \cos p_i t)$$

where A_i and B_i are arbitrary constants to be evaluated by boundary conditions, C_i and D_i are arbitrary constants to be evaluated by initial conditions, and p_i are the natural frequencies of the system. (See Problems 2.1-2.6.)

WAVE ELEMENTS

Plane acoustic waves are characterized by three important elements: particle displacement, acoustic pressure, and density change or condensation.

Particle displacements from their equilibrium positions are amplitudes of motion of small constant volume elements of the fluid medium possessing average identical properties, and can be expressed as

$$u(x,t) = Ae^{i(\omega t - kx)} + Be^{i(\omega t + kx)}$$

or

$$u(x,t) = A\cos(\omega t - kx) + B\cos(\omega t + kx)$$

Acoustic pressure p is the total instantaneous pressure at a point minus the static pressure. This is often referred to as excess pressure. The *effective sound pressure* p_{rms} at a point is the *root mean square* value of the instantaneous sound pressure over a complete cycle at that point. Thus

$$p = -\rho c^2 \frac{\partial u}{\partial x} = i\rho c\omega(Ae^{i(\omega t - kx)} - Be^{i(\omega t + kx)})$$

or

$$p = -\rho c\omega A \sin(\omega t - kx) + \rho c\omega B \sin(\omega t + kx)$$

Density change is the difference between the instantaneous density and the constant equilibrium density of the medium at any point, and is defined by the *condensation* s at such point as

$$s = \frac{\rho - \rho_0}{\rho_0} = -\frac{\partial u}{\partial x} = ikAe^{i(\omega t - kx)} - ikBe^{i(\omega t + kx)}$$

When plane acoustic waves are traveling in the positive x direction, it is clear that particle displacement lags particle velocity, condensation and acoustic pressure by 90°. On the other hand, when plane acoustic waves are traveling in the negative x direction, acoustic pressure and condensation lag particle displacement by 90° while particle velocity leads it by 90°. (See Problems 2.7-2.9.)

SPEED OF SOUND

The speed of sound is the speed of propagation of sound waves through the given medium. The speed of sound in air is

$$c = \sqrt{\gamma p/\rho} \text{ m/sec}$$

where γ is the ratio of the specific heat of air at constant pressure to that at constant volume, p is the pressure in newtons/m², and ρ is the density in kg/m³. At room temperature and standard atmospheric pressure, the speed of sound in air is 343 m/sec and increases approximately 0.6 m/sec for each degree centigrade rise. The speed of sound in air is independent of changes in barometric pressure, frequency and wavelength but is directly proportional to absolute temperature, i.e.

$$c_1/c_2 = \sqrt{T_1/T_2}$$

The speed of sound in solids having large cross-sectional areas is

$$c = \sqrt{\frac{Y(1-\mu)}{\rho(1+\mu)(1-2\mu)}} \text{ m/sec}$$

where Y is the Young's modulus of elasticity in nt/m², ρ the density in kg/m³, and μ Poisson's ratio. When the dimension of the cross section is small compared to the wavelength, the lateral effect considered in Poisson's ratio can be neglected and the speed of sound is simply

$$c = \sqrt{Y/\rho} \text{ m/sec}$$

The speed of sound in fluids is

$$c = \sqrt{B/\rho} \text{ m/sec}$$

where B is the bulk modulus in nt/m² and ρ is the density in kg/m³. (See Problems 2.10-2.13.)

ACOUSTIC INTENSITY

Acoustic intensity I of a sound wave is defined as the average power transmitted per unit area in the direction of wave propagation:

$$I = \frac{p_{\text{rms}}^2}{\rho c} \quad \text{watts/m}^2$$

where p_{rms} is the effective (root mean square) pressure in nt/m^2, ρ is the density in kg/m^3, and c is the speed of sound in m/sec.

At room temperature and standard atmospheric pressure, $p_{\text{rms}} = 0.00002$ nt/m^2, $\rho = 1.21$ kg/m^3, $c = 343$ m/sec, and so acoustic intensity for airborne sounds is approximately 10^{-12} watt/m^2. (See Problems 2.14-2.18.)

SOUND ENERGY DENSITY

Sound energy density is energy per unit volume in a given medium. Sound waves carry energy which is partly potential due to displacement of the medium and partly kinetic arising from the motion of the particles of the medium. If there are no losses, the sum of these two energies is constant. Energy losses are supplied from the sound source.

The instantaneous sound energy density is

$$E_{\text{ins}} = \rho\dot{x}^2 + \frac{p_0\dot{x}}{c} \quad \text{watt-sec/m}^3$$

and the average sound energy density is

$$E_{\text{av}} = \tfrac{1}{2}\rho\dot{x}^2 \quad \text{watt-sec/m}^3$$

where ρ is the instantaneous density in kg/m^3, p_0 is the static pressure in nt/m^2, $\dot{x}$ is particle velocity in m/sec, and c is the speed of sound in m/sec. (See Problems 2.19-2.20.)

SPECIFIC ACOUSTIC IMPEDANCE

Specific acoustic impedance z of a medium is defined as the ratio (real or complex) of sound pressure to particle velocity:

$$z = p/v \quad \text{kg/m}^2\text{-sec or rayls}$$

where p is sound pressure in nt/m^2, and v is particle velocity in m/sec.

For harmonic plane acoustic waves traveling in the positive x direction,

$$z = \frac{-\rho c\omega A}{-\omega A} = \rho c \quad \text{rayls}$$

and for harmonic plane acoustic waves traveling in the negative x direction,

$$z = \frac{-\rho c\omega A}{\omega A} = -\rho c \quad \text{rayls}$$

where ρ is the density in kg/m^3, c is the speed of sound in m/sec, and ρc is known as the *characteristic impedance* or *resistance* of the medium in *rayls*. At standard atmospheric pressure and 20°C, for example, the density of air is 1.21 kg/m^3, the speed of sound is 343 m/sec, and so the characteristic impedance of air is 1.21(343) or 415 rayls. For distilled water at standard atmospheric pressure and 20°C, the density is 998 kg/m^3 and the speed of sound is 1480 m/sec; hence its characteristic impedance is $1.48(10)^6$ rayls.

For standing waves, the specific acoustic impedance will vary from point to point in the x direction. In general, it is a complex ratio

$$z = r + ix \text{ rayls}$$

where r is the *specific acoustic resistance*, x is the *specific acoustic reactance* and $i = \sqrt{-1}$.

SOUND MEASUREMENTS

Because of the very wide range of sound power, intensity and pressure encountered in our acoustical environment, it is customary to use the logarithmic scale known as the *decibel scale* to describe these quantities, i.e. to relate the quantity logarithmically to some standard reference. *Decibel* (abbreviated db) is a dimensionless unit for expressing the ratio of two powers, which can be acoustical, mechanical, or electrical. The number of decibels is 10 times the logarithm to the base 10 of the power ratio. One *bel* is equal to 10 decibels. Thus *sound power level* PWL is defined as

$$\text{PWL} = 10 \log (W/W_0) \text{ db} \quad \text{re } W_0 \text{ watts}$$

where W is power in watts, W_0 is the reference power also in watts, and re = refer to the reference power W_0. For standard power reference $W_0 = 10^{-12}$ watt,

$$\text{PWL} = (10 \log W + 120) \text{ db}$$

The acoustical power radiated by a large rocket, for example, is approximately 10^7 watts or 190 db. For a very soft whisper, the acoustical power radiated is 10^{-10} watt or 20 db.

Sound intensity level IL is similarly defined as

$$\text{IL} = 10 \log (I/I_0) \text{ db} \quad \text{re } I_0 \text{ watts/m}^2$$

For standard sound intensity reference $I_0 = 10^{-12}$ watt/m^2,

$$\text{IL} = (10 \log I + 120) \text{ db}$$

Sound pressure level SPL is thus defined as

$$\text{SPL} = 20 \log (p/p_0) \text{ db} \quad \text{re } p_0 \text{ nt/m}^2$$

For standard sound pressure reference $p_0 = 2(10)^{-5}$ nt/m^2 or 0.0002 microbar,

$$\text{SPL} = (20 \log p + 94) \text{ db}$$

In vibration measurements, the *velocity level* VL is similarly defined as

$$\text{VL} = 20 \log (v/v_0) \text{ db} \quad \text{re } v_0 \text{ m/sec}$$

where $v_0 = 10^{-8}$ m/sec is the standard velocity reference. The *acceleration level* AL is

$$\text{AL} = 20 \log (a/a_0) \text{ db} \quad \text{re } a_0 \text{ m/sec}^2$$

where $a_0 = 10^{-5}$ m/sec^2 is the standard acceleration reference. (See Problems 2.21-2.29.)

RESONANCE OF AIR COLUMNS

Acoustic resonance of air columns is tuned response where the receiver is excited to vibrate by sound waves having the same frequency as its natural frequency. Resonant response depends on the distance between sound source and the receiver, and the coupling medium between them. It is, in fact, an exchange of energy of vibration between the source and the receiver.

The Helmholtz resonator makes use of the principle of air column resonance to detect a particular frequency of vibration to which it is accurately tuned. It is simply a spherical container filled with air, and having a large opening at one end and a much smaller one at the opposite end. The ear will hear amplified sound of some particular frequency from the small hole when sound is directed through the larger hole.

Half wavelength resonance of air columns will be observed when the phase change on reflection is the same at both ends of the tube, i.e. either two nodes or two antinodes. The effective lengths of air column and its resonant frequencies are

$$L = i\lambda/2, \quad f = c/\lambda = ic/2L, \quad i = 1, 2, \ldots$$

where λ is the wavelength and c is the speed of sound.

Quarter wavelength resonance of air columns will be observed when there is no change in phase at one end of a stationary wave but 180° phase change at the other end. The effective lengths of air column and its resonant frequencies are

$$L = \lambda(2i-1)/4, \quad f = c(2i-1)/4L, \quad i = 1, 2, 3, \ldots$$

In general, an open end of a tube of air is an antinode, and a closed end a node. (See Problems 2.30-2.37.)

DOPPLER EFFECT

When a source of sound waves is moving with respect to the medium in which waves are propagated, or an observer is moving with respect to the medium, or both the source and the observer have relative motion with respect to each other and to the medium, the frequency detected by the observer will be different from the actual frequency of the sound waves emitted by the source. This apparent change in frequency is known as the *Doppler effect.*

The observed frequency of a sound depends essentially on the number of sound waves reaching the ear per second, and is given by

$$f' = (c-v)f/(c-u) \text{ cyc/sec}$$

where f' is the observed frequency, c the speed of sound, v the speed of the observer relative to the medium, and u the speed of the source. When the source and observer are approaching each other, the observed frequency is increased; while if they are receding from each other, the observed frequency is lowered. (See Problems 2.38-2.41.)

Solved Problems

WAVE EQUATION

2.1. Derive the differential equation of motion for the free longitudinal elastic vibration of air columns and discuss its general solution.

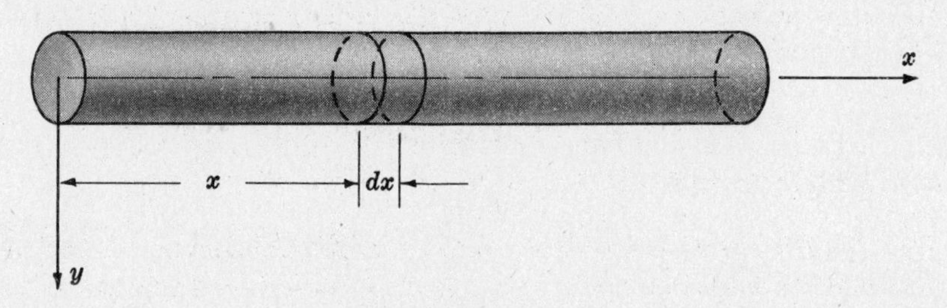

Fig. 2-1

An air column may be defined as a sample of air contained by a cylindrical tube of length L and of uniform cross-sectional area A. The tube is closed at both ends. Then the mass of the air column is $AL\rho$, where ρ is the density of air. Assume the temperature is constant throughout the tube, and also negligible air viscosity effects. In short, we have an ideal gas.

While the air column is vibrating, the density of the air in the neighborhood of any section changes with time. Also, at any instant, the density of the air varies from point to point along the column. Let u be the instantaneous displacement of any cross section dx of the air column as shown in Fig. 2-1. When the column of air is vibrating, the initial and instantaneous section dx and $(dx + du)$ will always contain the same mass of air, $A\rho\, dx$. Therefore we can write

$$A\rho\, dx = A(\rho + d\rho)(dx + du) \tag{1}$$

where $(\rho + d\rho)$ is the instantaneous density of air, and $(dx + du)$ is the instantaneous length of the section of air dx in question.

Expanding equation (1) and neglecting the higher order term $d\rho\, du$, we obtain

$$d\rho = -\rho\, du/dx \tag{2}$$

Now $dp = B\, d\rho/\rho$ is the change in pressure due to change of volume and B is the bulk modulus. We can write equation (2) as

$$dp = -B\, du/dx \tag{3}$$

While the air is vibrating, pressure changes indicated by (3) will exert forces on the section dx. Balancing the inertia force and the pressure forces on the section dx, we obtain

$$A\rho\, dx \frac{d^2u}{dt^2} = A\left(p - B\frac{du}{dx}\right) - A\left(p - B\frac{du}{dx} - B\frac{d^2u}{dx^2}dx\right)$$

Simplifying,

$$\frac{d^2u}{dt^2} = \frac{B}{\rho}\frac{d^2u}{dx^2} \tag{4}$$

Since u is a function of both x and t, we may use partial differentials to rewrite equation (4) as

$$\frac{\partial^2 u}{\partial t^2} = c^2 \frac{\partial^2 u}{\partial x^2} \tag{5}$$

where $c^2 = B/\rho$.

Equation (5) is therefore the differential equation of motion for the free longitudinal vibration of an air column inside a closed cylindrical tube and is commonly known as the one-dimensional wave equation. It has exactly the same form as the differential equation of motion for the free transverse vibration of strings and free longitudinal vibration of bars. (See Problems 1.16 and 1.21.) Hence all the theory discussed and problems solved in Chapter 1 for the free transverse vibration of strings and free longitudinal vibration of bars apply equally well for the vibration of air columns.

2.2. Prove that the following expressions are correct solutions for the one-dimensional wave equation:

(a) $u(x,t) = Ae^{i\omega t}\sin kx + Be^{i\omega t}\cos kx$

(b) $u(x,t) = (Ce^{ikx} + De^{-ikx})e^{i\omega t}$

(a) The one-dimensional wave equation is given by

$$\frac{\partial^2 u}{\partial t^2} = c^2 \frac{\partial^2 u}{\partial x^2} \tag{1}$$

Now

$$\frac{\partial u}{\partial x} = k(A\cos kx - B\sin kx)e^{i\omega t}, \qquad \frac{\partial^2 u}{\partial x^2} = -k^2(A\sin kx + B\cos kx)e^{i\omega t} \tag{2}$$

$$\frac{\partial u}{\partial t} = i\omega(A\sin kx + B\cos kx)e^{i\omega t}, \qquad \frac{\partial^2 u}{\partial t^2} = -\omega^2(A\sin kx + B\cos kx)e^{i\omega t} \tag{3}$$

Equating (2) and (3) gives $k^2 = \omega^2$. But $k = \omega/c$ is defined as the wave number. When this expression for the wave number is used, the required answer follows.

(b) If $u(x,t) = (Ce^{ikx} + De^{-ikx})e^{i\omega t}$, we proceed as in part (a):

$$\frac{\partial u}{\partial x} = (ikCe^{ikx} - ikDe^{-ikx})e^{i\omega t}, \qquad \frac{\partial^2 u}{\partial x^2} = -k^2(Ce^{ikx} + De^{-ikx})e^{i\omega t}$$

$$\frac{\partial u}{\partial t} = i\omega(Ce^{ikx} + De^{-ikx})e^{i\omega t}, \qquad \frac{\partial^2 u}{\partial t^2} = -\omega^2(Ce^{ikx} + De^{-ikx})e^{i\omega t}$$

and the wave equation becomes

$$-\omega^2(Ce^{ikx} + De^{-ikx})e^{i\omega t} = -c^2k^2(Ce^{ikx} + De^{-ikx})e^{i\omega t}$$

which again yields $k = \omega/c$ as in part (a). Therefore we conclude this is also a correct solution.

Since the wave equation for plane acoustic waves is linear, i.e. u and its coefficients never occur in any form other than that of the first degree, the principle of superposition can be applied to obtain solutions in series form. For example, if f_1 and f_2 are any two possible and correct solutions for the wave equation, $a_1f_1 + a_2f_2$ is also a possible and correct solution where a_1 and a_2 are two arbitrary constants. In short, the most general solution is in series form which is the sum of an arbitrary number of all possible solutions.

2.3. If $u(x,0) = U_0(x)$, $\dot{u}(x,0) = 0$ are the initial conditions, find the traveling-wave solution for the one-dimensional wave equation.

The traveling-wave solution for the one-dimensional wave equation can be written as

$$u(x,t) = f_1(x - ct) + f_2(x + ct)$$

where f_1 and f_2 are arbitrary functions.

From the given initial conditions,

$$u(x,0) = f_1(x) + f_2(x) = U_0(x) \tag{1}$$

$$\dot{u}(x,0) = -cf_1'(x) + cf_2'(x) = 0 \tag{2}$$

and from (2),

$$f_1'(x) = f_2'(x) \tag{3}$$

Integration of (3) gives

$$f_1(x) = f_2(x) + C \tag{4}$$

Substituting (4) into (1), we obtain

$$2f_2(x) + C = U_0(x) \quad \text{or} \quad f_2(x) = \tfrac{1}{2}[U_0(x) - C] \tag{5}$$

From (4),

$$f_1(x) = \tfrac{1}{2}[U_0(x) + C] \tag{6}$$

Substituting (5) and (6) into the traveling-wave solution,

$$u(x,t) = \tfrac{1}{2}[U_0(x-ct) + C] + [\tfrac{1}{2}U_0(x+ct) - C] = \tfrac{1}{2}[U_0(x-ct) + U_0(x+ct)]$$

2.4. Show that solutions to the one-dimensional wave equation can assume harmonic, complex exponential, hyperbolic and exponential forms.

Plane acoustic wave motion is governed by the one-dimensional wave equation

$$\frac{\partial^2 u}{\partial t^2} = c^2 \frac{\partial^2 u}{\partial x^2} \tag{1}$$

Let us look for a solution in the general form of $u(x,t) = X(x)\,T(t)$, where X and T are functions of x and t respectively. Substituting this expression for u into (1), we obtain

$$\frac{1}{X}\frac{d^2X}{dx^2} = \frac{1}{c^2T}\frac{d^2T}{dt^2} \tag{2}$$

Since the right-hand side of (2) is a function of t, and the left-hand side is a function of x alone, each side must be equal to the same constant. Let this constant be $-p^2$. This leads to the following ordinary differential equations

$$\frac{d^2X}{dx^2} + p^2X = 0 \quad \text{and} \quad \frac{d^2T}{dt^2} + c^2p^2T = 0$$

the solutions of which are

$$X(x) = A\cos px + B\sin px, \qquad T(t) = C\cos cpt + D\sin cpt$$

and so

$$u(x,t) = (A\cos px + B\sin px)(C\cos cpt + D\sin cpt) \tag{3}$$

or

$$X(x) = Ae^{ipx} + Be^{-ipx}, \qquad T(t) = Ce^{icpt} + De^{-icpt}$$

and

$$u(x,t) = (Ae^{ipx} + Be^{-ipx})(Ce^{icpt} + De^{-icpt}) \tag{4}$$

where A, B, C, D are arbitrary constants.

If we call the constant for equation (2) p^2, we obtain

$$\frac{d^2X}{dx^2} - p^2X = 0 \quad \text{and} \quad \frac{d^2T}{dt^2} - c^2p^2T = 0$$

the solutions of which are

$$X(x) = A\cosh px + B\sinh px, \qquad T(t) = C\cosh cpt + D\sinh cpt$$

and so

$$u(x,t) = (A\cosh px + B\sinh px)(C\cosh cpt + D\sinh cpt) \tag{5}$$

or

$$X(x) = Ae^{px} + Be^{-px}, \qquad T(t) = Ce^{cpt} + De^{-cpt}$$

and

$$u(x,t) = (Ae^{px} + Be^{-px})(Ce^{cpt} + De^{-cpt}) \tag{6}$$

Equations (3) to (6) represent the four different forms of solution for the one-dimensional wave equation. These forms of solution — the harmonic, the complex exponential, the hyperbolic and the exponential — are all interchangeable and will give rise to standing waves, formed by the superposition of two sets of waves equal in wavelength and amplitude but moving in opposite directions. (See Problem 2.3 for the progressive waves forms of solution for the one-dimensional wave equation.)

2.5. Show that the function $u = f(\omega t + kx)$ represents a progressive wave of fixed profile $f(kx)$ moving along the negative x axis with constant velocity $c = \omega/k$.

Since u is a linear, single-valued function of x and t, we may write

$$u = f(\omega t + kx) = kf(\omega t/k + x) = kf(x + ct)$$

where $c = \omega/k$.

Plotting the function u against x, the wave at time $t = 0$ is $u = kf(x)$ or $f(kx)$. As the wave is propagated without change of shape, the wave shape at a later time t will be identical to that at $t = 0$ except that the wave profile has moved a distance ct in the negative x direction.

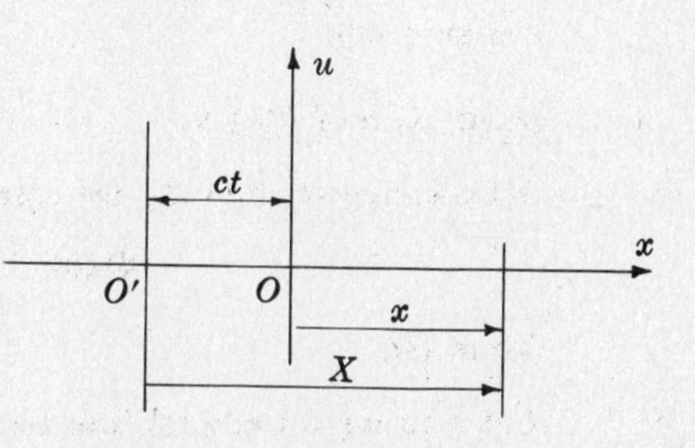

Fig. 2-2

Now O' is the new origin, and $x = X - ct$ as shown in Fig. 2-2. The equation of the wave profile referred to this new origin O' is

$$u = kf(X) = kf(x+ct)$$

Similarly, it can be shown that $u = f(\omega t - kx)$ or $u = kf(x - ct)$ represents a wave of fixed profile $f(kx)$ moving in the positive x direction with constant velocity $c = \omega/k$. If the wave profile is harmonic, we have free harmonic progressive waves, e.g. $A \sin(\omega t + kx)$, $A \cos k(x - ct)$, $Ae^{i(\omega t - kx)}$. A harmonic diverging spherical wave is therefore represented by $(A/r)\cos(\omega t - kx)$ or $(A/r)e^{i(\omega t - kx)}$ where its amplitude decreases with distance of propagation.

2.6. Use D'Alembert's method of integration to obtain the solution for the one-dimensional wave equation.

Let us introduce two new independent variables r and s such that

$$r = x - ct, \qquad s = x + ct$$

Then
$$\frac{\partial r}{\partial x} = 1, \quad \frac{\partial r}{\partial t} = -c, \quad \frac{\partial s}{\partial x} = 1, \quad \frac{\partial s}{\partial t} = c$$

Using the chain rule:

$$\frac{\partial u}{\partial x} = \frac{\partial u}{\partial r}\frac{\partial r}{\partial x} + \frac{\partial u}{\partial s}\frac{\partial s}{\partial x} = \frac{\partial u}{\partial r} + \frac{\partial u}{\partial s}$$

$$\frac{\partial^2 u}{\partial x^2} = \frac{\partial^2 u}{\partial r^2}\frac{\partial r}{\partial x} + \frac{\partial^2 u}{\partial r\,\partial s}\frac{\partial s}{\partial x} + \frac{\partial^2 u}{\partial s^2}\frac{\partial s}{\partial x} + \frac{\partial^2 u}{\partial r\,\partial s}\frac{\partial r}{\partial x} = \frac{\partial^2 u}{\partial r^2} + 2\frac{\partial^2 u}{\partial r\,\partial s} + \frac{\partial^2 u}{\partial s^2} \tag{1}$$

$$\frac{\partial u}{\partial t} = \frac{\partial u}{\partial r}\frac{\partial r}{\partial t} + \frac{\partial u}{\partial s}\frac{\partial s}{\partial t} = -c\frac{\partial u}{\partial r} + c\frac{\partial u}{\partial s}$$

$$\frac{\partial^2 u}{\partial t^2} = -c\frac{\partial^2 u}{\partial r^2}\frac{\partial r}{\partial t} - c\frac{\partial^2 u}{\partial r\,\partial s}\frac{\partial s}{\partial t} + c\frac{\partial^2 u}{\partial r\,\partial s}\frac{\partial r}{\partial t} + c\frac{\partial^2 u}{\partial s^2}\frac{\partial s}{\partial t}$$

$$= c^2\frac{\partial^2 u}{\partial r^2} - 2c^2\frac{\partial^2 u}{\partial r\,\partial s} + c^2\frac{\partial^2 u}{\partial s^2} \tag{2}$$

Substituting (1) and (2) into the wave equation $\dfrac{\partial^2 u}{\partial t^2} = c^2\dfrac{\partial^2 u}{\partial x^2}$ yields

$$\frac{\partial^2 u}{\partial r\,\partial s} = 0 \tag{3}$$

Integrating (3) first with respect to r gives

$$\partial u/\partial s = f_2'(s) \tag{4}$$

where $f_2'(s)$ is an arbitrary function of s. Integration of (4) with respect to s gives

$$u = \int f_2'(s)\,ds + f_1(r) = f_1(r) + f_2(s) \tag{5}$$

where $f_1(r)$ is an arbitrary function of r. Thus the general solution is

$$u(x,t) = f_1(x - ct) + f_2(x + ct)$$

where f_1 and f_2 are arbitrary functions.

WAVE ELEMENTS

2.7. For sinusoidal plane acoustic waves, show that the effective (root mean square) value of acoustic pressure $p_{\text{rms}} = p_{\text{peak}}/\sqrt{2}$. Find the intensity I of a plane acoustic wave having a peak acoustic pressure of 2 nt/m² at standard atmospheric pressure and temperature.

$$p_{\text{rms}} = \sqrt{\frac{1}{P}\int_0^P [p_{\text{peak}} \sin(\omega t - kx)]^2\,dt} = \sqrt{\frac{p_{\text{peak}}^2}{P}\left[\frac{1}{\omega}\left(\tfrac{1}{2}\omega t - \tfrac{1}{4}\sin 2\omega t\right)\right]_0^P}$$

Now the period $P = 2\pi/\omega$, then

$$p_{rms} = \sqrt{\tfrac{1}{2}p_{peak}^2} = p_{peak}/\sqrt{2}$$

and $I = p_{peak}^2/2\rho c = 2^2/[2(1.21)343] = 0.0024$ watt/m², where $\rho = 1.21$ kg/m³ is the density of air and $c = 343$ m/sec is the speed of sound in air.

Here we have ideal constant wave front propagation, i.e. intensity remains constant for any distance from the source because of plane acoustic waves. This is not true for spherical acoustic wave propagation.

2.8. For harmonic plane acoustic wave propagation in the positive x direction, show that particle velocity leads particle displacement by 90°. What is the phase relationship between acoustic pressure and particle displacement when the waves are traveling in the negative x direction?

For harmonic plane acoustic wave propagation in the positive x direction, particle displacement is expressed as

$$u(x,t) = Ae^{i(\omega t - kx)} \quad \text{or} \quad u(x,t) = A\cos(\omega t - kx)$$

Particle velocity
$$du/dt = i\omega Ae^{i(\omega t - kx)} = i\omega u$$

or
$$du/dt = -\omega A\sin(\omega t - kx) = \omega A\cos(\omega t - kx + 90°)$$

Thus the particle velocity du/dt leads the particle displacement u by 90°.

For harmonic acoustic wave propagation in the negative x direction,

$$u(x,t) = Ae^{i(\omega t + kx)}$$

Now acoustic pressure $p = -\rho c^2(du/dx) = -i\rho c\omega Ae^{i(\omega t + kx)} = -i\rho c\omega u$. Therefore the acoustic pressure p lags the particle displacement u by 90°.

2.9. Derive an expression for acoustic pressure p in a free progressive plane acoustic wave from measurement of particle velocity du/dt.

In the derivation of the wave equation for plane acoustic waves, the force acting is shown equal to the product of mass and acceleration, i.e.

$$-\partial p/\partial x = \rho(\partial^2 u/\partial t^2)$$

For steady state sinusoidal progressive wave motion, we can write particle displacement, velocity, and acceleration respectively as

$$u = Ae^{i(\omega t - kx)}, \quad du/dt = i\omega Ae^{i(\omega t - kx)}, \quad d^2u/dt^2 = -\omega^2 Ae^{i(\omega t - kx)} = i\omega(du/dt)$$

Substitute the above expression for the acceleration into the force equation and obtain

$$-\partial p/\partial x = \rho(i\omega)(du/dt) \quad \text{or} \quad \Delta p = -i\omega\Delta x\rho(du/dt) \text{ nt/m}^2$$

where $i = \sqrt{-1}$, ω is the frequency in rad/sec, ρ is the density in kg/m³, Δx is the particle displacement in m, and du/dt is the particle velocity in m/sec.

SPEED OF SOUND

2.10. Calculate the speed of sound in air at 20°C and standard atmospheric pressure.

$$c = \sqrt{\gamma p/\rho} = 343 \text{ m/sec}$$

where $\gamma = 1.4$ is the ratio of the specific heat of air at constant pressure to that at constant volume, $p = 1.01(10)^5$ nt/m² is the pressure, and $\rho = 1.21$ kg/m³ is the density of air.

2.11. The bulk modulus of water is $B = 2.1(10)^9$ nt/m². Find the speed of sound in water.

$$c = \sqrt{B/\rho} = \sqrt{2.1(10)^9/998} = 1450 \text{ m/sec}$$

where $\rho = 998$ kg/m³ is the density of water.

2.12. Young's modulus of copper is $12.2(10)^{10}$ nt/m², and the density of copper is 8900 kg/m³. Calculate the speed of sound in copper.

$$c = \sqrt{Y/\rho} = \sqrt{12.2(10)^{10}/8900} = 3700 \text{ m/sec}$$

2.13. Prove that the speed of sound in air is proportional to the square root of the absolute temperature.

The speed of sound in air at 0°C is given by

$$c_0 = \sqrt{\gamma p/\rho_0}$$

where γ is the ratio of the specific heat of air at constant pressure to that at constant volume, p is the effective pressure, and ρ_0 is the density at 0°C. Similarly, the speed of sound in air at t°C is

$$c_t = \sqrt{\gamma p/\rho_t}$$

where ρ_t is the density of air at t°C. But $\rho_0 = \rho_t(1+\alpha t) = \rho_t(T_t/T_0)$, where α is the coefficient of expansion of air, T_0 and T_t are absolute temperatures. Thus

$$c_t = \sqrt{\frac{\gamma p}{\rho_0/(T_t/T_0)}} = \sqrt{(\gamma p/\rho_0)(T_t/T_0)} = \sqrt{c_0^2 T_t/T_0} \quad \text{or} \quad c_t/c_0 = \sqrt{T_t/T_0}$$

INTENSITY AND ENERGY DENSITY

2.14. Derive a general expression for the intensity of harmonic progressive plane acoustic waves.

Acoustic intensity is the average rate of flow of sound energy through unit area, or the average of the instantaneous power flow through unit area. Instantaneous power per unit area is the product of instantaneous pressure p and instantaneous particle velocity v, and the average power per unit area or intensity is therefore given by

$$I = \frac{1}{P}\int_0^P pv\,dt = \frac{1}{P}\int_0^P [-\rho c\omega A \sin(\omega t - kx)][-\omega A \sin(\omega t - kx)]\,dt$$

where P is the period, ρ is the density, c is the speed of sound, $u = A\cos(\omega t - kx)$ is the harmonic progressive wave, $v = \partial u/\partial t = -\omega A \sin(\omega t - kx)$, and $p = -\rho c^2(\partial u/\partial x) = -\rho c\omega A\sin(\omega t - kx)$. Thus

$$\begin{aligned} I &= \frac{\rho c\omega^2 A^2}{P}\int_0^P [\sin(\omega t - kx)]^2\,dt \\ &= \frac{\rho c\omega^2 A^2}{P}\int_0^P (\cos^2 kx \sin^2 \omega t + \sin^2 kx \cos^2 \omega t - \tfrac{1}{2}\sin 2\omega t \sin 2kx)\,dt \\ &= \frac{\rho c\omega^2 A^2}{P}(P/2) = \tfrac{1}{2}\rho c\omega^2 A^2 \end{aligned}$$

Since $p = -\rho c\omega A\sin(\omega t - kx)$ and $p_{max} = -\rho c\omega A$, $p_{rms} = p_{max}/\sqrt{2}$, the general expression for acoustic intensity can be written as

$$I = p_{max}^2/2\rho c = p_{rms}^2/\rho c$$

2.15. Compare the intensities of sound in air and in water for (*a*) the same acoustic pressure, and (*b*) the same frequency and displacement amplitude.

(*a*) At standard atmospheric pressure and temperature, the density of air is $\rho = 1.21$ kg/m³ and the speed of sound in air is $c = 343$ m/sec. The characteristic impedance of air is $\rho c = 1.21(343) =$ 415 rayls. Similarly, the characteristic impedance of distilled water is $\rho c = 998(1480) =$ $1.48(10)^6$ rayls.

Intensity $I = p^2_{rms}/\rho c$ and so the ratio is

$$\frac{I_{air}}{I_{water}} = \frac{p^2_{rms}/(\rho c)_{air}}{p^2_{rms}/(\rho c)_{water}} = \frac{1.48(10)^6}{415} = 3560$$

This indicates that for the same acoustic pressure, the acoustic intensity in air is 3560 times that in water.

(*b*)
$$\frac{I_{water}}{I_{air}} = \frac{\frac{1}{2}(\rho c\omega^2 A^2)_{water}}{\frac{1}{2}(\rho c\omega^2 A^2)_{air}} = \frac{(\rho c)_{water}}{(\rho c)_{air}} = \frac{1.48(10)^6}{415} = 3560$$

For the same frequency and displacement amplitude, the acoustic intensity in water is 3560 times that in air.

2.16. A plane acoustic wave in air has an intensity of 10 watts/m². Calculate the force on a wall of area 10 m² due to the impact of the wave at right angles to the surface of the wall.

Acoustic intensity is defined as power per unit area, and power is the product of force and velocity. Acoustic intensity can be expressed as

$$I = pc \text{ watts/m}^2$$

where p is the acoustic pressure in nt/m², and c is the velocity of sound wave in air. Thus

$$p = I/c = 10/343 = 0.0292 \text{ nt/m}^2$$

where $c = 343$ m/sec for air at room temperature and pressure. The force on the wall is therefore

$$F = pA = (0.0292)(10) = 0.292 \text{ nt}$$

2.17. Compute the intensity and acoustic pressure of a plane acoustic wave having an intensity level of 100 db re 10^{-12} watt/m².

From the definition of sound intensity level, we have

$$\text{IL} = 10 \log I + 120 \text{ db re } 10^{-12} \text{ watt/m}^2 \quad \text{or} \quad 100 = 10 \log I + 120$$

from which $\log I = -2$ and $I = 0.01$ watt/m².

$$\text{Acoustic pressure } p = \sqrt{I\rho c} = \sqrt{0.01(1.21)343} = 2.04 \text{ nt/m}^2$$

where $\rho = 1.21$ kg/m³ is the density of air, and $c = 343$ m/sec is the speed of sound in air.

If the sound pressure level is assumed equal to the intensity level (see Problem 2.27), then

$$\text{SPL} = 20 \log p + 94 \text{ db re } 2(10)^{-5} \text{ nt/m}^2 \quad \text{or} \quad 100 = 20 \log p + 94$$

from which $\log p = 0.3$ and $p = 2.00$ nt/m².

2.18. What is the acoustic intensity in water produced by a free progressive plane acoustic wave having a sound pressure level of 100 db re 1 microbar? Find also the ratio of sound pressures produced if an identical sound wave of equal intensity is propagated through air and water.

The sound pressure level SPL $= 20 \log (p/p_0) = 20 \log (p/0.1) = 100$ db re 1 microbar $=$ 0.1 nt/m². The effective pressure of the given wave is

$$p_{rms} = 0.1 \text{ antilog } 5 = 10^4 \text{ nt/m}^2$$

Since acoustic intensity $I = p_{rms}^2/\rho c$ where $\rho = 998$ kg/m³ is the density of water, $c = 1480$ m/sec is the speed of sound in water, then

$$I = (10^4)^2/998(1480) = 77.6 \text{ watts/m}^2$$

For sound waves of equal intensities,

$$\frac{I_{water}}{I_{air}} = \frac{(p_{rms}^2/\rho c)_{water}}{(p_{rms}^2/\rho c)_{air}} = \frac{(p_{rms}^2)_{water}/1{,}480{,}000}{(p_{rms}^2)_{air}/415} = 1.0$$

where ρc is the characteristic impedance. Thus

$$p_{water}^2/1{,}480{,}000 = p_{air}^2/415 \quad \text{or} \quad p_{water}/p_{air} = 60$$

Sound pressure in water is therefore 60 times greater than sound pressure in air for waves of equal intensities.

2.19. Find the sound energy density in air and in water of a free progressive plane acoustic wave having an intensity level of 80 db re 10^{-12} watt/m².

Wave in air:

Intensity level $\text{IL} = 10 \log (I/I_0)$ where $I_0 = 10^{-12}$ watt/m² is the reference intensity. Thus $80 = 10 \log I + 120$ or $I = 10^{-4}$ watt/m². The sound energy density is

$$I/c = 10^{-4}/343 = 2.9(10)^{-7} \text{ joules/m}^3$$

where $c = 343$ m/sec is the speed of sound in air.

Wave in water:

The sound intensity is the same but the speed of sound is different. The sound energy density is therefore

$$I/c = 10^{-4}/1480 = 6.7(10)^{-8} \text{ joules/m}^3$$

where $c = 1480$ m/sec is the speed of sound in water.

2.20. Derive an expression for the sound energy density of a harmonic plane acoustic wave.

The sound energy density associated with a medium at any instant is the sum of the kinetic and potential energies per unit volume. The kinetic energy is $\frac{1}{2}\rho V\dot{x}^2$, where ρ is the average density, V the volume of the medium, $\dot{x}$ the average particle velocity over the volume. The potential energy is determined as follows:

The potential energy is equal to the work done by the sound pressure and change in volume of the medium, i.e. $W = -\int p'\,dV'$ where p' is the instantaneous pressure and V' the instantaneous volume. But $dV' = -V\,dp'/B$ where $B = \rho c^2$ is the bulk modulus of the medium. Let p_0 be the static sound pressure; then

$$W = V/B \int_{p_0}^{p'} p'\,dp' = V/2B[(p')^2 - p_0^2] \quad \text{or} \quad W = (V/2B)(2p_0 p + p^2)$$

where $p = p' - p_0$ is the excessive pressure.

For harmonic plane acoustic progressive waves, $p = \rho c\dot{x}$; hence the total sound energy is the sum of kinetic and potential energies,

$$E = \tfrac{1}{2}\rho V\dot{x}^2 + (V/2B)(2p_0 p + p^2) = V(\rho\dot{x}^2 + p_0\dot{x}/c)$$

Then the instantaneous sound energy density is

$$E_{ins} = E/V = \rho\dot{x}^2 + p_0\dot{x}/c \text{ watt-sec/m}^3$$

and the average sound energy density is therefore given by

$$E_{av} = 1/P \int_0^P (\rho\dot{x}^2 + p_0\dot{x}/c)\,dt$$

averaging over a complete cycle of period P. If $x(t) = A\cos(\omega t - kx)$, then $\dot{x} = -\omega A \sin(\omega t - kx)$, and the above expression will yield

$$E_{av} = \tfrac{1}{2}\rho\dot{x}^2 \quad \text{or} \quad \tfrac{1}{2}\rho\omega^2 A^2 \text{ watt-sec/m}^3$$

SOUND MEASUREMENTS

2.21. The power output from a loudspeaker is raised from 5 to 50 watts. What is the change in sound power level?

Sound power level is PWL $= 10 \log (W/W_0)$ db re W_0 watts, where W_0 is the reference power in watts. Thus

$$(\text{PWL})_1 = 10 \log (5/W_0) \text{ db}, \qquad (\text{PWL})_2 = 10 \log (50/W_0) \text{ db}$$

and
$$\Delta\text{PWL} = (\text{PWL})_2 - (\text{PWL})_1 = 10 \log (50/W_0) - 10 \log (5/W_0)$$
$$= 10 \log \frac{50/W_0}{5/W_0} = 10 \log 10 = 10 \text{ db}$$

Conversely, if the power output is lowered from 50 to 5 watts, the change in power level would be -10 db.

2.22. Show that the ratio of the acoustic powers of two sounds in decibels is equal to the difference of their power levels.

Let W_1 and W_2 be the acoustic powers of two sounds. The ratio of the powers is W_1/W_2, and in decibels this ratio becomes $10 \log (W_1/W_2)$ db.

Now the sound power levels are

$$(\text{PWL})_1 = 10 \log (W_1/W_0) \text{ db}, \qquad (\text{PWL})_2 = 10 \log (W_2/W_0) \text{ db}$$

where W_0 is the reference power.

The difference in sound power level is given by

$$\Delta\text{PWL} = (\text{PWL})_1 - (\text{PWL})_2 = 10 \log (W_1/W_0) - 10 \log (W_2/W_0)$$
$$= 10 \log \frac{W_1/W_0}{W_2/W_0} = 10 \log (W_1/W_2) \text{ db}$$

2.23. Determine the acoustic intensity level at a distance of 10 m from a source which radiates 1 watt of acoustic power. Use reference intensities of (*a*) 100, (*b*) 1, (*c*) 10^{-12} and (*d*) 10^{-13} watts/m^2.

The acoustic intensity level is defined as IL $= 10 \log (I/I_0)$ db re I_0 watts/m^2, where I_0 is the reference intensity.

First calculate the sound intensity at 10 m from the source:

$$\text{Power radiated } W = (\text{intensity})(\text{area}) = 4\pi r^2 I$$

(Here we assume spherical wave propagation.) Then $I = W/A = 1/4(3.14)100 = 0.00079$ watt/m^2.

(*a*) $\text{IL} = 10 \log (0.00079/100) = 10 \log 0.00079 - 10 \log 100 = -51$ db re 100 watts/m^2

(*b*) $\text{IL} = 10 \log (0.00079/1.0) = 10(-3.1) = -31$ db re 1 watt/m^2

(*c*) $\text{IL} = 10 \log (0.00079/10^{-12}) = 89$ db re 10^{-12} watt/m^2

(*d*) $\text{IL} = 10 \log (0.00079/10^{-13}) = -31 + 130 = 99$ db re 10^{-13} watt/m^2

In general, the acoustic intensity level of a sound source at a given distance is given in the number of decibels, omitting the reference intensity which is commonly accepted as 10^{-12} watt/m^2.

2.24. An air-conditioning unit operates with a sound intensity level of 73 db. If it is operated in a room with an ambient sound intensity level of 68 db, what will be the resultant intensity level?

$$(\text{IL})_1 = 10 \log (I_1/I_0) = 73 \text{ db} \quad \text{or} \quad I_1 = I_0 \text{ antilog } 7.3 = 4.77(10)^7 I_0 \text{ watts/m}^2$$
$$(\text{IL})_2 = 10 \log (I_2/I_0) = 68 \text{ db} \quad \text{or} \quad I_2 = I_0 \text{ antilog } 6.8 = 0.9(10)^7 I_0 \text{ watts/m}^2$$

The total sound intensity $I = I_1 + I_2 = 5.67(10)^7 I_0$ watts/m^2 and the resultant intensity level is

$$\text{IL} = 10 \log (I/I_0) = 10 \log 5.67(10)^7 = 73.69 \text{ db}$$

2.25. Calculate the sound pressure level for a sound wave having an effective pressure of 3.5 nt/m². Use reference pressures of (*a*) 10, (*b*) 1, (*c*) 10^{-4} and (*d*) $2(10)^{-4}$ microbars.

The sound pressure level SPL $= 20 \log(p/p_0)$ db re p_0 microbars, where 1 microbar = 0.1 nt/m².

(*a*) SPL $= 20 \log(35/10) = 10.8$ db re 10 microbars

(*b*) SPL $= 20 \log 35 = 30.8$ db re 1 microbar

(*c*) SPL $= 20 \log(35/10^{-4}) = 110.8$ db re 10^{-4} microbar

(*d*) SPL $= 20 \log(35/0.0002) = 104.8$ db re $2(10)^{-4}$ microbar

In general, reference pressure of 1 microbar is commonly used for underwater sound. For audible sound, reference pressure of 0.0002 microbar is being used.

2.26. If sound pressure is doubled, find the increase in sound pressure level.

Let p be the initial sound pressure. Then $(\text{SPL})_1 = 20 \log(p/p_0)$ db and similarly $(\text{SPL})_2 = 20 \log(2p/p_0)$ db. Thus

$$\Delta\text{SPL} = (\text{SPL})_2 - (\text{SPL})_1 = 20 \log \frac{2p/p_0}{p/p_0} = 20 \log 2 = 6 \text{ db}$$

2.27. For plane acoustic waves, express the intensity level in terms of the sound pressure level.

The intensity level is defined as IL $= 10 \log(I/I_0)$ db where I is the intensity and I_0 is the reference intensity. Now $I = p^2/\rho c$ and $I_0 = p^2/(\rho c)_0$ where $p = p_{\text{rms}}$ = effective pressure. Thus

$$\begin{aligned} \text{IL} &= 10 \log I - 10 \log I_0 = 10 \log(p^2/\rho c) - 10 \log(p^2/\rho_0 c_0) \\ &= 10 \log p^2 - 10 \log \rho c - 10 \log p_0^2 + 10 \log(\rho c)_0 \\ &= 10 \log(p^2/p_0^2) + 10 \log(\rho_0 c_0/\rho c) = \text{SPL} + 10 \log(\rho_0 c_0/\rho c) \end{aligned}$$

If the measured characteristic impedance ρc is equal to the reference characteristic impedance $(\rho c)_0$ (e.g. measurements are made in the same medium under identical environment), intensity level IL will be equal to the sound pressure level SPL.

2.28. Two sound sources S_1 and S_2 are radiating sound waves of different frequencies. If their sound pressure levels recorded at position S as shown in Fig. 2-3 are 75 and 80 db respectively, find the total sound pressure level at S due to the two sources together.

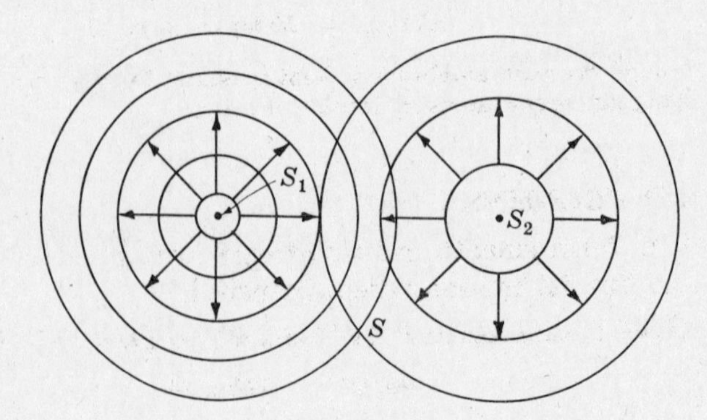

Fig. 2-3

By definition, sound pressure level SPL $= 20 \log(p/p_0)$ db. Then

$$(\text{SPL})_1 = 20 \log(p_1/p_0) = 75 \quad \text{or} \quad p_1 = 5.6 \times 10^3 p_0 \text{ nt/m}^2$$

$$(\text{SPL})_2 = 20 \log(p_2/p_0) = 80 \quad \text{or} \quad p_2 = 10^4 p_0 \text{ nt/m}^2$$

Thus the total sound pressure at S is $p = p_1 + p_2 = 15.6 \times 10^3 p_0$ nt/m² and the total sound pressure level is

$$(\text{SPL})_{\text{total}} = 20 \log (p/p_0) = 20 \log (15{,}600 p_0/p_0) = 20(4.195) = 83.9 \text{ db}$$

The total sound pressure level is not at all equal to the arithmetic sum of the individual sound pressure levels. It is not necessary to determine the actual sound pressure in the computation of total sound pressure level.

On the other hand, if the two sound sources are radiating sound waves of the same frequency, the total sound pressure level at S will be different from the one calculated above.

$$(\text{SPL})_1 = 20 \log p_1 + 94 = 75 \quad \text{or} \quad p_1 = 0.11 \text{ nt/m}^2$$

$$(\text{SPL})_2 = 20 \log p_2 + 94 = 80 \quad \text{or} \quad p_2 = 0.2 \text{ nt/m}^2$$

and the total sound pressure $p = \sqrt{p_1^2 + p_2^2} = \sqrt{(0.11)^2 + (0.2)^2} = 0.23$ nt/m². Thus

$$(\text{SPL})_{\text{total}} = 20 \log 0.23 + 94 = 92.7 \text{ db}$$

2.29. The pressure amplitude of a plane acoustic wave is kept constant while the temperature increases from 0°C to 20°C. Find (*a*) the percent change in sound intensity, (*b*) the change in sound intensity level, and (*c*) the change in sound pressure level.

(*a*) Sound intensity is $I = p^2/2\rho c$, where p is the pressure amplitude in nt/m², ρ is the density of air in kg/m³, and c is the speed of sound in air in m/sec.

Let the sound intensity at 0°C be $I_{(0)} = p^2/2(1.3)332 = p^2/862$ watts/m² and the sound intensity at 20°C be $I_{(20)} = p^2/2(1.2)343 = p^2/824$ watts/m². Then

$$\Delta I = I_{(20)} - I_{(0)} = p^2/824 - p^2/862$$

where p is the constant pressure amplitude. Hence the percent change in sound intensity is given by

$$\frac{\Delta I}{I_{(0)}} = \frac{p^2/824 - p^2/862}{p^2/862} = 0.05 \text{ or } 5\%$$

(*b*) The sound intensity level is $\text{IL} = 10 \log I - 10 \log I_0$ db where I is the sound intensity and I_0 is the reference intensity. At 0°C, we have

$$\text{IL}_{(0)} = 10 \log (p^2/862) - 10 \log I_0 \text{ db}$$

and at 20°C,

$$\text{IL}_{(20)} = 10 \log (p^2/824) - 10 \log I_0 \text{ db}$$

Then

$$\text{IL}_{(20)} - \text{IL}_{(0)} = 10 \log 862 - 10 \log 824 = 10(2.936 - 2.916) = 0.2 \text{ db}$$

(*c*) The sound pressure level is $\text{SPL} = 20 \log (p/p_0)$ db where p is the pressure amplitude and p_0 is the reference sound pressure amplitude. At 0°C, we have

$$\text{SPL}_{(0)} = 20 \log (p_{(0)}/p_0) \text{ db}$$

and at 20°C,

$$\text{SPL}_{(20)} = 20 \log (p_{(20)}/p_0) \text{ db}$$

But since the sound pressure amplitude is kept constant, i.e. $p_{(0)} = p_{(20)} = p$, $\text{SPL}_{(0)} = \text{SPL}_{(20)}$. We find no change in sound pressure level.

RESONANCE OF AIR COLUMNS

2.30. A rigid tube of uniform smooth cross-sectional area is closed at both ends. If the tube contains air, find its motion when disturbed.

The one-dimensional wave equation for harmonic progressive plane acoustic wave is (see Problem 2.1)

$$\partial^2 u/\partial t^2 = c^2(\partial^2 u/\partial x^2) \tag{1}$$

where $c = \sqrt{B/\rho}$ is the speed of sound, B the bulk modulus and ρ the density. The general solution is

$$u(x, t) = \sum_{i=1,2,\ldots}^{\infty} (A_i \cos p_i t + B_i \sin p_i t)\left(C_i \cos \frac{p_i}{c} x + D_i \sin \frac{p_i}{c} x\right) \tag{2}$$

where A_i, B_i are arbitrary constants to be determined by initial conditions, C_i, D_i are arbitrary constants to be determined by boundary conditions, and p_i are the natural frequencies of the system.

The boundary conditions are $u(0,t) = 0$ and $u(L,t) = 0$ where L is the length of the tube. From the first boundary condition,

$$C_i(A_i \cos p_it + B_i \sin p_it) = 0 \quad \text{or} \quad C_i = 0$$

and from the second boundary condition,

$$D_i \sin(p_iL/c)(A_i \cos p_it + B_i \sin p_it) = 0$$

Because D_i cannot equal zero all the time, $\sin(p_iL/c)$ must equal zero. Therefore $\sin(p_iL/c) = 0$ and $p_i = i\pi c/L$, $i = 1, 2, \ldots$ are the natural frequencies of the system.

The normal modes of vibration are given by $X_i(x) = \sin(i\pi x/L)$ and the general motion of the air inside the tube is

$$u(x,t) = \sum_{i=1,2,\ldots}^{\infty} \sin(i\pi x/L)\,(A_i' \cos p_it + B_i' \sin p_it) \tag{3}$$

where $A_i' = A_iD_i$, $B_i' = B_iD_i$.

The analysis and results of this problem are exactly the same as for the transverse vibration of a uniform string fixed at both ends and the longitudinal vibration of a uniform bar fixed at both ends. (See Problems 1.16 and 1.22.) This is because their differential equations of motion are mathematically similar; they are thus equivalent to one another. As a result, there are almost complete analogies between the wave motion of uniform strings and plane acoustic waves. The analogy between longitudinal vibration of a bar and plane acoustic waves in air columns is almost complete except that the bar is not a three-dimensionally infinite solid of the same physical constituent. As the outer surface of the bar is free, any longitudinal elongation of the bar will result in a transverse linear dilatation $-\mu\epsilon$, where μ is the Poisson's ratio for the material of the bar. This will in turn affect the value for Young's modulus which is one of the two factors governing the speed of wave propagation.

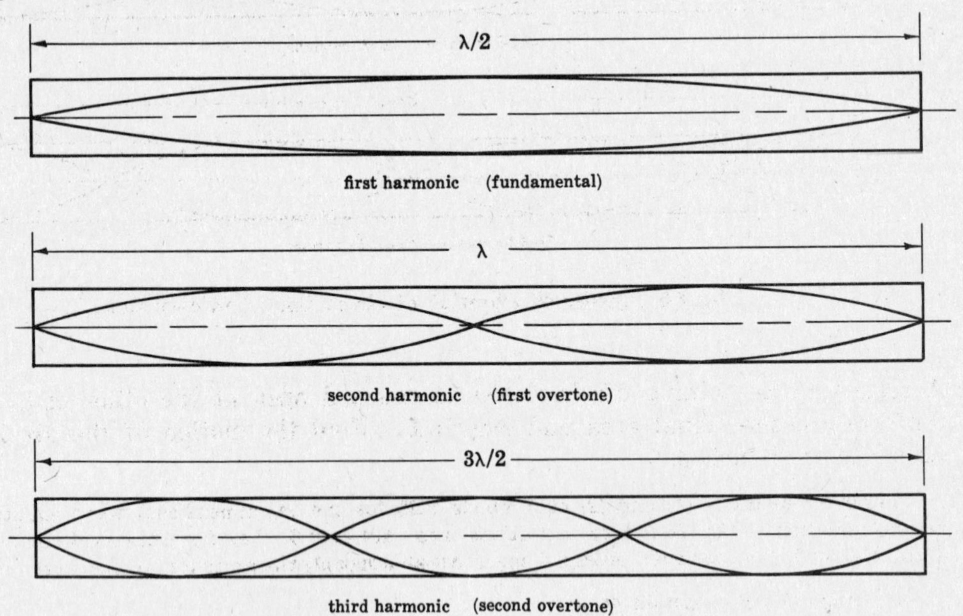

Fig. 2-4. Modes of vibration of air column in a closed tube.

2.31. A rigid tube of uniform cross-sectional area and length L is opened at both ends. Investigate the motion of plane acoustic waves inside the tube.

Refer to equations (1) and (2) of Problem 2.30 for the one-dimensional wave equation and its general solution.

The boundary conditions are $du/dx = -dp/B = 0$ at $x = 0$ and $x = L$, i.e. the acoustic pressure at both ends of the tube must equal atmospheric pressure. From the first boundary condition,

$$D_i(p_i/c)(A_i \cos p_it + B_i \sin p_it) = 0 \quad \text{or} \quad D_i = 0$$

and from the second boundary condition,

$$-C_i(p_i/c)\sin(p_iL/c)\,(A_i\cos p_it + B_i\sin p_it) = 0 \quad\text{or}\quad \sin(p_iL/c) = 0$$

Hence $p_i = i\pi c/L,\ i = 1, 2, \ldots$.

The motion for plane acoustic waves inside a tube open at both ends is therefore given by

$$u(x,t) = \sum_{i=1,2,\ldots}^{\infty} \cos(i\pi x/c)\,(A'_i\cos p_it + B'_i\sin p_it)$$

where $A'_i = A_iC_i$ and $B'_i = B_iC_i$ are arbitrary constants to be evaluated by initial conditions, and p_i are the natural frequencies.

The motion is equivalent to the free longitudinal vibration of a uniform bar free at both ends. (See Problem 1.23.)

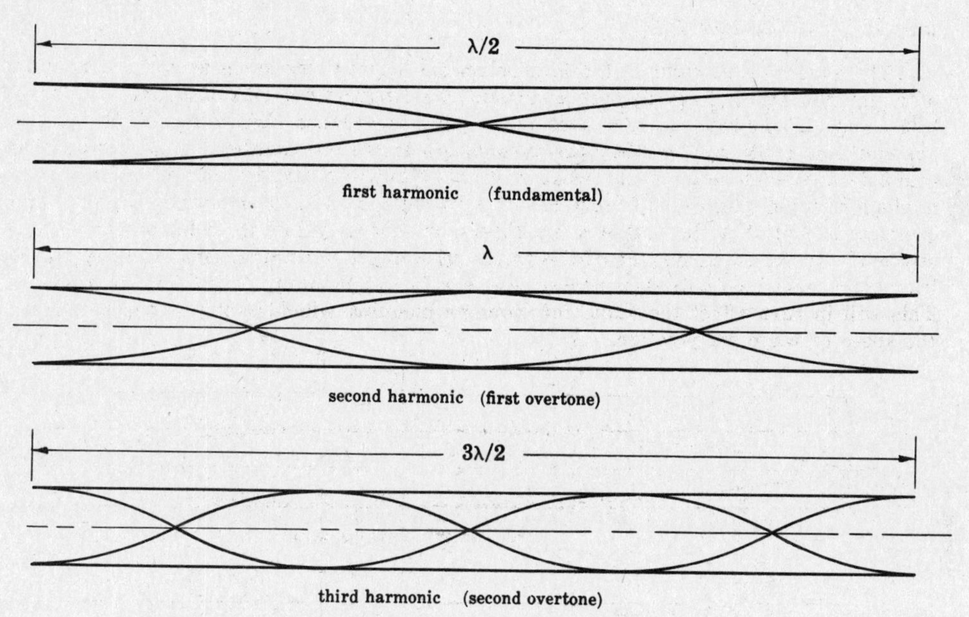

Fig. 2-5. Modes of vibration of air column in an open tube.

2.32. A rigid tube containing air is closed at one and open at the other end. It has a uniform cross-sectional area and length L. Find the motion of the air inside the tube if it is disturbed.

Refer to equations (*1*) and (*2*) of Problem 2.30 for the one-dimensional wave equation and its general solution. The boundary conditions are $u(0,t) = 0$, i.e. no motion at the closed end. $du(L,t)/dx = 0$, i.e. acoustic pressure must equal atmospheric pressure at the open end.

From boundary condition at $x = 0$,

$$C_i(A_i\cos p_it + B_i\sin p_it) = 0 \quad\text{or}\quad C_i = 0$$

and from boundary condition at $x = L$,

$$D_i(p_i/c)\cos(p_iL/c)\,(A_i\cos p_it + B_i\sin p_it) = 0 \quad\text{or}\quad \cos(p_iL/c) = 0$$

Hence $p_i = i\pi c/2L,\ i = 1, 3, \ldots$.

The motion of air inside a tube open at one end and closed at the other is therefore given by

$$u(x,t) = \sum_{i=1,3,\ldots}^{\infty} \sin\frac{i\pi x}{2L}\,(A'_i\cos p_it + B'_i\sin p_it)$$

where $A'_i = A_iD_i$ and $B'_i = B_iD_i$ are arbitrary constants to be evaluated by initial conditions, and p_i are the natural frequencies.

The motion is equivalent to the free longitudinal vibration of a uniform bar fixed at one end and free at the other. (See Problem 1.24.)

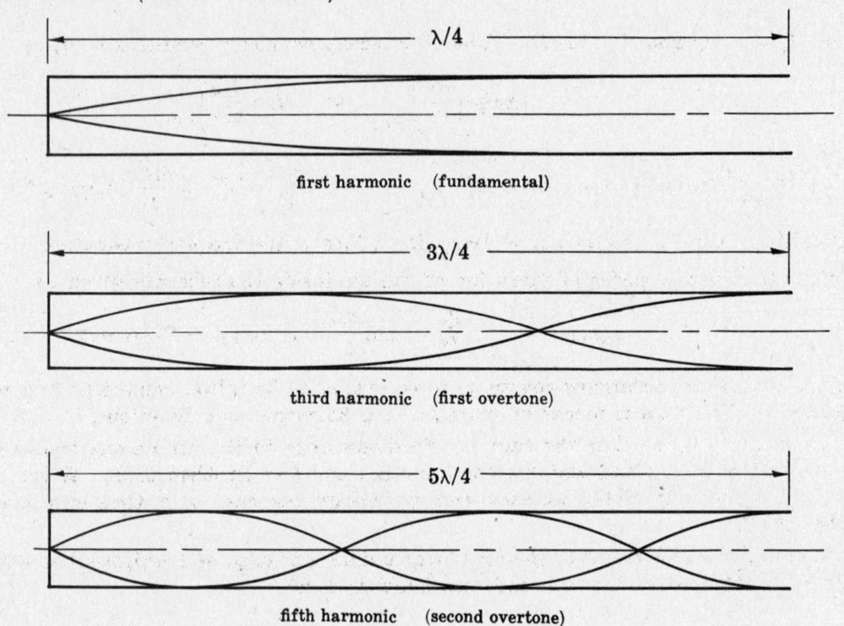

Fig. 2-6. Modes of vibration of air column in a tube open at one end and closed at the other end.

2.33. A rigid tube of uniform cross-sectional area is closed at one end by a rigid boundary and at the other end by a mass M_0 free to move along the tube as shown in Fig. 2-7. If the tube contains air, find the normal modes of vibration of the air inside the tube.

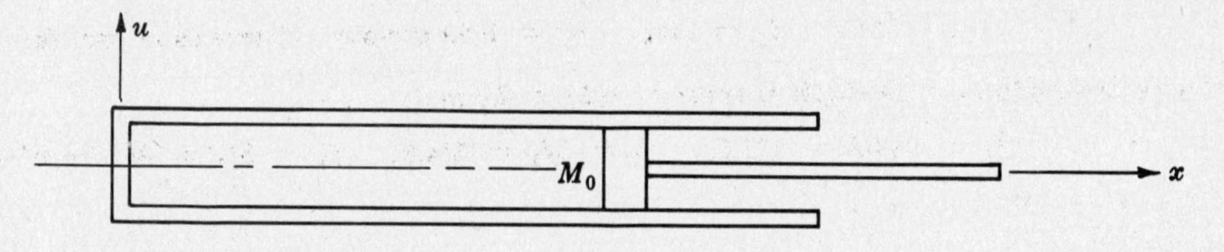

Fig. 2-7

Refer to equations (*1*) and (*2*) of Problem 2.30 for the one-dimensional wave equation and its general solution.

Let the fixed boundary be taken as $x = 0$, and the normal equilibrium position of the movable mass M_0 be at $x = L$. The boundary conditions are

$$u(0,t) = 0, \quad A(p - p_0) = M_0\left[\frac{\partial^2 u}{\partial t^2}\right]_{x=L}$$

i.e. at $x = 0$ the wave motion of the air is zero, and at $x = L$ the force on the surface of the mass M_0 due to the excessive pressure inside the tube causes the acceleration of the mass M_0. A is the area of the surface of the mass M_0.

From the first boundary condition,

$$C_i(A_i \cos p_i t + B_i \sin p_i t) = 0 \quad \text{or} \quad C_i = 0$$

and from the second boundary condition,

$$p - p_0 = dp = -B(du/dx) = -\rho c^2(du/dx)$$

where $\left[\dfrac{du}{dx}\right]_{x=L} = \dfrac{p_i}{c}(A_i \cos p_i t + B_i \sin p_i t)\left(D_i \cos\dfrac{p_i L}{c}\right)$. Then

$$\left[\frac{d^2u}{dt^2}\right]_{x=L} = -p_i^2 \sin\frac{p_iL}{c}(A_i' \cos p_it + B_i' \sin p_it)$$

where $A_i' = A_iD_i$ and $B_i' = B_iD_i$. Thus the second boundary condition becomes

$$-Ac^2\rho\left(\frac{du}{dx}\right)_{x=L} = M_0\left[\frac{d^2u}{dt^2}\right]_{x=L}$$

or
$$-Ac^2\rho\left[\frac{p_i}{c}\cos\frac{p_iL}{c}(A_i' \cos p_it + B_i' \sin p_it)\right] = M_0\left[-p_i^2 \sin\frac{p_iL}{c}(A_i' \cos p_it + B_i' \sin p_it)\right]$$

and finally we obtain $\tan(p_iL/c) = Ac\rho/p_iM_0$ which is the frequency equation.

Thus the normal modes of vibration of the air inside the tube are given by

$$u(x,t) = \sum_{i=1,2,\ldots}^{\infty} \sin\frac{p_i}{c}x\,(A_i' \cos p_it + B_i' \sin p_it)$$

where A_i' and B_i' are arbitrary constants to be evaluated by initial conditions and p_i are the natural frequencies. The normal modes of vibration are harmonic sine functions.

When $M_0 = 0$, so that the tube is effectively open to the air at one end, we obtain the case of a tube closed at one end and open at the other end (see Problem 2.32). When $M_0 = \infty$, so that the tube is effectively closed at each end, we obtain the case of a tube closed at both ends (see Problem 2.30).

The motion is equivalent to the free longitudinal vibration of a uniform bar with a concentrated heavy mass attached at the free end (see Problem 1.25).

2.34. Calculate the three lowest frequencies of (*a*) closed tube, (*b*) open tube and (*c*) closed-open tube, each of length 0.5 m and at standard atmospheric pressure and temperature.

(*a*) Wavelength λ_1 = 2(length of tube) = 2(.5) = 1.0 m

$$f_1 = c/\lambda_1 = 343/1.0 = 343, \quad f_2 = 2f_1 = 686, \quad f_3 = 3f_1 = 1029 \text{ cyc/sec}$$

(*b*) Wavelength λ_1 = 2(length of tube) = 1.0 m

$$f_1 = 343, \quad f_2 = 686, \quad f_3 = 1029 \text{ cyc/sec} \quad \text{(same as in part (a))}$$

(*c*) Wavelength λ_1 = 4(length of tube) = 4(.5) = 2.0 m

$$f_1 = 343/2 = 171.5, \quad f_2 = 3f_1 = 514.5, \quad f_3 = 5f_1 = 857.5 \text{ cyc/sec}$$

2.35. A resonance tube (a tube open at one end and closed at the other) is employed to find the frequency of a tuning fork. If resonance is obtained when the length of air column is 0.52 and 2.25 m, what is the frequency of the tuning fork? What is the end correction factor for this resonance tube?

Assume the two measurements of air column represent the shortest and the next shortest lengths for resonance as shown in Fig. 2-8. The lengths of air column plus the same end correction factor e is equal to a quarter and three quarters of a wavelength respectively; or the difference of their sums equals one half wavelength, i.e.

$$(2.25 + e) - (0.52 + e) = \tfrac{1}{2}\lambda \quad \text{or} \quad \lambda = 3.46 \text{ m}$$

Since $\lambda = cf$, where λ is the wavelength, $c = 343$ m/sec is the speed of sound in air and f is the frequency of the tuning fork, then

$$f = c/\lambda = 343/3.46 = 99 \text{ cyc/sec}$$

To find the end correction factor, we write

$$0.52 + e = \tfrac{1}{4}\lambda, \quad 2.25 + e = 3\lambda/4$$

from which $e = 0.23/0.67 = 0.34$ m.

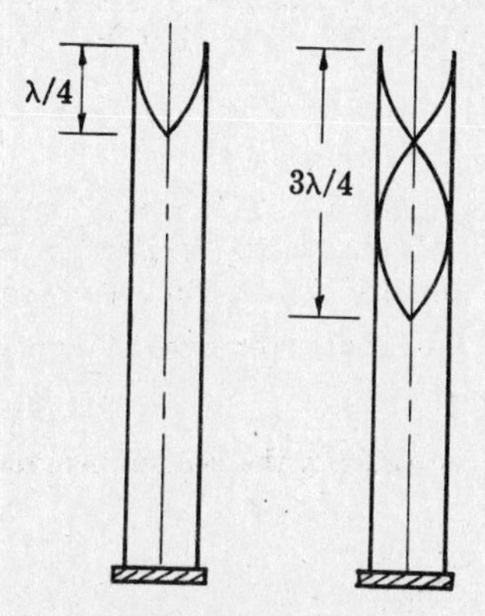

Fig. 2-8

2.36. In order to determine the speed of sound at room temperature, a resonance tube is used. A tuning fork of frequency $f_1 = 200$ cyc/sec causes it to resonate when the water level is 0.344 m below the reference mark. A second tuning fork of frequency $f_2 = 400$ cyc/sec obtains resonance when the water level is 0.136 m below the reference mark. What is the speed of sound in air?

As shown in Problem 2.32, the shortest length for resonance for a tube open at one end and closed at the other is equal to one quarter wavelength. Thus

$$L + 0.344 = \lambda_1/4, \qquad L + 0.136 = \lambda_2/4$$

where L is the distance from the open end of the tube to the reference mark, $\lambda_1 = c/f_1 = c/200$ and $\lambda_2 = c/f_2 = c/400$ are the wavelengths and c is the speed of sound in air at room temperature. Substituting these values in the above equations and solving, we obtain $c = 334$ m/sec.

2.37. An air column 0.8 m long resonates in a closed cylindrical tube of diameter 0.1 m with an unmarked tuning fork. Calculate the frequency of vibration of the unmarked tuning fork.

Resonance of air column is an exchange of energy of vibration between a tuning fork and a closed air column whose natural frequency can be adjusted to that of the tuning fork. This is also the maximum acoustic response obtainable.

From Problem 2.32, for a closed tube the wave length $\lambda = 4L$ where L is the effective length of the resonant air column. The effective length of the resonant air column is equal to the actual length of the air column plus a correction. This correction, found by experiments to be equal to $0.3d_0$ where d_0 is the diameter of the tube, is due to the spherical spread of the reflected plane acoustic waves at the open end of the tube. Thus we have

$$L = 0.8 + (0.3)(0.1) = 0.83 \text{ m} \qquad \text{and} \qquad \lambda = 4L = 3.32 \text{ m}$$

Now $f = c/\lambda$ where $c = 343$ m/sec is the speed of sound in air. Thus

$$f = 343/3.32 = 103 \text{ cyc/sec}$$

DOPPLER EFFECT

2.38. Develop an expression for the Doppler effect, i.e. the apparent change in frequency due to relative motion of the sound-producing source and the sound receiver.

We have previously shown that the speed at which sound waves propagate in a medium is independent of the source producing it. If the source is moving relative to the medium, the speed of sound is unchanged, but the wavelength and the frequency as observed by a stationary receiver will be changed.

For example, take a square wave whose source is moving toward the stationary receiver R with velocity u as shown in Fig. 2-9.

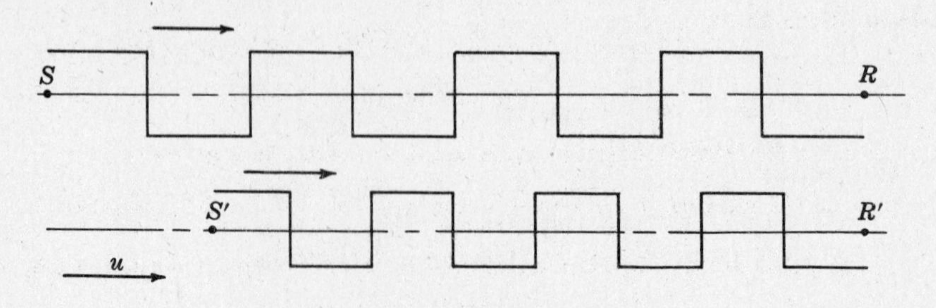

Fig. 2-9

First assume the source S is fixed. The sound waves will fill the distance SR between the fixed source S and stationary receiver R in a certain time Δt. Now let the source S move toward the receiver R with velocity u. Then in the same time interval Δt, the same sound waves will be compressed into the distance $S'R' = u\,\Delta t$ which is the distance covered by the source S in Δt. Now $SR - S'R' = u\,\Delta t$ or $\lambda f\,\Delta t - \lambda' f\,\Delta t = u\,\Delta t$, from which $\lambda' = (f\lambda - u)/f$ where f is the frequency of the emitted sound waves, λ the original wavelength and λ' the apparent wavelength. Since $c = f\lambda = f'\lambda'$, we obtain

$$f' = f\lambda/\lambda' = \frac{c}{c-u}f \tag{1}$$

Similarly, if the source is fixed while the receiver is moving in a straight line with velocity v, the apparent frequency is given by

$$f' = \frac{c-v}{c}f \tag{2}$$

When both the source and receiver are moving along the same straight line with velocities u and v respectively, the apparent frequency becomes

$$f' = \frac{c-v}{c-u}f \tag{3}$$

For general plane motion of the source and receiver relative to the medium as shown in Fig. 2-10, the apparent frequency observed by the receiver is

$$f' = \frac{c - v\cos(\gamma - \beta)}{c - u\cos\alpha}f \tag{4}$$

Expression (*4*) will reduce to (*1*), (*2*) and (*3*) under identical conditions.

If the medium through which sound waves travel moves with respect to some inertial reference with velocity w, expression (*4*) becomes

$$f' = \frac{c - v + w}{c - u + w}f \tag{5}$$

where velocities u, v and w are in the x direction.

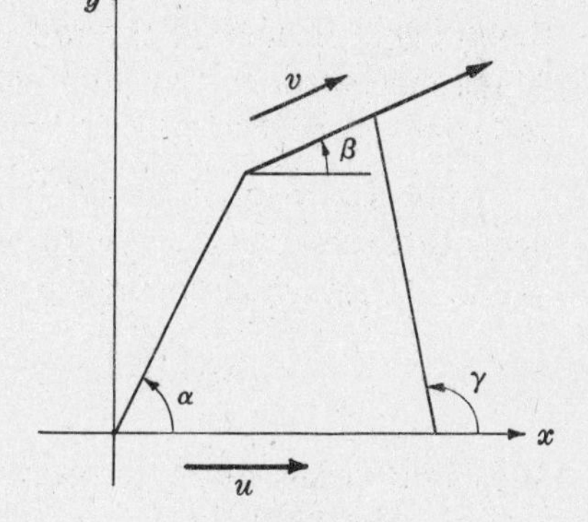

Fig. 2-10

To summarize, we have

(1) If the speed of the source $u = c$ while the receiver is at rest, $f' = \infty$, and all the sound waves travel with the source and reach the stationary receiver together. If $u > c$, the sound waves emitted are being received in the reversed order. If $u \gg c$ and the receiver is stationary, the Doppler effect develops into what is commonly known as *sonic boom.* The boom is heard on the ground when an aircraft in the vicinity exceeds the speed of sound.

(2) If the receiver has the same speed as sound waves, i.e. $v = c$, the apparent frequency f' is zero. If $v = 2c$ and source is fixed, equation (*2*) gives $f' = -f$, which indicates that the receiver will hear the sounds in correct time and tune but backward. If $v > c$ and the source is stationary, $f' \to -\infty$; this means that sound waves produced after the motion of the receiver has begun will never reach the receiver (the person does not hear anything). But for sound waves propagated before the motion of the receiver, he will gradually overtake the sound waves and hear them in the reverse of the natural order. Finally, if $v \to -\infty$, $f' \to \infty$ and the receiver is approaching the source with great speed.

(3) If the medium in which sound waves are propagated is moving with velocity w with respect to some inertial reference, this will be the same as if the medium were at rest while the source and receiver have a common velocity w relative to the medium. If $u = v$, then $f' = f$. This implies the velocity of the medium has no effect at all on the observed frequency.

(4) From equation (*2*), f' will be greater than f when source and receiver approach each other; and f' will be less than f when source and receiver separate from each other. This explains the fact that the whistle of a locomotive is heard high as it approaches, and low as it moves away from a stationary observer at the railway station, changing rather abruptly at the moment of passage. If the relative velocity is not in a straight line joining the source and receiver, the change in apparent frequency is more gradual, from $cf/(c-u)$ to $(c-u)f/c$.

2.39. An automobile emitting sound at a frequency of 100 cyc/sec moves away from a stationary observer towards a rigid flat wall with velocity of 10 m/sec. How many beats/sec will be heard by the observer?

The stationary observer hears sound of apparent frequency f_1' from the moving source directly, and also sound of apparent frequency f_2' from the waves reflected by the wall. Now

$$f_1' = cf/(c-v) = 343(100)/(343-10) = 103 \text{ cyc/sec}$$

and

$$f_2' = cf/(c+v) = 343(100)/(343+10) = 97.2 \text{ cyc/sec}$$

where $c = 343$ m/sec is the speed of sound in air, $v = 10$ m/sec is the velocity of the source and $f = 100$ cyc/sec is the frequency of the source.

The beat frequency is $f_{\text{beat}} = f_1' - f_2' = 103 - 97.2 = 5.8$ beats/sec.

2.40. Train A travels at 50 m/sec in still air while its whistle emits sound of frequency 600 cyc/sec. (*a*) What are the frequencies of the emitted sound observed by a stationary receiver in front of and behind the train? (*b*) Another train B is passing train A at 100 m/sec. What are the frequencies of the emitted sound from the whistle of train A as observed by passengers in train B before and after they pass train A? (*c*) For a wind velocity of 20 m/sec in the direction of the motion of the trains, calculate the results of parts (*a*) and (*b*).

(*a*) For a moving source and a stationary receiver, the apparent frequency as given by equation (*1*) of Problem 2.38 is

$$f'_{\text{front}} = \frac{c}{c-u}f = \frac{343}{343-50}600 = 700 \text{ cyc/sec}$$

i.e. the apparent frequency will be greater than the actual frequency when source and receiver approach each other.

$$f'_{\text{behind}} = \frac{343}{343-(-50)}600 = 520 \text{ cyc/sec}$$

i.e. the apparent frequency will be less than the actual frequency when source and receiver separate from each other.

Thus the whistle of the train is heard high ($f'_{\text{front}} = 700$ cyc/sec) as it approaches and low ($f'_{\text{behind}} = 520$ cyc/sec) as it moves away from a stationary observer at the railway station, changing rather abruptly (from 700 to 520 cyc/sec) at the moment of passage.

(*b*) For both moving source and receiver, the apparent frequencies are given by equation (*3*) of Problem 2.38,

$$f'_{\text{before}} = \frac{c-v}{c-u}f = \frac{343-100}{343-50}600 = 497 \text{ cyc/sec}$$

$$f'_{\text{after}} = \frac{343-100}{343-(-50)}600 = 370 \text{ cyc/sec}$$

(*c*) When the air moves, the apparent frequencies are given by equation (*5*) of Problem 2.38. Thus

(*a*) $$f'_{\text{front}} = \frac{343+20}{343-50+20}600 = 697 \text{ cyc/sec}$$

$$f'_{\text{behind}} = \frac{343-20}{343+50-20}600 = 521 \text{ cyc/sec}$$

(*b*) $$f'_{\text{front}} = \frac{343-100+20}{343-50+20}600 = 503 \text{ cyc/sec}$$

$$f'_{\text{behind}} = \frac{343-100+20}{343+50-20}600 = 360 \text{ cyc/sec}$$

2.41. Considering the same relative velocity in the Doppler effect, we obtain different apparent frequencies according as the source or the observer is in motion relative to the medium. Prove that this statement is correct.

Let the given relative velocity of approach be w. If the observer is approaching the stationary source, we obtain

$$f_1' = \frac{c+w}{c} f \qquad (1)$$

where f' is the apparent frequency, c the speed of sound and f the actual frequency of the source.

If the source is approaching the stationary observer,

$$f_2' = \frac{c}{c-w} f \qquad (2)$$

Thus

$$f_1'/f_2' = \left(\frac{c+w}{c} f\right) \Big/ \left(\frac{c}{c-w} f\right) = 1 - w^2/c^2 \qquad (3)$$

Equation (3) shows that unless the relative velocity of approach w is equal to the speed of sound c, the two apparent frequencies will not be the same.

If the observer is moving away from the stationary source with the same velocity w,

$$f_3' = \frac{c-w}{c} f \qquad (4)$$

and if the source is also moving away from the stationary observer with velocity w,

$$f_4' = \frac{c}{c+w} f \qquad (5)$$

and

$$f_3'/f_4' = 1 - w^2/c^2 \quad \text{(as in (3))} \qquad (6)$$

Supplementary Problems

WAVE EQUATION

2.42. Prove that $u(x,t) = A(ct-x)^{-B(ct-x)}$ is a possible solution for the one-dimensional wave equation.

2.43. Use the Fourier transform to obtain the solution for the one-dimensional wave equation.

2.44. Show that the one-dimensional wave equation may be expressed in polar coordinates as

$$\frac{1}{c^2}\frac{\partial^2 u}{\partial t^2} = \frac{1}{r}\frac{\partial}{\partial r}\left(r\frac{\partial u}{\partial r}\right) + \frac{1}{r^2}\frac{\partial^2 u}{\partial \theta^2}$$

2.45. Prove that the following expression is a possible general solution for the one-dimensional wave equation.

$$u(x,t) = \sum_{i=0,1,2,\ldots}^{\infty} A_i \cos(ix + \theta_i)\, e^{-i^2c^2t}$$

2.46. For one-dimensional wave propagation, find the initial conditions such as to cause only a wave traveling in the negative x direction. *Ans.* $u(x,0) = 0$, $\dot{u}(x,0) = c\,du/dx$

WAVE ELEMENTS

2.47. Show that the maximum particle displacement and maximum pressure at a given point do not occur simultaneously in a sound wave.

2.48. Show that the kinetic and potential energies of a free progressive plane acoustic wave are equal.

2.49. Show that the kinetic and potential energies of stationary sound waves in a rectangular room have a constant sum.

2.50. The pressure amplitude of a plane acoustic wave is kept constant while the temperature rises from 0°C to 80°C. Find the percent change in sound intensity and the intensity level.
Ans. 14%, 0.7 db

SPEED OF SOUND

2.51. Find the speed of sound wave propagation in an aluminum bar. *Ans.* $c = 5100$ m/sec

2.52. The planet Jupiter has an atmosphere of methane at a temperature of −130°C. Find the speed of sound there. *Ans.* $c = 310$ m/sec

2.53. A blow is made by a hammer on a steel rail 1 km from a listener who puts one ear to the rail and hears two sounds. Calculate the time interval between the arrivals of the sounds.
Ans. $t = 2.85$ sec

ACOUSTIC INTENSITY AND ENERGY DENSITY

2.54. Prove that intensity at any distance from the sound source for a one-dimensional cylindrical wave is inversely proportional to the first power of the radius. (A *one-dimensional cylindrical wave* is a wave radiated outward from the longitudinal axis of a long cylinder expanding and contracting radially.)

2.55. Show that $I = 2\pi^2 f^2 A^2 \rho c$ watts/m² is a correct expression for acoustic intensity of a plane wave.

2.56. Compute the intensity of a plane acoustic wave in air at standard atmospheric pressure and temperature if its frequency is 1000 cyc/sec and its displacement amplitude is 10^{-5} m.
Ans. $I = 0.82$ watt/m²

2.57. Show that the average sound energy density for a standing wave is twice that for a free progressive plane wave and is equal to $p^2/\rho c$.

SPECIFIC ACOUSTIC IMPEDANCE

2.58. Calculate the characteristic impedances of hydrogen at 0°C and steam at 100°C.
Ans. 114, 242 rayls

2.59. Prove that the characteristic impedance of a gas is inversely proportional to the square root of its absolute temperature.

SOUND MEASUREMENTS

2.60. Two electric motors have intensity levels of 58 and 60 db respectively. Find the total sound intensity level if both motors run simultaneously. *Ans.* 62.1 db

2.61. What will be the total sound pressure level of two typewriters if each has sound pressure level 70 db? *Ans.* 76 db

2.62. The sound pressure levels of three machines are respectively 90, 93 and 95 db. Determine the total sound pressure level if all the machines are turned on. *Ans.* 97.8 db

2.63. At standard atmospheric pressure and temperature, show that SPL = IL + 0.2 db.

2.64. What is the power level of 0.02 watts of power? *Ans.* PWL = 103 db

2.65. The power levels of two engines are 90 and 100 db respectively. Find the combined power level.
Ans. 100.4 db

RESONANCE OF AIR COLUMNS

2.66. If two parallel reflecting surfaces are 10 m apart, find the lowest frequency for resonant standing waves that can exist between the surfaces. *Ans.* 172 cyc/sec

2.67. A resonance box is to be made for use with a tuning fork of frequency 472 cyc/sec. Find the shortest length of the box if it is closed at one end. *Ans.* 0.18 m

2.68. A vertical tube of length 5 m is filled with water. A tuning fork of frequency 589 cyc/sec is held over the open top end of the tube while water is running out gradually from the bottom of the tube. Find the maximum number of times that resonance can occur. *Ans.* 3 times

2.69. A closed tube of length 0.25 m and an open tube of length 0.3 m, both made of the same material and same diameter, are each sounding its first overtone. What is the end correction for these tubes? *Ans.* $e = 0.05$ m

2.70. Show that $f_i = (2i-1)f_1$, $i = 1, 2, \ldots,$ where f_1 is the fundamental frequency for resonant tubes open at one end.

2.71. A cylindrical tube of length 0.2 m and closed at one end is found to be at resonance when a tuning fork of frequency 900 cyc/sec is sounded over the open end. Find the end correction.
Ans. $e = 0.036$ m

DOPPLER EFFECT

2.72. An automobile traveling at 50 m/sec emits sound at a frequency of 450 cyc/sec. Determine the apparent frequency as the automobile is approaching a stationary observer.
Ans. $f' = 526$ cyc/sec

2.73. The frequency of a car is observed to drop from 272 to 256 cyc/sec as the car passes an observation post. What is the speed of the car? *Ans.* 23 mph

2.74. A locomotive is passing by a stationary observer at a railway station with speed v, and is sounding a whistle of frequency f. Determine the change in pitch heard by the observer.
Ans. $f = 2cfv/(c^2 - v^2)$

2.75. Two observers A and B carry identical sound sources of frequency 1000 cyc/sec. If A is stationary while B moves away from A at a speed of 10 m/sec, how many beats/sec are heard by A and B?
Ans. A, 2.8; B, 3.0 beats/sec

Chapter 3

Spherical Acoustic Waves

NOMENCLATURE

a = radius, m
A = area, m^2
B = bulk modulus, nt/m^2
c = speed of sound in air, m/sec
D = directivity factor
d_r = directivity index, db
D_r = directivity ratio
E_d = energy density, joules/m^3
f = frequency, cyc/sec
I = acoustic intensity, watts/m^2
J_1 = Bessel function of the first kind of order one
k = wave number; spring constant, nt/m
k_i = constant
KE = kinetic energy, joules
m = mass, kg
p = acoustic pressure, nt/m^2
P = period, sec
PE = potential energy, joules
Q = source strength, m^3/sec
r = radial distance, m
R_m = dissipation coefficient, nt-sec/m
R_r = radiation resistance, kg/sec
s = condensation
u = particle displacement, m; component velocity, m/sec
v = particle velocity, m/sec
V = volume, m^3
w = component velocity, m/sec
W = power, watts
X_r = radiation reactance, kg/sec
z = specific acoustic impedance, rayls
z_m = mechanical impedance, rayls
z_r = radiation impedance, rayls
ω = circular frequency, rad/sec
λ = wavelength, m
ρ = density, kg/m^3

INTRODUCTION

When the surface of a pulsating sphere expands and contracts radially about its mean position, a force will be exerted on the fluid medium in contact with the surface. The fluid is hence disturbed from its equilibrium position. As a result, a disturbance is produced and propagated away from the sphere uniformly in all directions as *spherical waves.* If the fluid medium is air, we have *spherical acoustic waves.*

Though the spherical wave moves outward with a spherical wavefront in a three-dimensional homogeneous medium, it is one-dimensional since all points of the wave can be related to one distance – the radial distance r of the wavefront from the center of the sphere.

Spherical acoustic waves do not change shape as they spread out, and resemble circular waves on a membrane in that they have infinite value at $r = 0$. Although the wavefront of spherical acoustic waves can be assumed plane at great distances from the source, many acoustical problems are concerned with diverging spherical acoustic waves radiated from a simple source rather than plane acoustic waves.

WAVE EQUATION

The *three-dimensional wave equation* in rectangular coordinates is

$$\frac{\partial^2 p}{\partial x^2} + \frac{\partial^2 p}{\partial y^2} + \frac{\partial^2 p}{\partial z^2} = \frac{1}{c^2}\frac{\partial^2 p}{\partial t^2}$$

where p is acoustic pressure, $c = \sqrt{B/\rho}$ is the speed of sound, B is the bulk modulus, and ρ is the density. The general solution can be expressed in *progressive waves* form as

$$p(x,y,z,t) = f(lx + my + nz - ct) + g(lx + my + nz + ct)$$

where f and g are arbitrary functions, and $l^2 + m^2 + n^2 = 1$. In *standing waves form,* the general solution can be written as

$$p(x,y,z,t) = [(A_1 \sin ck_1t + B_1 \cos ck_1t)(A_2 \sin k_2x + B_2 \cos k_2x)$$
$$(A_3 \sin k_3y + B_3 \cos k_3y)(A_4 \sin k_4z + B_4 \cos k_4z)]$$

where A_1 and B_1 are arbitrary constants to be evaluated by initial conditions, and $A_2, B_2, A_3, B_3, A_4, B_4$ are arbitrary constants to be evaluated by boundary conditions.

The three-dimensional wave equation can be written in spherical coordinates as

$$\frac{\partial^2(rp)}{\partial t^2} = c^2\frac{\partial^2(rp)}{\partial r^2}$$

with solution

$$p(r,t) = \frac{1}{r}f(ct - r) + \frac{1}{r}g(ct + r)$$

where r is the radial distance from the source to the wavefront, and f and g are arbitrary functions. (See Problems 3.1-3.8.)

WAVE ELEMENTS

For harmonic progressive spherical acoustic waves, we have

$$\text{particle displacement } u = -\left(\frac{1}{r} + ik\right)\frac{p}{\omega^2\rho}$$

$$\text{particle velocity } v = \left(\frac{1}{r} + ik\right)\frac{p}{i\omega\rho}$$

$$\text{condensation } s = p/\rho c^2$$

where p is acoustic pressure and $i = \sqrt{-1}$. (See Problems 3.12-3.13.)

ACOUSTIC INTENSITY AND ENERGY DENSITY

Acoustic intensity is the average rate of flow of sound energy through unit area. For spherical acoustic waves this becomes

$$I = \tfrac{1}{2}p_0 v_0 \cos\theta = \tfrac{1}{2}p_0^2/\rho c \text{ watts/m}^2$$

where p_0 is the amplitude of acoustic pressure in nt/m^2, v_0 is the velocity amplitude in m/sec, ρ is the density in kg/m^3, and c is the speed of sound in m/sec.

Energy density of a spherical acoustic wave at any instant is the sum of kinetic and potential energies per unit volume.

$$E_d = \tfrac{1}{4}(\rho v_0^2 + p_0^2/\rho c^2) \text{ joules/m}^3$$

(See Problems 3.9-3.11.)

SPECIFIC ACOUSTIC IMPEDANCE

Specific acoustic impedance has been defined as the ratio of acoustic pressure over velocity at any point in the wave. For harmonic progressive spherical acoustic waves, the specific acoustic impedance is given by

$$z = \rho ckr\left[\frac{kr}{1+k^2r^2} + i\frac{1}{1+k^2r^2}\right] \text{ rayls}$$

where the real part is known as the *specific acoustic resistance* while the imaginary part is called the *specific acoustic reactance*. r is the distance from the source to the wavefront, and $k = \omega/c$ is the wave number. (See Problems 3.14-3.15.)

RADIATION OF SOUND

If waves radiated outward from a sound source are symmetric and uniform in all directions, the source is an *isotropic radiator*. The simplest isotropic radiator is a *pulsating sphere,* which is a uniform and homogeneous sphere whose surface expands and contracts radially and sinusoidally with time. If the dimensions of a radiator are small compared with the wavelength of the sound radiated, the radiator can be approximated by a pulsating sphere.

Sound waves produced by the vibration of an extended surface such as a diaphragm will not have the symmetric spherical radiation pattern characteristic of an isotropic radiator. However, the radiation produced at any point by such a source can be assumed equal to the sum of the radiation produced by an equivalent array of isotropic radiators.

In general, sound waves produced by most sources have pronounced directional effects known as the *directivity* of the source. This is due to the following factors: (1) size and shape of source, (2) radiation impedance, (3) mode of vibration of the surface of the radiator, and (4) reaction of the fluid medium on the surface of the radiator. The presence of any large rigid surface known as *infinite baffle* near the vicinity of a sound source not only confines the radiation to one side of the surface but also affects the directivity of the source.

The *directivity pattern* of a sound source is therefore a graphical description of the response of a radiator as a function of the direction of the transmitted waves in a specified plane for a specified frequency.

The directivity of a sound source is described by the *directivity factor D*,

$$D = \frac{2J_1(ka \sin \theta)}{ka \sin \theta}$$

where J_1 is the Bessel function of the first kind of order one, k is the wave number, a is the radius of the source, and θ is the directional angle from the axial direction of the source. Hence a plot of the directivity factor in decibels will yield the relative values of acoustic pressure and intensity at points equidistant from the source but at different angles from the axial direction of the source.

Radiation of sound will be found equal to zero at certain angles from the axial direction of the source, beyond which it will reach a maximum, and so on. The second maximum, called the *side or minor lobe,* is usually much weaker than the first maximum at an earlier angle.

The *directivity ratio* $D_r = I_0/I_{ref}$ is the ratio of the intensity at any point on the axis of the sound source to the intensity that would be produced at the same point by a simple source of equal strength. The *directivity index* or *gain* $d_r = 10 \log D_r$ db is simply the decibel expression for the directivity ratio. *Beam width* is defined as the angle at which sound intensity drops down to one half of its value at the axial direction of the source. (See Problems 3.16-3.20.)

SOURCE STRENGTH

Source strength is the product of the surface area and velocity amplitude v_0 of a pulsating sphere, i.e. $Q = 4\pi a^2 v_0$ where a is the radius of the sphere. A hemispherical source mounted in an infinite baffle, for example, has half the strength of a similar spherical source having the same radius and velocity amplitude.

Acoustic doublet is an arrangement of two simple sound sources of identical strength and frequency. The directivity pattern of this array of sound sources depends on the distance between the two sources and the phase between them. (See Problems 3.21-3.23.)

RADIATION IMPEDANCE

Radiation impedance $z_r = f/v$ kg/sec is defined as the ratio of the force f in newtons exerted by the radiator on the medium to the velocity v in m/sec of the radiator. The force is due to the reaction acting on the radiator given by $\int p\, dA$, where p is acoustic pressure acting on the surface A of the radiator.

The total impedance acting on the radiator is therefore the sum of its mechanical impedance $z_m = R_m + i(\omega m - k/\omega)$ and the radiation impedance z_r as defined above. Since these impedances are functions of frequency ω, the velocity amplitude $v_0 = f/(z_m + z_r)$ will not remain constant as the frequency is varied. (See Problem 3.16.)

Solved Problems

WAVE EQUATION

3.1. Derive the general three-dimensional acoustic wave equation.

The derivation of the general acoustic wave equation in a form valid for discussing any three-dimensional type of nondissipative progressive wave is based on the following assumptions and procedure.

(1) The medium is assumed to be continuous and homogeneous, (2) the process is adiabatic, (3) a completely elastic medium, and (4) small amplitudes of particle displacements and velocities, as well as small changes in pressure and density.

(*a*) Develop the equation of continuity, (*b*) derive the dynamic equations from elastic properties and force equations, and (*c*) combine the three dynamic equations to form the general wave equation.

Consider a small element $dx\,dy\,dz$ of the fluid as having equilibrium coordinates x, y, z as shown in Fig. 3-1. Let u, v, w be the components of the particle velocity in the x, y, z directions respectively and ρ the density of the element. Then the mass flow of fluid through the left surface of this element will be

$$\left[\rho u - \frac{\partial}{\partial x}(\rho u)\frac{dx}{2}\right] dy\,dz$$

while the mass flow through the right surface is

$$\left[\rho u + \frac{\partial}{\partial x}(\rho u)\frac{dx}{2}\right] dy\,dz$$

The resultant flow in the x direction is therefore equal to the difference of these two flows,

$$\frac{\partial}{\partial x}(\rho u)\,dx\,dy\,dz$$

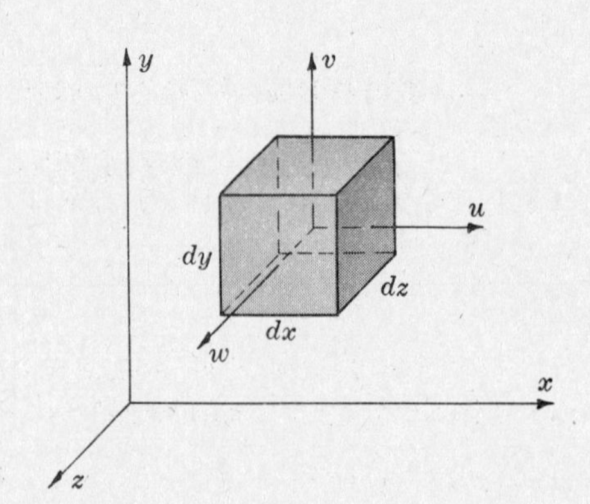

Fig. 3-1

Similarly, the resultant flows in the y and z directions are

$$\frac{\partial}{\partial y}(\rho v)\,dx\,dy\,dz, \qquad \frac{\partial}{\partial z}(\rho w)\,dx\,dy\,dz$$

so that the net flow through the entire element is

$$\left[\frac{\partial}{\partial x}(\rho u) + \frac{\partial}{\partial y}(\rho v) + \frac{\partial}{\partial z}(\rho w)\right] dx\,dy\,dz$$

Thus the equation of continuity is given by equating the net flow per unit mass to the time rate change of density

$$\frac{\partial}{\partial x}(\rho u) + \frac{\partial}{\partial y}(\rho v) + \frac{\partial}{\partial z}(\rho w) \;=\; -\frac{\partial \rho}{\partial t}$$

To obtain the dynamic equation in the x direction, let p be the pressure at the center of the element. Then the pressures at the left and right faces of the element are respectively

$$p - \frac{\partial p}{\partial x}\left(\frac{dx}{2}\right), \qquad p + \frac{\partial p}{\partial x}\left(\frac{dx}{2}\right)$$

Hence the net force acting on the element in the x direction is

$$\left\{\left[p - \frac{\partial p}{\partial x}(dx/2)\right] - \left[p + \frac{\partial p}{\partial x}(dx/2)\right]\right\} dy\,dz \;=\; -(\partial p/\partial x)\,dx\,dy\,dz$$

For small amplitudes of particle displacement and velocity, the mass of this element can be expressed as $\rho\,dx\,dy\,dz$, and the velocity throughout the element in the x direction is u. From Newton's second law $\sum F = \frac{d}{dt}(mv)$, we have

$$-\frac{\partial p}{\partial x}\,dx\,dy\,dz \;=\; \frac{\partial}{\partial t}(\rho u\,dx\,dy\,dz)$$

or

$$-\frac{\partial p}{\partial x} \;=\; \frac{\partial}{\partial t}(\rho u) \tag{1}$$

Similar dynamic equations in the y and z directions are

$$-\frac{\partial p}{\partial y} = \frac{\partial}{\partial t}(\rho v) \tag{2}$$

$$-\frac{\partial p}{\partial z} = \frac{\partial}{\partial t}(\rho w) \tag{3}$$

Now differentiate equations (1), (2), (3) respectively with respect to x, y, z:

$$-\frac{\partial^2 p}{\partial x^2} = \frac{\partial^2}{\partial t\, \partial x}(\rho u) \tag{4}$$

$$-\frac{\partial^2 p}{\partial y^2} = \frac{\partial^2}{\partial t\, \partial y}(\rho v) \tag{5}$$

$$-\frac{\partial^2 p}{\partial z^2} = \frac{\partial^2}{\partial t\, \partial z}(\rho w) \tag{6}$$

Adding equations (4), (5), (6) yields

$$-\left(\frac{\partial^2 p}{\partial x^2} + \frac{\partial^2 p}{\partial y^2} + \frac{\partial^2 p}{\partial z^2}\right) = \frac{\partial}{\partial t}\left[\frac{\partial}{\partial x}(\rho u) + \frac{\partial}{\partial y}(\rho v) + \frac{\partial}{\partial z}(\rho w)\right] \tag{7}$$

or

$$\frac{\partial^2 p}{\partial x^2} + \frac{\partial^2 p}{\partial y^2} + \frac{\partial^2 p}{\partial z^2} = \frac{\partial^2 \rho}{\partial t^2} \tag{8}$$

Now $\rho = \rho_0(1+s)$ and $p = Bs$. Then $\dfrac{\partial^2 \rho}{\partial t^2} = \rho_0 \dfrac{\partial^2 s}{\partial t^2}$, $\dfrac{\partial^2 s}{\partial t^2} = \dfrac{1}{B}\dfrac{\partial^2 p}{\partial t^2}$ where ρ_0 is the static density, s is the condensation, and B is bulk modulus. Equation (8) can be written as

$$\frac{\partial^2 p}{\partial x^2} + \frac{\partial^2 p}{\partial y^2} + \frac{\partial^2 p}{\partial z^2} = \frac{\rho_0}{B}\frac{\partial^2 p}{\partial t^2} \tag{9}$$

Equation (9) is then the three-dimensional wave equation with acoustic pressure p as the variable.

3.2. Obtain solutions for the general two-dimensional wave equation in rectangular coordinates.

The general two-dimensional wave equation in rectangular coordinates is

$$\frac{\partial^2 p}{\partial x^2} + \frac{\partial^2 p}{\partial y^2} = \frac{1}{c^2}\frac{\partial^2 p}{\partial t^2} \tag{1}$$

where p is the acoustic pressure and c the speed of sound.

(a) As in the case of the one-dimensional wave equation, we can write the solution in progressive waves form as

$$p(x, y, t) = f(mx + ny - ct) + g(mx + ny + ct), \qquad m^2 + n^2 = 1 \tag{2}$$

which represents waves of the same shape moving in opposite directions along x and y axes with velocity c. This can be verified by differentiating equation (2) and substituting into (1).

(b) Let us next look for solutions in standing waves form which is represented by $p = X(x)\,Y(y)\,T(t)$ where X, Y, and T are functions of x, y and t respectively. Substitute this expression for p into (1) to obtain

$$\frac{1}{X}\frac{d^2X}{dx^2} + \frac{1}{Y}\frac{d^2Y}{dy^2} = \frac{1}{c^2 T}\frac{d^2T}{dt^2} \tag{3}$$

Since the right-hand side of (3) is a function of t alone, and the left-hand side a function of x and y, each side must be equal to the same constant. Let this constant be $-p^2$. This leads to the following two differential equations:

$$\frac{d^2T}{dt^2} + c^2p^2T = 0 \tag{4}$$

with solution $$T(t) = A \sin cpt + B \cos cpt \tag{5}$$

or $$T(t) = Ae^{icpt} + Be^{-icpt} \tag{6}$$

and $$-\frac{1}{X}\frac{d^2X}{dx^2} - p^2 = \frac{1}{Y}\frac{d^2Y}{dy^2} \tag{7}$$

Using the same argument as before, we see that both sides of equation (7) equal the same constant, $-p^2 + q^2$. So we have

$$\frac{d^2X}{dx^2} + q^2X = 0 \tag{8}$$

with solution $$X(x) = A \cos qx + B \sin qx \tag{9}$$

or $$X(x) = Ae^{iqx} + Be^{-iqx} \tag{10}$$

where the A's and B's are arbitrary constants. Similarly,

$$\frac{d^2Y}{dy^2} - (q^2 - p^2)Y = 0 \tag{11}$$

with solution $$Y(y) = A \cosh \sqrt{q^2 - p^2}\, y + B \sinh \sqrt{q^2 - p^2}\, y \tag{12}$$

or $$Y(y) = Ae^{\sqrt{q^2 - p^2}\, y} + Be^{-\sqrt{q^2 - p^2}\, y} \tag{13}$$

where the A's and B's are arbitrary constants.

(c) If we replace the constant $-p^2$ in equation (3) by p^2, we obtain

$$\frac{d^2T}{dt^2} - c^2p^2T = 0 \tag{14}$$

with solution $$T(t) = A \cosh cpt + B \sinh cpt \tag{15}$$

or $$T(t) = Ae^{cpt} + Be^{-cpt} \tag{16}$$

Similarly, if we replace the constant $(q^2 - p^2)$ in equation (7) by $-(q^2 + p^2)$, we obtain

$$\frac{d^2X}{dx^2} - q^2X = 0 \tag{17}$$

with solution $$X(x) = A \cosh qx + B \sinh qx \tag{18}$$

or $$X(x) = Ae^{qx} + Be^{-qx} \tag{19}$$

Also, $$\frac{d^2Y}{dy^2} + (q^2 + p^2)Y = 0 \tag{20}$$

with solution $$Y(y) = A \cos \sqrt{q^2 + p^2}\, y + B \sin \sqrt{q^2 + p^2}\, y \tag{21}$$

or $$Y(y) = Ae^{i\sqrt{p^2 + q^2}\, y} + Be^{-i\sqrt{p^2 + q^2}\, y} \tag{22}$$

The complete solution is $$p(x, y, t) = X(x)\, Y(y)\, T(t)$$

which is expressed in harmonic terms by equations (5), (9) and (21); in complex exponential terms by equations (6), (10) and (22); in hyperbolic terms by equations (12), (15) and (18); and in exponential terms by equations (13), (16) and (19).

The theory and solution carried out here for the two-dimensional wave equation can be applied to the three-dimensional wave equation. Although there are four possible forms of solution available for the wave equations, the harmonic form of solution is widely employed. To account for the change of phase, it is advantageous to use the complex exponential form of solution.

3.3. Transform the two-dimensional wave equation $\dfrac{\partial^2 u}{\partial x^2} + \dfrac{\partial^2 u}{\partial y^2} = \dfrac{1}{c^2}\dfrac{\partial^2 u}{\partial t^2}$ into polar coordinates.

In polar coordinates r and θ, we have

$$x^2 + y^2 = r^2, \qquad \theta = \tan^{-1}\frac{y}{x}$$

or $2x = 2r\dfrac{\partial r}{\partial x}, \quad \dfrac{\partial r}{\partial x} = \dfrac{x}{r}; \quad \dfrac{\partial \theta}{\partial x} = \dfrac{-(y/x^2)}{1+(y/x)^2} = -y/r^2.$

Using the chain rule,

$$\frac{\partial u}{\partial x} = \frac{\partial u}{\partial r}\frac{\partial r}{\partial x} + \frac{\partial u}{\partial \theta}\frac{\partial \theta}{\partial x}$$

$$\frac{\partial^2 u}{\partial x^2} = \frac{\partial^2 u}{\partial r\,\partial x}\frac{\partial r}{\partial x} + \frac{\partial u}{\partial r}\frac{\partial^2 r}{\partial x^2} + \frac{\partial^2 u}{\partial \theta\,\partial x}\frac{\partial \theta}{\partial x} + \frac{\partial u}{\partial \theta}\frac{\partial^2 \theta}{\partial x^2}$$

$$\frac{\partial^2 r}{\partial x^2} = \frac{r - x(\partial r/\partial x)}{r^2} = y^2/r^3$$

$$\frac{\partial^2 \theta}{\partial x^2} = -y(-2/r^3)\frac{\partial r}{\partial x} = 2xy/r^4$$

$$\frac{\partial^2 u}{\partial x\,\partial r} = \frac{\partial^2 u}{\partial x^2}\frac{\partial r}{\partial x} + \frac{\partial^2 u}{\partial r\,\partial \theta}\frac{\partial \theta}{\partial x}$$

$$\frac{\partial^2 u}{\partial x\,\partial \theta} = \frac{\partial^2 u}{\partial \theta\,\partial r}\frac{\partial r}{\partial x} + \frac{\partial^2 u}{\partial \theta^2}\frac{\partial \theta}{\partial x}$$

Hence
$$\frac{\partial^2 u}{\partial x^2} = \frac{x^2}{r^2}\frac{\partial^2 u}{\partial r^2} - 2\frac{xy}{r^3}\frac{\partial^2 u}{\partial r\,\partial \theta} + \frac{y^2}{r^4}\frac{\partial^2 u}{\partial \theta^2} + \frac{y^2}{r^3}\frac{\partial u}{\partial r} + 2\frac{xy}{r^4}\frac{\partial u}{\partial \theta}$$

A similar expression can be obtained for $\dfrac{\partial^2 u}{\partial y^2}$ as

$$\frac{\partial^2 u}{\partial y^2} = \frac{y^2}{r^2}\frac{\partial^2 u}{\partial r^2} + 2\frac{xy}{r^3}\frac{\partial^2 u}{\partial r\,\partial \theta} + \frac{x^2}{r^4}\frac{\partial^2 u}{\partial \theta^2} + \frac{x^2}{r^3}\frac{\partial u}{\partial \theta} - 2\frac{xy}{r^4}\frac{\partial u}{\partial \theta}$$

The wave equation becomes

$$\frac{\partial^2 u}{\partial x^2} + \frac{\partial^2 u}{\partial y^2} = \frac{\partial^2 u}{\partial r^2} + \frac{1}{r}\frac{\partial u}{\partial r} + \frac{1}{r^2}\frac{\partial^2 u}{\partial \theta^2} = \frac{1}{c^2}\frac{\partial^2 u}{\partial t^2}$$

3.4. Find a solution for the general three-dimensional acoustic wave equation in rectangular coordinates.

The general three-dimensional acoustic wave equation in rectangular coordinates is

$$\frac{\partial^2 p}{\partial x^2} + \frac{\partial^2 p}{\partial y^2} + \frac{\partial^2 p}{\partial z^2} = \frac{1}{c^2}\frac{\partial^2 p}{\partial t^2} \tag{1}$$

where p is the acoustic pressure, and $c = \sqrt{B/\rho}$ is the speed of sound waves.

Let us look for a solution in the form of $p = X(x)\,Y(y)\,Z(z)\,T(t)$ where X, Y, Z, T are functions of x, y, z, t respectively. Substituting this expression for p into equation (1), we obtain

$$\frac{1}{X}\frac{d^2X}{dx^2} + \frac{1}{Y}\frac{d^2Y}{dy^2} + \frac{1}{Z}\frac{d^2Z}{dz^2} = \frac{1}{c^2T}\frac{d^2T}{dt^2} \tag{2}$$

Now the right-hand side of (2) is a function of t alone, and the left-hand side a function of $x, y,$ and z. Each side must be equal to a constant. Let this constant be $-k_1^2$. This leads to the following equations:

$$\frac{d^2T}{dt^2} + c^2k_1^2T = 0 \tag{3}$$

with solution $T(t) = A_1 \sin ck_1t + B_1 \cos ck_1t$ and

$$\frac{1}{X}\frac{d^2X}{dx^2} + \frac{1}{Y}\frac{d^2Y}{dy^2} + \frac{1}{Z}\frac{d^2Z}{dz^2} = -k_1^2 \tag{4}$$

Equation (4) can be rewritten as

$$\frac{1}{Y}\frac{d^2Y}{dy^2} + \frac{1}{Z}\frac{d^2Z}{dz^2} = -k_1^2 - \frac{1}{X}\frac{d^2X}{dx^2} = -k_1^2 + k_2^2 \tag{5}$$

where k_2 is another constant.

Using the same argument as before, we see that both sides of equation (5) must equal the same constant, $-k_1^2 + k_2^2$, so we have

$$\frac{d^2X}{dx^2} + k_2^2 X = 0 \tag{6}$$

with solution $X(x) = A_2 \sin k_2 x + B_2 \cos k_2 x$ and

$$\frac{1}{Y}\frac{d^2Y}{dy^2} + \frac{1}{Z}\frac{d^2Z}{dz^2} = -k_1^2 + k_2^2 \tag{7}$$

We can rewrite (7) as

$$\frac{1}{Z}\frac{d^2Z}{dz^2} = -k_1^2 + k_2^2 - \frac{1}{Y}\frac{d^2Y}{dy^2} = -k_1^2 + k_2^2 + k_3^2 \tag{8}$$

where k_3 is an arbitrary constant.

From (8) we obtain
$$\frac{d^2Y}{dy^2} + k_3^2 Y = 0 \tag{9}$$

with solution $Y(y) = A_3 \sin k_3 y + B_3 \cos k_3 y$ and

$$\frac{d^2Z}{dz^2} + (k_1^2 - k_2^2 - k_3^2)Z = 0 \quad \text{or} \quad \frac{d^2Z}{dz^2} + k_4^2 Z = 0 \tag{10}$$

with solution $Z(z) = A_4 \sin k_4 z + B_4 \cos k_4 z$, $k_4^2 = k_1^2 - k_2^2 - k_3^2$.

The general solution for the three-dimensional wave equation is therefore given by

$$p(x,y,z,t) = [(A_1 \sin ck_1 t + B_1 \cos ck_1 t)(A_2 \sin k_2 x + B_2 \cos k_2 x) \times (A_3 \sin k_3 y + B_3 \cos k_3 y)(A_4 \sin k_4 z + B_4 \cos k_4 z)]$$

where A's and B's are arbitrary constants.

3.5. A rectangular room has rigid walls of lengths L_1, L_2 and L_3 as shown in Fig. 3-2. Find the normal modes of acoustic wave oscillation.

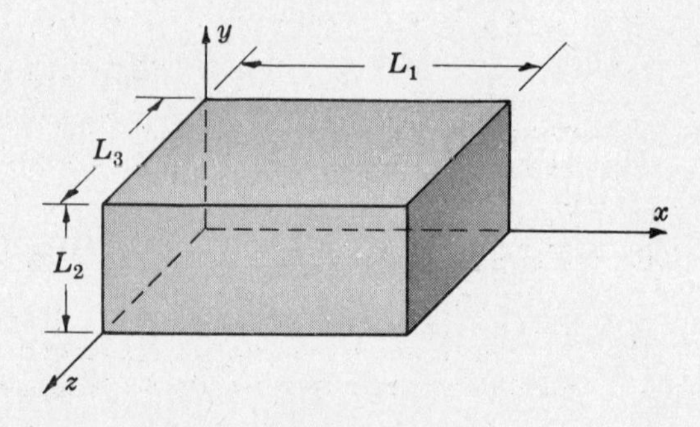

Fig. 3-2

The general three-dimensional acoustic wave equation is given by

$$\frac{\partial^2 p}{\partial x^2} + \frac{\partial^2 p}{\partial y^2} + \frac{\partial^2 p}{\partial z^2} = \frac{1}{c^2}\frac{\partial^2 p}{\partial t^2}$$

where p is the acoustic pressure, $c = \sqrt{B/\rho}$ is the speed of sound waves. The general solution is

$$p(x,y,z,t) = [(A_1 \sin ck_1 t + B_1 \cos ck_1 t)(A_2 \sin k_2 x + B_2 \cos k_2 x) \times (A_3 \sin k_3 y + B_3 \cos k_3 y)(A_4 \sin k_4 z + B_4 \cos k_4 z)]$$

The boundary conditions are that the particle velocities normal to any wall surface must be zero, i.e.

$$V_x = 0 \quad \text{at } x = 0 \text{ and } x = L_1 \tag{1}$$

$$V_y = 0 \quad \text{at } y = 0 \text{ and } y = L_2 \tag{2}$$

$$V_z = 0 \quad \text{at } z = 0 \text{ and } z = L_3 \tag{3}$$

But $-\frac{\partial p}{\partial x} = \rho\frac{\partial V_x}{\partial t}$, $-\frac{\partial p}{\partial y} = \rho\frac{\partial V_y}{\partial t}$, $-\frac{\partial p}{\partial z} = \rho\frac{\partial V_z}{\partial t}$, and so the boundary conditions (1), (2), and (3) become

$$\frac{\partial p}{\partial x} = 0 \quad \text{at } x = 0 \text{ and } x = L_1 \tag{4}$$

$$\frac{\partial p}{\partial y} = 0 \quad \text{at } y = 0 \text{ and } y = L_2 \tag{5}$$

$$\frac{\partial p}{\partial z} = 0 \quad \text{at } z = 0 \text{ and } z = L_3 \tag{6}$$

Now $(\partial p/\partial x)_{x=0} = 0$ or

$$A_2k_2(A_1 \sin ck_1t + B_1 \cos ck_1t)(A_3 \sin k_3y + B_3 \cos k_3y)(A_4 \sin k_4z + B_4 \cos k_4z) = 0$$

Since k_2 cannot always be zero, $A_2 = 0$ for the above expression equals zero. Similarly, $A_3 = 0$ from boundary condition $(\partial p/\partial y)_{y=0} = 0$; and $A_4 = 0$ from boundary condition $(\partial p/\partial z)_{z=0} = 0$. Then the general solution becomes

$$p(x, y, z, t) = (A_1 \sin ck_1t + B_1 \cos ck_1t)(B_2 \cos k_2x)(B_3 \cos k_3y)(B_4 \cos k_4z)$$

or $$p(x, y, z, t) = (\cos k_2x)(\cos k_3y)(\cos k_4z)(C_1 \sin ck_1t + C_2 \cos ck_1t)$$

where $C_1 = A_1B_2B_3B_4$ and $C_2 = B_1B_2B_3B_4$.

The second parts of boundary conditions (4), (5) and (6) yield

$$\left(\frac{\partial p}{\partial x}\right)_{x=L_1} = -k_2(\sin k_2L_1)(\cos k_3y)(\cos k_4z)(C_1 \sin ck_1t + C_2 \cos ck_1t) = 0$$

or $$\sin k_2L_1 = 0, \quad k_2 = l\pi/L_1 \quad \text{where} \quad l = 0, 1, 2, \ldots \tag{7}$$

$$\left(\frac{\partial p}{\partial y}\right)_{y=L_2} = -k_3(\cos k_2x)(\sin k_3L_2)(\cos k_4z)(C_1 \sin ck_1t + C_2 \cos ck_1t) = 0$$

or $$\sin k_3L_2 = 0, \quad k_3 = m\pi/L_2 \quad \text{where} \quad m = 0, 1, 2, \ldots \tag{8}$$

$$\left(\frac{\partial p}{\partial z}\right)_{z=L_3} = -k_4(\cos k_2x)(\cos k_3y)(\sin k_4L_3)(C_1 \sin ck_1t + C_2 \cos ck_1t) = 0$$

or $$\sin k_4L_3 = 0, \quad k_4 = n\pi/L_3 \quad \text{where} \quad n = 0, 1, 2, \ldots \tag{9}$$

The natural frequencies of the system are given by

$$\omega = ck_1 = c\sqrt{k_2^2 + k_3^2 + k_4^2}$$

and the normal modes of vibration are

$$X(x)\,Y(y)\,Z(z) = \cos k_2x \cos k_3y \cos k_4z \tag{10}$$

which has the same form as the free transverse vibration of a uniform rectangular membrane fixed at the edges. (See Problem 1.27.)

3.6. Write the general acoustic wave equation in cylindrical coordinates and find its solution.

The general acoustic wave equation in rectangular coordinates for any three-dimensional space is

$$\frac{\partial^2 p}{\partial x^2} + \frac{\partial^2 p}{\partial y^2} + \frac{\partial^2 p}{\partial z^2} = \frac{1}{c^2}\frac{\partial^2 p}{\partial t^2} \tag{1}$$

where p is acoustic pressure and $c = \sqrt{B/\rho}$ the speed of sound in air.

In cylindrical coordinates, a point A in space is described by the three coordinates r, θ and z as shown in Fig. 3-3 where

$$x = r\cos\theta, \quad y = r\sin\theta, \quad z = z \tag{2}$$

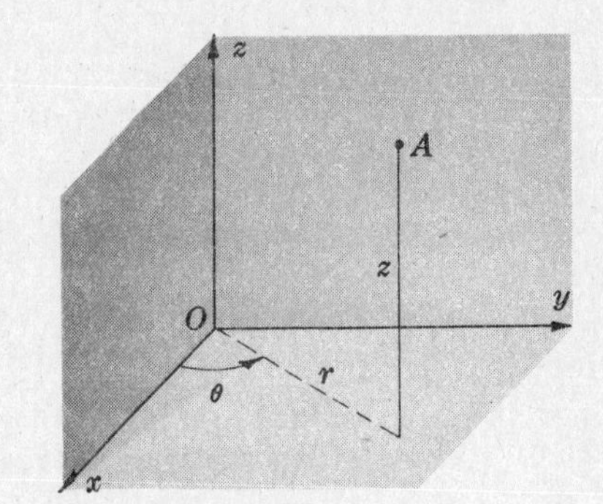

Fig. 3-3

Differentiating acoustic pressure p with respect to r, θ and z and transforming, we obtain

$$\frac{\partial^2 p}{\partial x^2} + \frac{\partial^2 p}{\partial y^2} + \frac{\partial^2 p}{\partial z^2} = \frac{\partial^2 p}{\partial r^2} + \frac{1}{r}\frac{\partial p}{\partial r} + \frac{1}{r^2}\frac{\partial^2 p}{\partial \theta^2} + \frac{\partial^2 p}{\partial z^2} \tag{3}$$

and so (1) becomes $$\frac{\partial^2 p}{\partial r^2} + \frac{1}{r}\frac{\partial p}{\partial r} + \frac{1}{r^2}\frac{\partial^2 p}{\partial \theta^2} + \frac{\partial^2 p}{\partial z^2} = \frac{1}{c^2}\frac{\partial^2 p}{\partial t^2} \tag{4}$$

A solution of the following form can be found by the method of separation of variables,

$$p(r, \theta, z, t) = R(r)\,\Theta(\theta)\,Z(z)\,T(t) \tag{5}$$

where R, Θ, Z and T are functions of r, θ, z and t respectively. Substituting (5) into (4), we obtain

$$\Theta ZT\frac{d^2R}{dr^2} + \frac{\Theta ZT}{r}\frac{dR}{dr} + \frac{RTZ}{r^2}\frac{d^2\Theta}{d\theta^2} + R\Theta T\frac{d^2Z}{dz^2} = \frac{R\Theta Z}{c^2}\frac{d^2T}{dt^2} \tag{6}$$

and dividing by $R\Theta ZT$,

$$\frac{1}{R}\frac{d^2R}{dr^2} + \frac{1}{rR}\frac{dR}{dr} + \frac{1}{r^2\Theta}\frac{d^2\Theta}{d\theta^2} + \frac{1}{Z}\frac{d^2Z}{dz^2} = \frac{1}{c^2T}\frac{d^2T}{dt^2} \tag{7}$$

Now the right-hand side of (7) is a function of t alone, and the left-hand side a function of r, θ and z. Each side must equal a constant. Let this same constant be $-k_1^2$. This leads to the following equations:

$$\frac{d^2T}{dt^2} + c^2k_1^2T = 0 \tag{8}$$

with solution $T(t) = A_1 \sin ck_1t + B_1 \cos ck_1t$, and

$$\frac{1}{R}\frac{d^2R}{dr^2} + \frac{1}{rR}\frac{dR}{dr} + \frac{1}{r^2\Theta}\frac{d^2\Theta}{d\theta^2} + \frac{1}{Z}\frac{d^2Z}{dz^2} = -k_1^2 \tag{9}$$

Rewrite (9) as
$$\frac{1}{R}\frac{d^2R}{dr^2} + \frac{1}{rR}\frac{dR}{dr} + \frac{1}{r^2\Theta}\frac{d^2\Theta}{d\theta^2} = -k_1^2 - \frac{1}{Z}\frac{d^2Z}{dz^2} = -k_1^2 + k_2^2 \tag{10}$$

where k_2 is another arbitrary constant. Using the same argument as before, we see that both sides of (10) equal the same constant, $-k_1^2 + k_2^2$, so we obtain

$$\frac{d^2Z}{dz^2} + k_2^2Z = 0 \tag{11}$$

with solution $Z(z) = A_2 \sin k_2z + B_2 \cos k_2z$, and

$$\frac{1}{R}\frac{d^2R}{dr^2} + \frac{1}{rR}\frac{dR}{dr} = -k_1^2 + k_2^2 - \frac{1}{r^2\Theta}\frac{d^2\Theta}{d\theta^2} \tag{12}$$

Multiplying (12) by r^2 and rearranging,

$$\frac{r^2}{R}\frac{d^2R}{dr^2} + \frac{r}{R}\frac{dR}{dr} + (k_1^2 - k_2^2)r^2 = -\frac{d^2\Theta}{d\theta^2} = +k_3^2\Theta \tag{13}$$

where k_3 is another arbitrary constant.

By the same argument as before, we have

$$\frac{d^2\Theta}{d\theta^2} + k_3^2\Theta = 0 \tag{14}$$

with solution $\Theta(\theta) = A_3 \sin k_3\theta + B_3 \cos k_3\theta$, and

$$\frac{d^2R}{dr^2} + \frac{1}{r}\frac{dR}{dr} - \frac{k_3^2}{r^2}R + (k_1^2 - k_2^2)R = 0 \tag{15}$$

Equation (15) is Bessel's equation of order k_3, with solutions $J_{k_3}(r\sqrt{k_1^2 - k_2^2})$ and $Y_{k_3}(r\sqrt{k_1^2 - k_2^2})$. J_{k_3} is finite and Y_{k_3} is infinite when $r = 0$, so we usually require only the J_{k_3} solutions.

The final form for the general acoustic wave equation in cylindrical coordinates is therefore given by the solutions of equations (8), (11), (14) and (15),

$$p(r,\theta,z,t) = J_{k_3}(A_1 \sin ck_1t + B_1 \cos ck_1t)(A_2 \sin k_2z + B_2 \cos k_2z)(A_3 \sin k_3\theta + B_3 \cos k_3\theta) \tag{16}$$

where the A's and B's are arbitrary constants.

3.7. Prove $\dfrac{\partial^2 u}{\partial x^2} + \dfrac{\partial^2 u}{\partial y^2} + \dfrac{\partial^2 u}{\partial z^2} = \dfrac{\partial^2 u}{\partial r^2} + \dfrac{2}{r}\dfrac{\partial^2 u}{\partial r^2}$ in the general three-dimensional wave equation.

Using $r^2 = x^2 + y^2 + z^2$, $\partial r/\partial x = x/r$ and $\dfrac{\partial u}{\partial x} = \dfrac{\partial u}{\partial r}\dfrac{\partial r}{\partial x} = \dfrac{x}{r}\dfrac{\partial u}{\partial r}$, we have

$$\frac{\partial^2 u}{\partial x^2} = \frac{1}{r}\frac{\partial u}{\partial r} - \frac{x}{r^2}\frac{\partial u}{\partial r}\frac{\partial r}{\partial x} + \frac{x}{r}\frac{\partial^2 u}{\partial r^2}\frac{\partial r}{\partial x} = \left[\frac{y^2 + z^2}{r^3}\right]\frac{\partial u}{\partial r} + \frac{x^2}{r^2}\frac{\partial^2 u}{\partial r^2}$$

Similarly,

$$\frac{\partial^2 u}{\partial y^2} = \left[\frac{z^2+x^2}{r^3}\right]\frac{\partial u}{\partial r} + \frac{y^2}{r^2}\frac{\partial^2 u}{\partial r^2} \quad \text{and} \quad \frac{\partial^2 u}{\partial z^2} = \left[\frac{y^2+x^2}{r^3}\right]\frac{\partial u}{\partial r} + \frac{z^2}{r^2}\frac{\partial^2 u}{\partial r^2}$$

Thus

$$\frac{\partial^2 u}{\partial x^2} + \frac{\partial^2 u}{\partial y^2} + \frac{\partial^2 u}{\partial z^2} = \frac{2}{r}\frac{\partial u}{\partial r} + \frac{\partial^2 u}{\partial r^2}$$

3.8. **Determine the general solutions for the three-dimensional wave equation in spherical coordinates for (*a*) waves having spherical symmetry, (*b*) waves having circular symmetry, and (*c*) waves having no symmetry.**

The general three-dimensional wave equation in rectangular coordinates is

$$\frac{\partial^2 p}{\partial x^2} + \frac{\partial^2 p}{\partial y^2} + \frac{\partial^2 p}{\partial z^2} = \frac{1}{c^2}\frac{\partial^2 p}{\partial t^2} \tag{1}$$

where p is the acoustic pressure, and $c = \sqrt{B/\rho}$ is the speed of sound.

In spherical coordinates, a point in space is described by the three coordinates r, θ and ϕ as shown in Fig. 3-4, where

$$x = r \sin\theta \cos\phi$$
$$y = r \sin\theta \sin\phi$$
$$z = r \cos\theta$$

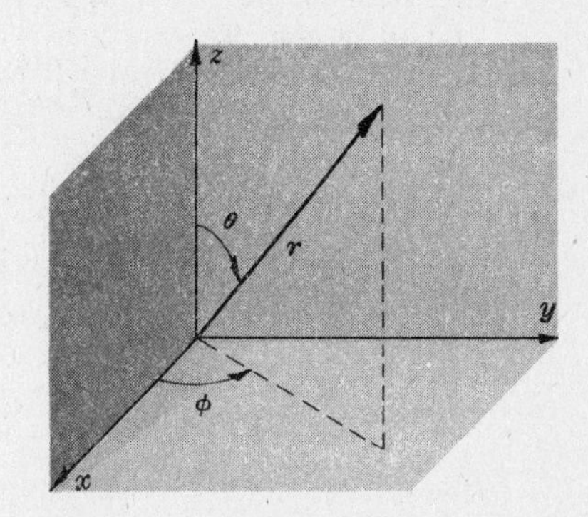

Fig. 3-4

It can be shown that the general three-dimensional wave equation in spherical coordinates is given by

$$\frac{\partial^2 p}{\partial r^2} + \frac{1}{r^2}\frac{\partial^2 p}{\partial \theta^2} + \frac{1}{r^2 \sin^2\theta}\frac{\partial^2 p}{\partial \phi^2} + \frac{2}{r}\frac{\partial p}{\partial r} + \frac{1}{r^2 \tan\theta}\frac{\partial p}{\partial \theta} = \frac{1}{c^2}\frac{\partial^2 p}{\partial t^2} \tag{2}$$

(*a*) For waves having spherical symmetry, acoustic pressure $p = p(r, t)$ is a function of radial distance r and time t. Equation (*1*) reduces to

$$\frac{\partial^2 p}{\partial r^2} + \frac{2}{r}\frac{\partial p}{\partial r} = \frac{1}{r}\frac{\partial^2(rp)}{\partial r^2} = \frac{1}{c^2}\frac{\partial^2 p}{\partial t^2} \tag{3}$$

or

$$\frac{\partial^2 p}{\partial t^2} = \frac{c^2}{r}\frac{\partial^2(rp)}{\partial r^2} \tag{4}$$

Since we assume spherical symmetry here, we could derive the wave equation (*4*) from (*1*) directly as follows:

$$r^2 = x^2 + y^2 + z^2, \qquad \partial r/\partial x = x/r$$

and so

$$\frac{\partial p}{\partial x} = \frac{\partial p}{\partial r}\frac{\partial r}{\partial x} = \frac{x}{r}\frac{\partial p}{\partial r}$$

Differentiating the last expression with respect to x,

$$\frac{\partial^2 p}{\partial x^2} = \frac{1}{r}\frac{\partial p}{\partial r} - \frac{x}{r^2}\frac{\partial p}{\partial r}\frac{\partial r}{\partial x} + \frac{x}{r}\frac{\partial^2 p}{\partial r^2}\frac{\partial r}{\partial x} = \frac{x^2}{r^2}\frac{\partial^2 p}{\partial r^2} + \left(\frac{1}{r} - \frac{x^2}{r^3}\right)\frac{\partial p}{\partial r} = \frac{x^2}{r^2}\frac{\partial^2 p}{\partial r^2} + \frac{y^2+z^2}{r^3}\frac{\partial p}{\partial r}$$

$$\frac{\partial^2 p}{\partial y^2} = \frac{y^2}{r^2}\frac{\partial^2 p}{\partial r^2} + \frac{z^2+x^2}{r^3}\frac{\partial p}{\partial r} \quad \text{and} \quad \frac{\partial^2 p}{\partial z^2} = \frac{z^2}{r^2}\frac{\partial^2 p}{\partial r^2} + \frac{x^2+y^2}{r^3}\frac{\partial p}{\partial r}$$

$$\frac{\partial^2 p}{\partial t^2} = c^2\left[\frac{\partial^2 p}{\partial x^2} + \frac{\partial^2 p}{\partial y^2} + \frac{\partial^2 p}{\partial z^2}\right] = c^2\left(\frac{\partial^2 p}{\partial r^2} + \frac{2}{r}\frac{\partial p}{\partial r}\right) = \frac{c^2}{r}\frac{\partial^2(rp)}{\partial r^2}$$

Since r is an independent variable which is not a function of time t, we can write the above expression as

$$\frac{\partial^2(rp)}{\partial t^2} = r\frac{\partial^2 p}{\partial t^2} \quad \text{or} \quad \frac{\partial^2(rp)}{\partial t^2} = c^2\frac{\partial^2(rp)}{\partial t^2} \tag{5}$$

If the term (rp) is considered a single term, the wave equation in spherical coordinates for any three-dimensional space is of the same form as the plane wave equation derived earlier in Chapter 2. The general solution is therefore

$$rp(r,t) \quad = \quad f_1(ct-r) \;+\; f_2(ct+r)$$

or

$$p(r,t) \quad = \quad \frac{1}{r}f_1(ct-r) \;+\; \frac{1}{r}f_2(ct+r) \tag{6}$$

where the first term $\frac{1}{r}f_1(ct-r)$ represents a spherical wave diverging from the origin of the coordinate with a velocity c, and the second term $\frac{1}{r}f_2(ct+r)$ similarly represents a wave converging on the origin with velocity c. Both waves diminish in amplitude as the distance from the source increases. The converging wave has little application in acoustics while the diverging wave is frequently produced by a small pulsating sphere completely isolated from reflecting surfaces, and has many uses. If the pulsation of the sphere is sinusoidal, the resulting waves are diverging harmonic spherical waves designated by

$$p(r,t) \quad = \quad \frac{A}{r}\,e^{i(\omega t - kr)} \tag{7}$$

where A is an arbitrary constant (real or complex), ω is the frequency, and $k = \omega/c$ is the wave number.

(*b*) If we assume that the waves have circular symmetry, then acoustic pressure $p = p(r, \theta, t)$ is function of r, θ and t. The general wave equation is reduced to

$$\frac{\partial^2 p}{\partial r^2} + \frac{1}{r}\frac{\partial p}{\partial r} + \frac{1}{r^2}\frac{\partial^2 p}{\partial \theta^2} \quad = \quad \frac{1}{c^2}\frac{\partial^2 p}{\partial t^2} \tag{8}$$

which we solve by the method of separation of variables. First we assume a solution of the form

$$p(r,\theta,t) \quad = \quad R(r)\,\Theta(\theta)\,T(t)$$

where R, Θ and T are functions of r, θ and t respectively. Substituting into (*8*) yields

$$\Theta T\frac{d^2R}{dr^2} + \frac{\Theta T}{r}\frac{dR}{dr} + \frac{RT}{r^2}\frac{d^2\Theta}{d\theta^2} \quad = \quad \frac{\Theta R}{c^2}\frac{d^2T}{dt^2}$$

and dividing through by $R\Theta T$,

$$\frac{1}{R}\frac{d^2R}{dr^2} + \frac{1}{Rr}\frac{dR}{dr} + \frac{1}{r^2\Theta}\frac{d^2\Theta}{d\theta^2} \quad = \quad \frac{1}{c^2T}\frac{d^2T}{dt^2} \tag{9}$$

Now the right-hand side of (*9*) is a function of t alone, and the left-hand side a function of r and θ. Each side must equal the same constant. Let this constant be $-k_1^2$. This leads to the following two equations:

$$\frac{d^2T}{dt^2} + c^2k_1^2T \quad = \quad 0 \tag{10}$$

with solution $T(t) \;=\; A_1 \sin ck_1t + B_1 \cos ck_1t$, and

$$\frac{r^2}{R}\frac{d^2R}{dr^2} + \frac{r}{R}\frac{dR}{dr} + \frac{1}{\Theta}\frac{d^2\Theta}{d\theta^2} \quad = \quad -k_1^2r^2$$

or

$$\frac{r^2}{R}\frac{d^2R}{dr^2} + \frac{r}{R}\frac{dR}{dr} \quad = \quad -k_1^2r^2 - \frac{1}{\Theta}\frac{d^2\Theta}{d\theta^2} \quad = \quad -k_1^2r^2 + k_2^2 \tag{11}$$

where k_2 is another arbitrary constant.

Using the same argument as before, we see that both sides of (*11*) equal the same constant $(-k_1^2r^2 + k_2^2)$, and so we obtain

$$\frac{d^2\Theta}{d\theta^2} + k_2^2\Theta \quad = \quad 0 \tag{12}$$

with solution $\Theta(\theta) \;=\; A_2 \sin k_2\theta + B_2 \cos k_2\theta$ and

$$r^2\frac{d^2R}{dr^2} + r\frac{dR}{dr} + k_1^2r^2R - k_2^2R = 0$$

or

$$\frac{d^2R}{dr^2} + \frac{1}{r}\frac{dR}{dr} - (k_2^2/r^2)R + k_1^2R = 0 \qquad (13)$$

which is Bessel's equation of order k_2, with solutions $J_{k_2}(rk_1)$ and $Y_{k_2}(rk_1)$. J_{k_2} is finite and Y_{k_2} is infinite when $r = 0$, so we usually require only the J_{k_2} solutions.

The final form of solution for the general acoustic wave equation in plane polar coordinates is therefore given by the solutions of (10), (12) and (13),

$$p(r, \theta, t) = J_{k_2}(A_1 \sin ck_1t + B_1 \cos ck_1t)(A_2 \sin k_2\theta + B_2 \cos k_2\theta) \qquad (14)$$

where the A's and B's are arbitrary constants.

(c) If we assume the waves have no symmetry, then the acoustic pressure $p = p(r, \theta, \phi, t)$ is a function of r, θ, ϕ and t. We assume the following form of solution and then solve by the method of separation of variables.

$$p(r, \theta, \phi, t) = R(r)\,\Theta(\theta)\,\Phi(\phi)\,T(t) \qquad (15)$$

where R, Θ, Φ and T are functions of r, θ, ϕ and t respectively. Now (2) can be rewritten as

$$\frac{\partial^2p}{\partial r^2} + \frac{2}{r}\frac{\partial p}{\partial r} + \frac{1}{r^2 \sin\theta}\frac{\partial}{\partial\theta}\left(\sin\theta\frac{\partial p}{\partial\theta}\right) + \frac{1}{r^2\sin^2\theta}\frac{\partial^2p}{\partial\phi^2} = \frac{1}{c^2}\frac{\partial^2p}{\partial t^2} \qquad (16)$$

Substituting (15) into (16),

$$\Phi T\Theta\frac{d^2R}{dr^2} + \frac{2\Theta T\Phi}{r}\frac{dR}{dr} + \frac{R\Phi T}{r^2\sin\theta}\frac{d}{d\theta}\left(\sin\theta\frac{d\Theta}{d\theta}\right) + \frac{RT\Theta}{r^2\sin^2\theta}\frac{d^2\Phi}{d\phi^2} = \frac{\Theta R\Phi}{c^2}\frac{d^2T}{dt^2}$$

and dividing by $R\Theta T\Phi$,

$$\frac{1}{R}\frac{d^2R}{dr^2} + \frac{2}{Rr}\frac{dR}{dr} + \frac{1}{\Theta r^2\sin\theta}\frac{d}{d\theta}\left(\sin\theta\frac{d\Theta}{d\theta}\right) + \frac{1}{\Phi r^2\sin^2\theta}\frac{d^2\Phi}{d\phi^2} = \frac{1}{Tc^2}\frac{d^2T}{dt^2} \qquad (17)$$

The right-hand side of (17) is a function of t alone, while the left-hand side is a function of r, θ and ϕ. Each side must equal the same constant. Let this constant be $-k_1^2$. This leads to the following two equations:

$$\frac{d^2T}{dt^2} + c^2k_1^2T = 0 \qquad (18)$$

with solution $T(t) = A_1 \sin ck_1t + B_1 \cos ck_1t$ and

$$\frac{1}{R}\frac{d^2R}{dr^2} + \frac{2}{Rr}\frac{dR}{dr} + \frac{1}{\Theta r^2\sin\theta}\frac{d}{d\theta}\left(\sin\theta\frac{d\Theta}{d\theta}\right) = -k_1^2 - \frac{1}{\Phi r^2\sin^2\theta}\frac{d^2\Phi}{d\phi^2} \qquad (19)$$

Multiplying (19) by $r^2\sin^2\theta$ and rearranging,

$$r^2\sin^2\theta\left[\left(\frac{1}{R}\frac{d^2R}{dr^2} + \frac{2}{Rr}\frac{dR}{dr} + \frac{1}{\Theta r^2\sin\theta}\frac{d}{d\theta}\left(\sin\theta\frac{d\Theta}{d\theta}\right) + k_1^2\right]\right. = -\frac{1}{\Phi}\frac{d^2\Phi}{d\phi^2} = k_2^2 \qquad (20)$$

where k_2 is another arbitrary constant.

Using the same argument as before, we see that both sides of (20) equal to the same constant k_2^2, so we have

$$\frac{d^2\Phi}{d\phi^2} + k_2^2\Phi = 0 \qquad (21)$$

with solution $\Phi(\phi) = A_2 \sin k_2\phi + B_2 \cos k_2\phi$ and

$$\frac{1}{\Theta\sin\theta}\frac{d}{d\theta}\left(\sin\theta\frac{d\Theta}{d\theta}\right) - \frac{k_2^2}{\sin^2\theta} = -\frac{r^2}{R}\frac{d^2R}{dr^2} - \frac{2r}{R}\frac{dR}{dr} - k_1^2r^2 \qquad (22)$$

As before, if we let both sides of (22) equal the same constant $-k_3(k_3+1)$, then we obtain the following two equations:

$$\frac{1}{\sin\theta}\frac{d}{d\theta}\left(\sin\theta\frac{d\Theta}{d\theta}\right) + [k_3(k_3+1) - (k_2^2/\sin^2\theta)]\Theta = 0 \qquad (23)$$

$$\frac{d^2R}{dr^2} + \frac{2}{r}\frac{dR}{dr} + [k_1^2 - k_3(k_3+1)/r^2]R = 0 \qquad (24)$$

Equation (*23*) is the generalized Legendre equation with solutions

$$\Theta(\theta) = P_n^m(\cos\theta) \tag{25}$$

where $k_2 = m$ and $k_3 = n$.

To solve (*24*) we make the substitution $R(r) = r^{-1/2}R'(r)$ and obtain

$$\frac{d^2R'}{dr^2} + \frac{1}{r}\frac{dR'}{dr} + [k_1^2 - (k_3 + \tfrac{1}{2})^2/r^2]R' = 0 \tag{26}$$

which is the Bessel equation with solution

$$R'(r) = J_{n+1/2}(pr) \quad \text{or} \quad Y_{n+1/2}(pr)$$

The final form of solution for the general acoustic wave equation in spherical coordinates is therefore given by the solutions of (*18*), (*21*), (*24*) and (*26*),

$$p(r,\theta,\phi,t) = P_n^m(\cos\theta)r^{-1/2}[A_3J_{n+1/2}(pr) + B_3Y_{n+1/2}(pr)](A_2\sin m\phi + B_2\cos m\phi)(A_1\sin cpt + B_1\cos cpt)$$

where the A's and B's are arbitrary constants.

ACOUSTIC INTENSITY AND ENERGY DENSITY

3.9. Derive an expression for the acoustic intensity of harmonic diverging spherical waves.

From Problem 2.14, page 48,

$$I = \frac{1}{P}\int_0^P pv\,dt = \frac{1}{P}\int_0^P p_0\cos(\omega t - kx)v_0\cos(\omega t - kx - \theta)\,dt = \tfrac{1}{2}p_0v_0\cos\theta \tag{1}$$

where P = period, p_0 = pressure amplitude, v_0 = particle velocity amplitude, k = wave number, and $\theta = \cos^{-1}(kr/\sqrt{1+k^2r^2})$ is the phase angle between acoustic pressure and particle velocity. Since acoustic pressure $p = p_0/\sqrt{2}$ and particle velocity $v = v_0/\sqrt{2}$, (*1*) can be written as

$$I = pv\cos\theta \tag{2}$$

Since specific acoustic impedance $z = p/v = \rho c\cos\theta$ for harmonic diverging spherical waves, (*2*) becomes

$$I = \rho cv^2\cos^2\theta = \frac{\rho cv^2k^2r^2}{1+k^2r^2} \tag{3}$$

where ρ is the density and c the speed of wave propagation.

3.10. Derive an expression for the energy density of a harmonic diverging spherical acoustic wave.

The energy density of a sound wave at any instant is the sum of kinetic and potential energies per unit volume. Now

$$\text{KE} = \tfrac{1}{2}\rho V_0v^2 = \tfrac{1}{4}\rho V_0v_0^2 \tag{1}$$

where ρ is the average density, V_0 the average volume, v the average particle velocity over unit volume, and v_0 the amplitude of particle velocity.

The potential energy is equal to the work done by pressure and change in volume of the medium, i.e.

$$\text{PE} = -\int p\,dV \tag{2}$$

where p is the instantaneous pressure and V the instantaneous volume. Since $V = V_0(1 - p/\rho c^2)$ and $dV = -pV_0\,dp/\rho c^2$,

$$\text{PE} = V_0/\rho c^2\int_0^p p\,dp = p^2V_0/2\rho c^2 = p_0^2V_0/4\rho c^2 \tag{3}$$

For harmonic diverging spherical waves,

$$p = (A/r)\cos(\omega t - kr) = p_0\cos(\omega t - kr) \tag{4}$$

and $$v = p(1/r + ik)/i\rho\omega = (A\sqrt{1+k^2r^2}/\rho ckr^2)\cos(\omega t - kx - \theta) \quad (5)$$

which yields $$v_0 = A\sqrt{1+k^2r^2}/\rho ckr^2 \quad (6)$$

The expression for energy density becomes

$$E_d = (\text{KE}+\text{PE})/V_0 = \tfrac{1}{4}(\rho v_0^2 + p_0^2/\rho c^2) \quad (7)$$

Substituting expressions for p_0 and v_0 from (4) and (6) into (7) gives

$$E_d = \frac{1}{4}\left[\frac{A^2(1+k^2r^2)}{\rho c^2 r^4 k^2} + \frac{A^2}{\rho c^2 r^2}\right] = \frac{p_0^2}{2\rho c^2}\left(1+\frac{1}{2k^2r^2}\right) \quad (8)$$

where c is the speed of sound, $k = \omega/c$ is the wave number, and r is the distance from the source to the point of interest in the wave.

3.11. A diverging spherical wave has a peak acoustic pressure of 2 nt/m² at a distance of 1 m from the source at standard atmospheric pressure and temperature. What is its intensity at a distance of 10 m from the source?

Assume the source is emitting a constant amount of energy to the sound waves. For diverging spherical waves, the area of the wavefront increases as the waves are traveling farther and farther from the source. Hence intensity of such waves diminishes with distance of propagation.

At a distance of 1 m from the source,

$$I = p^2/2\rho c = 2^2/[2(1.21)343] = 0.0048 \text{ watt/m}^2$$

where $\rho = 1.21$ kg/m³ is the density of air, and $c = 343$ m/sec is the speed of sound in air at standard atmospheric pressure and temperature.

At a distance of 10 m from the source, the effective sound pressure will change but the power radiated will remain the same.

$$W = 4\pi r^2 I = 4(3.14)(1)^2(0.0048) = 0.062 \text{ watt}$$

Thus $$I = W/4\pi r^2 = 0.062/4(3.14)100 = 0.000048 \text{ watt/m}^2$$

3.12. A simple sound source radiates harmonic diverging spherical waves into free space with 10 watts of acoustic power at a frequency of 500 cyc/sec. Find the (*a*) intensity, (*b*) acoustic pressure, (*c*) particle velocity, (*d*) particle displacement, (*e*) energy density, (*f*) condensation and (*g*) sound pressure level at a radial distance of 1 m from the source.

(*a*) Intensity $I = W/4\pi r^2 = 10/4(3.14)(1)^2 = 0.8$ watt/m²

(*b*) Acoustic pressure $p = \sqrt{2\rho c I} = \sqrt{2(1.21)(343)(0.8)} = 25.8$ nt/m²

(*c*) Particle velocity $v = p/\rho c \cos\theta = 0.062$ m/sec

where $p = 25.8$ nt/m² is acoustic pressure, $\rho = 1.21$ kg/m³ is density of air, $c = 343$ m/sec is speed of sound in air, $\cos\theta = kr/\sqrt{1+k^2r^2} = 0.99$ and $kr = \omega r/c = 2(3.14)(500)(1.0)/343 = 9.18$.

(*d*) Particle displacement $u = v/\omega = 0.062/6.28(500) = 1.97(10)^{-5}$ m

(*e*) Energy density (see Problem 3.10)

$$E_d = \frac{p_0^2}{2\rho c^2}\left(1+\frac{1}{2k^2r^2}\right) = \frac{25.8^2}{2(1.21)(343)^2}\left(1+\frac{1}{2(9.18)^2}\right) = 2.34(10)^{-3} \text{ watt-sec/m}^3$$

(*f*) Condensation $s = p/\rho c^2 = 25.8/(1.21)(343)^2 = 1.8(10)^{-4}$

(*g*) Sound pressure level SPL $= 20\log p + 94 = 20\log 25.8 + 94 = 122.3$ db

3.13. Calculate the amplitude of particle displacement u_0 and the amplitude of particle velocity v_0 of spherical acoustic waves in air at standard atmospheric pressure and temperature. The pressure amplitude at a distance 0.01 m from the source is 10 nt/m², and the frequency of the wave is 25 cyc/sec.

For spherical acoustic waves, acoustic pressure is

$$p_0 = \rho c v \cos\theta = \rho c v k r/\sqrt{1+k^2r^2} \text{ nt/m}^2$$

where $\rho c = 415$ rayls is the characteristic impedance of air at standard atmospheric pressure and temperature. Now $\cos\theta \doteq \omega r/c = 6.28(25)(0.01)/343 = 0.0046$, and so

$$v_0 = p_0/\rho c\cos\theta = 10/[(415)(0.0046)] = 5.23 \text{ m/sec}$$

$$u_0 = v_0/\omega = 5.23/[(6.28)(25)] = 0.033 \text{ m}$$

These values are much greater than the corresponding values for plane acoustic waves under similar conditions.

SPECIFIC ACOUSTIC IMPEDANCE

3.14. Derive an expression for the specific acoustic impedance of a harmonic diverging spherical wave.

Specific acoustic impedance is defined as the ratio of pressure over velocity at any point in the wave. For harmonic diverging spherical waves, we have

$$p = \frac{A}{r}e^{i(\omega t - kr)} \qquad (1)$$

$$-\frac{\partial p}{\partial x} = \rho\frac{\partial V_x}{\partial t}, \quad -\frac{\partial p}{\partial y} = \rho\frac{\partial V_y}{\partial t}, \quad -\frac{\partial p}{\partial z} = \rho\frac{\partial V_z}{\partial t} \qquad (2)$$

where k is the wave number, V_x, V_y, V_z are velocity components in the x, y, z directions, and ρ is the density. From equations (1) and (2) we obtain

$$-\frac{\partial p}{\partial r} = \rho\frac{\partial v}{\partial t} \qquad (3)$$

which shows that the radial pressure gradient is directly proportional to the radial acceleration. Integrating (3), we obtain the radial velocity

$$v = -\frac{1}{\rho}\int\frac{\partial p}{\partial r}\,dt = -\frac{1}{i\rho\omega}\frac{\partial p}{\partial r} = \left(\frac{1}{r}+ik\right)\frac{p}{i\rho\omega} \qquad (4)$$

Hence the specific acoustic impedance is given by

$$z = \frac{p}{v} = \frac{i\rho\omega}{(1/r+ik)} = \frac{\rho c k^2 r^2}{(1+k^2r^2)} + i\frac{\rho c k r}{(1+k^2r^2)} \qquad (5)$$

which consists of the real part known as the specific acoustic resistance, and the imaginary part known as the specific acoustic reactance. From equation (5),

$$|z| = \rho c k r/\sqrt{1+k^2r^2} \qquad (6)$$

3.15. Spherical acoustic waves of frequency 125 cyc/sec are emitted from a small source. At a radial distance of 1.5 m from the source, what is the phase angle between acoustic pressure and particle velocity? Find the magnitude of the specific acoustic impedance at this point.

For harmonic diverging spherical waves, acoustic pressure and particle velocity may be written as

$$p = (A/r)e^{i(\omega t - kr)}, \qquad v = kp/\rho\omega + ip/\rho r\omega$$

The phase angle is found from

$$p/v = \frac{c\rho r^2k^2 + icr\rho k}{1+r^2k^2} = \frac{\rho c r^2 k^2}{1+r^2k^2} + \frac{i\rho c r k}{1+r^2k^2}$$

or $\theta = \tan^{-1}(1/kr) = \tan^{-1}(1/3.42) = 16.2°$

where $k = \omega/c = 125(6.28)/343 = 2.28$, $kr = 2.28(1.5) = 3.42$.

The magnitude of the specific acoustic impedance is given by equation (6) of Problem 3.14,

$$z = \rho ckr/\sqrt{1+k^2r^2} = 1.21(343)(3.4)/\sqrt{1+3.4^2} = 397 \text{ rayls}$$

RADIATION OF SOUND

3.16. A small circular piston of mass 0.01 kg and radius 0.05 m radiates sound at a frequency of 1000 cyc/sec. It is mounted in an infinite baffle; the stiffness of the suspension is 1000 nt/m and the mechanical resistance is 10 kg/m. If the effective driving force is 1 newton, determine (*a*) the relative acoustic pressure at a point equidistant from the piston but at an angle of 30° from the axis of the piston, (*b*) the beam width 3 db down, (*c*) the power output, (*d*) the directivity factor, and (*e*) the directivity index.

(*a*) $$\frac{p(30°)}{p(0°)} = \frac{2J_1(ka \sin\theta)}{ka \sin\theta} = \frac{2(0.22)}{0.46} = 0.98 \text{ or } -0.4 \text{ db}$$

where $k = \omega/c = 1000(6.28)/343 = 18.3$ is the wave number, $ka = 18.3(0.05) = 0.92$, $ka \sin 30° = 0.46$, and $J_1(0.46) = 0.22$ is the Bessel function of the first kind of order one.

(*b*) To compute the angle $\theta°$ at which the intensity is 3 db less than the axial intensity at equidistance, we write

$$20 \log \frac{p(\theta°)}{p(0°)} = -3 \text{ db} \quad \text{or} \quad \frac{p(\theta°)}{p(0°)} = 0.707 = \frac{2J_1(ka \sin\theta)}{ka \sin\theta}$$

from which we obtain $ka \sin\theta = 1.6$, or $\sin\theta = 1.6/0.92 = 1.74$ which is greater than unity. This indicates that there is no angle at which the fall-off in acoustic intensity from the axial direction is as great as 3 db.

(*c*) Power output is given by $W = v^2R_r$, where v is the particle velocity and $R_r = \rho c\pi a^2 R_1$ is the radiation resistance.

In order to determine R_r, we have

$$R_1(2ka) = 1 - 2J_1(2ka)/2ka, \quad R_1(1.83) = 0.38$$

where $2ka = 2(18.3)(0.05) = 1.83$. Hence

$$R_r = \rho c\pi a^2 R_1 = 1.21(343)(3.14)(0.05)^2(0.38) = 12.3$$

Now $$X_1(x) = \frac{4}{\pi}\left[\frac{x}{3} - \frac{x^3}{3^2(5)} + \frac{x^5}{3^2(5)^2(7)} - \cdots\right]$$

and $X_1(1.83) = 0.62$, $X_r = \rho c\pi a^2 X_1 = 1.21(343)(0.05)^2(3.14)(0.62) = 20.2$

Then $z_r = R_r + iX_r$ or $|z_r| = \sqrt{(12.3)^2 + (20.2)^2} = 23.3$ acoustic ohms.

Mechanical impedance $z_m = R_m + i(\omega m - k/\omega) = 10 + i(62.8 - 0.16)$

or $|z_m| = \sqrt{(10)^2 + (62.6)^2} = 63.8$ ohms where $R_m = 10$ kg/m, $\omega_m = 6280(0.01) = 62.8$, and $k/\omega = 1000/6280 = 0.16$.

The total impedance of the system is

$$z = |z_r| + |z_m| = 23.3 + 63.8 = 87.1 \text{ ohms}$$

from which $$v = F_0/z = 1/87.1 = 0.0115 \text{ m/sec}$$

and finally $$W = v^2R_r = (0.0115)^2(12.3) = 1.62(10)^{-3} \text{ watts}$$

(*d*) $$\text{Directivity factor } D = \frac{k^2a^2}{1 - 2J_1(2ka)/2ka} = 2.4$$

where $2ka = 1.83$ and $J_1(1.83) = 0.582$.

(*e*) $$\text{Directivity index } d = 10 \log D = 10 \log 2.4 = 3.8 \text{ db}$$

3.17. Investigate the radiation pattern of a square plane rigid piston of sides L mounted flush in an infinite plane baffle as shown in Fig. 3-5.

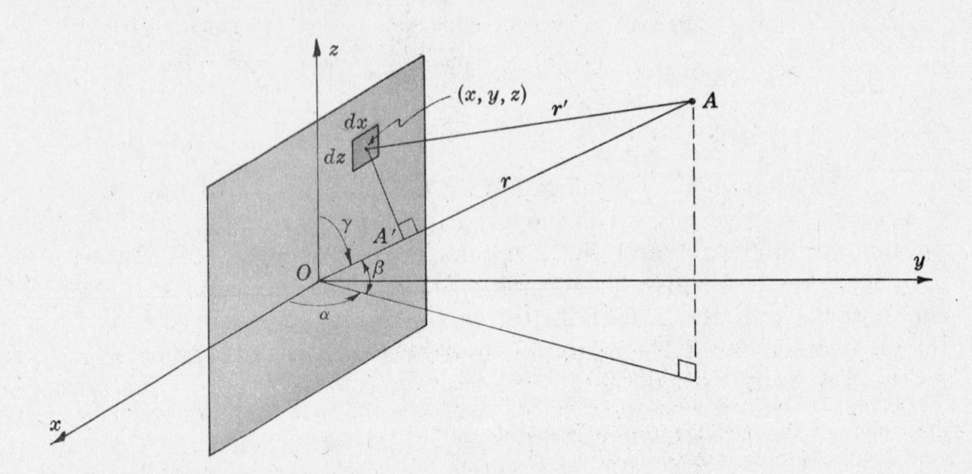

Fig. 3-5

In general, the radiation produced by the vibration of extended surfaces, such as pistons, diaphragms, or cones, do not have symmetrical spherical radiation patterns characteristic of a simple source. It is to be expected that these sources will have definite directional characteristics if their linear dimensions are comparable to the wavelength. The radiation produced by these sources can, however, be found by considering them to be assemblages of simple sources whose pressure at a point is given by

$$p = (i\rho c k v_0/4\pi r')e^{i(\omega t - kr')}$$

where r' is the distance from point A to the source, and v_0 is the velocity amplitude of the surface of the piston.

An elementary area of the surface of the piston, $dx\,dz$, can be considered to be a simple point source radiating into half of the infinite space to the right of the baffle. This amounts to twice the effect of the same source radiating into a free space. Then

$$dp = (i\rho c k v_0/2\pi r')e^{i(\omega t - kr')}\,dx\,dz$$

where r' is now the distance from point A to the $dx\,dz$ element. The total pressure at A due to the vibration of the entire piston is therefore found by integrating the above expression over the surface of the piston. Now

$$OA' = x\cos\alpha + z\cos\gamma$$

$$r' = r - OA' = r - (x\cos\alpha + z\cos\gamma)$$

and at great distance from the piston, $r = r'$, so we have

$$dp = \frac{i\rho c k v_0}{2\pi r}e^{i(\omega t - kr)}\int_{-L}^{L} dz \int_{-L}^{L} e^{ik(x\cos\alpha + z\cos\gamma)}\,dx$$

from which
$$p = \frac{2i\rho c k v_0 L^2}{\pi r}e^{i(\omega t - kr)}\left[\frac{\sin(ka\cos\alpha)}{ka\cos\alpha}\right]\left[\frac{\sin(ka\cos\gamma)}{ka\cos\gamma}\right]$$

The radiation in the yz plane can be determined by putting $\cos\gamma = \sin\beta$, and $\dfrac{\sin(ka\cos\alpha)}{ka\cos\alpha} = 1$ as α approaches 90°,

$$p = \frac{2i\rho c k v_0 L^2}{\pi r}e^{i(\omega t - kr)}\left[\frac{\sin(ka\sin\beta)}{ka\sin\beta}\right]$$

where $\dfrac{\sin(ka\sin\beta)}{ka\sin\beta}$ is known as the *directivity function* which determines the directional characteristics of the radiation of the source.

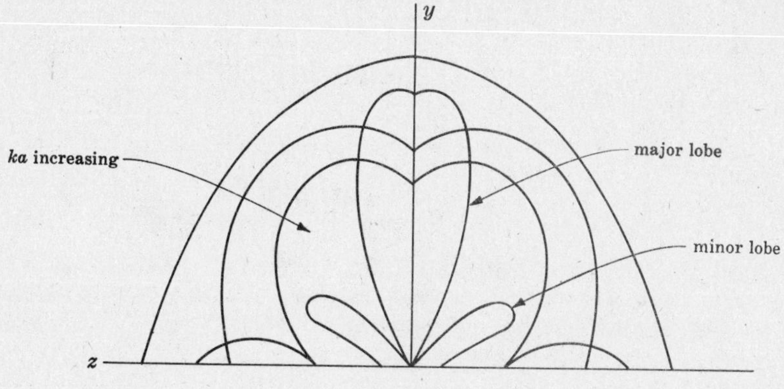

Fig. 3-6

The directivity function is plotted for increasing values of ka as shown in Fig. 3-6 above. It is clear that the directional pattern becomes more pronounced with higher frequencies or with increasing dimensions of the arrangement of sources. In other words, the greater the line dimensions of the radiator, the more pronounced the directivity will be. At the same time, minor lobes develop in addition to major lobes as the dimensions of the piston are increased.

Similar analysis can be applied to any other extended vibrating bodies in space. In general, pronounced directional effects will be observed when the frequencies are high enough so that wavelengths are comparable to the dimensions of the radiator.

3.18. A dynamic loudspeaker cone of diameter 0.2 m is mounted in an infinite baffle. Find the frequency at which the pressure amplitude along the wall is equal to one half of its axial value.

The loudspeaker cone may be regarded as a rigid circular piston of the same radius. From Problem 3.16, acoustic pressure amplitude ratio is given by

$$\frac{2J_1(ka \sin\theta)}{ka \sin\theta} = \frac{1}{2}$$

For $\theta = 90°$, we have

$$\frac{2J_1(ka)}{ka} = \frac{1}{2} \quad \text{or} \quad ka = 2$$

where J_1 is Bessel function of the first kind of order one. Thus

$$\omega = kc = (2/0.2)343 = 3430 \text{ rad/sec} \quad \text{or} \quad f = 546 \text{ cyc/sec}$$

3.19. A pulsating sphere of radius a is vibrating with a surface velocity amplitude v_0. Obtain expressions for the radiation resistance and radiation reactance acting upon the surface of the sphere.

For harmonic diverging spherical waves, the specific acoustic impedance is given by equation (5) of Problem 3.14,

$$z = \frac{\rho c k^2 r^2}{1 + k^2 r^2} + i\frac{\rho c k r}{1 + k^2 r^2}$$

where the real part is known as the specific acoustic resistance and the imaginary part is known as the specific acoustic reactance.

The radiation resistance acting upon the surface of the sphere is equal to the product of the area of the sphere and acoustic resistance of the medium in contact with the surface of the sphere,

$$R_r = 4\pi a^2\left(\frac{\rho c k^2 r^2}{1 + k^2 r^2}\right) = \frac{4\pi\rho c k^2 a^4}{1 + k^2 a^2} \text{ acoustic ohms}$$

Similarly the radiation reactance acting upon the surface of the sphere is given by

$$X_r = \frac{4\pi\rho c k a^3}{1 + k^2 a^2} \text{ acoustic ohms}$$

3.20. A pulsating sphere of radius 0.2 m is submerged in water. It radiates 200 watts of acoustic energy at a frequency of 1000 cyc/sec. Find the velocity amplitude of the sphere at the surface.

For harmonic diverging spherical waves, acoustic intensity at the surface of the sphere is given by equation (*3*) of Problem 3.9,

$$I_a = \frac{\rho c k^2 a^2 v_a^2}{2(1+k^2a^2)} \text{ watts/m}^2$$

where $\rho = 998 \text{ kg/m}^3$ is the density of water, $c = 1480$ m/sec is the speed of sound in water, $k = \omega/c = 1000(6.28)/1480 = 4.23$ is the wave number, $a = 0.2$ m is the radius of the sphere, and v_a is the velocity amplitude of the sphere at the surface.

Now acoustic power output at the surface of the sphere is

$$W_a = 4\pi a^2 I_a = \frac{2\pi\rho c k^2 a^4 v_a^2}{1+k^2a^2} \text{ watts}$$

Thus $$v_a = \sqrt{\frac{(1+k^2a^2)W_a}{2\pi\rho c k^2 a^4}} = \sqrt{\frac{(1+0.72)200}{6.28(998)(1480)(4.23)^2(0.2)^4}} = 0.115 \text{ m/sec}$$

SOURCE STRENGTH

3.21. Derive expressions for acoustic intensity and power radiated by a harmonic diverging spherical wave in terms of source strength.

Source strength of a pulsating sphere is defined as the product of its surface area and the velocity amplitude at its surface, i.e.

$$Q = 4\pi a^2 v_a \text{ m}^3\text{/sec} \tag{1}$$

where a is the radius of the sphere in meters, and v_a is the velocity amplitude at the surface in m/sec.

From equation (*3*) of Problem 3.9,

$$I_a = \frac{\rho c v_0^2 k^2 a^2}{2(1+k^2a^2)} \tag{2}$$

Because of the continuity of velocity at the boundary, $v = v_a$ and so

$$I_a = \frac{\rho c k^2 Q^2}{32\pi^2 a^2(1+k^2a^2)} \tag{3}$$

or in general $$I_r = \frac{a^2}{r^2} I_a = \frac{\rho c k^2 Q^2}{32\pi^2 r^2(1+k^2a^2)} \tag{4}$$

If the pulsating sphere is small compared with the wavelength, k^2a^2 is negligible and (*4*) becomes

$$I_r = \frac{\rho c k^2 Q^2}{32\pi^2 r^2} \text{ watts/m}^2 \tag{5}$$

The power radiated equals the product of the area of the surface and the intensity,

$$W = 4\pi r^2 I_r = \rho c k^2 Q^2/8\pi \text{ watts} \tag{6}$$

3.22. A hemispherical sound source of radius 0.2 m is mounted in an infinite baffle and radiates harmonic diverging spherical waves into water at a frequency of 500 cyc/sec. If the sound pressure level at a distance of 4 m from the source is 50 db re 2 microbars, determine the surface displacement amplitude of the source.

Sound pressure level SPL $= 20 \log (p/p_0) = 50$ db where p is the effective pressure and $p_0 = 2$ microbars or 0.2 nt/m^2 is the reference pressure. Then $20 \log p = 50 + 20 \log 0.2$ and $p = 63 \text{ nt/m}^2$.

From equation (*1*) of Problem 3.21,

$$Q = 4\pi a^2 v_a = 4\pi a^2 u_a \omega \text{ m}^3\text{/sec} \tag{1}$$

where $u_a = v_a/\omega$ is the surface displacement amplitude of the source. And from equation (*5*) of Problem 3.21,

$$I_a = \frac{\rho c k^2 Q^2}{32\pi^2 a^2} = \frac{p^2}{2\rho c} \tag{2}$$

Substituting equation (1) into (2) and solving for u_a, we obtain

$$u_a = \frac{p}{a\rho c k\omega} = \frac{p}{a\rho\omega^2} = \frac{63}{0.2(998)(500)^2(6.28)^2} = 3.2(10)^{-8}\text{ m}$$

where $k = \omega/c$ is the wave number and $\rho = 998\text{ kg/m}^3$ is the density of water.

3.23. Two simple sound sources S_1 and S_2 spaced a half wavelength apart radiate harmonic diverging waves of equal magnitudes uniformly in all directions. If the radiation of the sources are in phase with each other, study the sound radiation pattern of this arrangement.

Let the midpoint between S_1 and S_2 shown in Fig. 3-7 be the reference point O for the radiation pattern. Acoustic pressure at point A_1, a great distance from the sources, will be the vector sum of pressures radiated from S_1 and S_2.

For harmonic diverging spherical acoustic waves,

$$p = \frac{A}{r}e^{i(\omega t - kr)} = \frac{A}{r}e^{i(\omega t - 2\pi r/\lambda)}$$

which shows that the phase angle of acoustic pressure decreases linearly with the radial distance from the source.

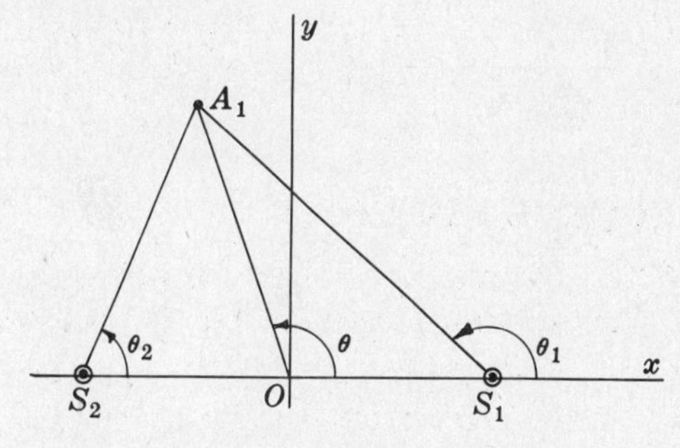

Fig. 3-7

Now $\theta_1 \doteq \theta_2 = \theta$, and sound waves from S_1 travel $\frac{1}{2}\lambda\cos\theta$ farther than waves from S_2 in reaching A_1. There will be a phase difference of $\frac{1}{2}\lambda(\cos\theta)(2\pi/\lambda)$ or $\pi\cos\theta$ rad between the waves. In other words, the wave from S_1 lags that from S_2 by $\pi\cos\theta$ rad. Acoustic pressure at A_1 becomes

$$p = \frac{A}{r}e^{i\omega t} + \frac{A}{r}e^{i(\omega t - \pi\cos\theta)}$$

When $\theta = 0$, we have two sound waves of equal magnitude but 180° out of phase with each other; hence $p = 0$. When $\theta = 90°$, we have two sound waves of same magnitude and phase; hence $p_0 = 2A/r$.

Continuing in this fashion with a locus of points equidistant from the reference point O, we obtain a polar plot of pressure versus angular displacements as shown in Fig. 3-8, which is the radiation pattern or directivity of this particular arrangement of two simple sound sources.

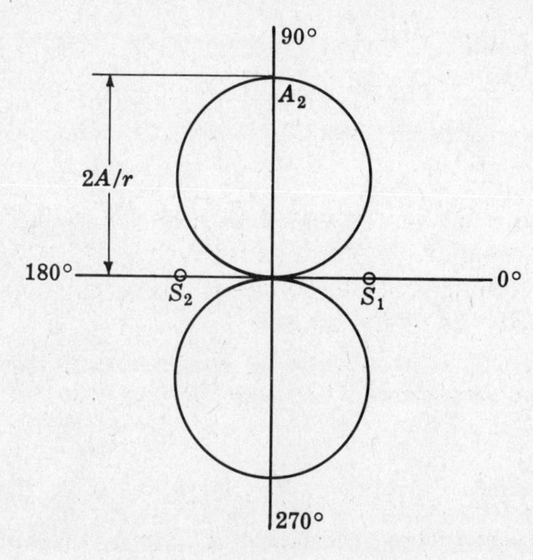

Fig. 3-8

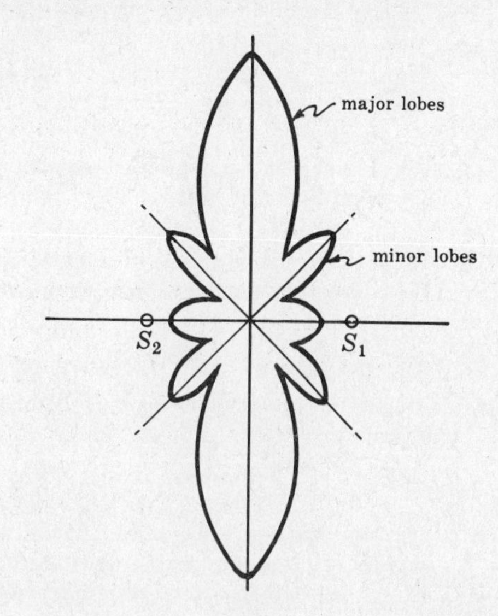

Fig. 3-9

The magnitude of the acoustic pressure at any point in this two-dimensional plot is given by the radial distance from the origin O to the point in question, e.g. at A_2, the pressure is $2A/r$, given by the line OA_2.

In general, the larger the extent of the radiator (here we mean the spacing between the sources), the sharper will be the major lobe and the greater the number of side lobes. The greater the number of the sources, the smaller will be the side lobes as shown in Fig. 3-9 above.

Practically all sound radiators have pronounced directional effects. This is particularly true when the source is radiating sound waves at high frequencies. The analysis and polar plot are similar.

Supplementary Problems

WAVE EQUATION

3.24. Obtain an expression for a two-dimensional wave traveling in the xy plane with velocity c in a direction at an angle θ to the x axis. *Ans.* $\phi(x,y,t) = f(x\cos\theta + y\sin\theta - ct)$

3.25. Show that $p = f(lx+my+nz-ct) + g(lx+my+nz+ct)$ represents the standing waves form of solution for equation (*9*) of Problem 3.1, page 68.

3.26. Prove that $p = \dfrac{A}{r}\cos(\omega t - kr)$ is a possible solution for the spherical acoustic wave equation (equation (*4*) of Problem 3.8, page 75).

3.27. Show that $p = \dfrac{A}{r}(ct - r)$ is a possible solution for equation (*4*) of Problem 3.8.

3.28. Compute the three lowest frequencies of a rectangular room of dimensions $10 \times 15 \times 20$ meters. *Ans.* 28, 37, 50 cyc/sec

WAVE ELEMENTS

3.29. Show that the velocity amplitude of a harmonic diverging spherical acoustic wave is not inversely proportional to the distance of the wave from the source.

3.30. For plane and spherical acoustic waves of the same frequency, find the ratios of their particle velocity amplitudes and particle displacement amplitudes. *Ans.* 1/110

3.31. Show that acoustic pressure and particle velocity of harmonic diverging spherical waves are essentially in phase at great distances from the source.

3.32. What is the phase angle between particle displacement and particle velocity of harmonic diverging spherical acoustic waves? *Ans.* 90°

3.33. Prove that the maxima of kinetic and potential energies at any point in a harmonic diverging spherical acoustic wave are equal.

ACOUSTIC INTENSITY AND ENERGY DENSITY

3.34. A simple underwater sound source radiates 10 watts of acoustic power at a frequency of 500 cyc/sec. Find the intensity and acoustic pressure at a distance of 5 m from the source. *Ans.* $I = 0.032$ watt/m², $p = 22$ nt/m²

3.35. The minimum audible sound wave (assume harmonic diverging spherical waves) at the frequency to which the human ear is most sensitive is 3500 cyc/sec at an effective pressure of $8(10)^{-6}$ nt/m². Find the corresponding intensity, velocity amplitude, and displacement amplitude.
Ans. $I = 1.55(10)^{-8}$ watt/m², $v_0 = 2.74(10)^{-8}$ m/sec, $u_0 = 1.25(10)^{-12}$ m

3.36. An isolated point source of sound of strength Q_0 radiates harmonic diverging spherical waves into free space. Find the average energy radiated and the specific acoustic impedance.
Ans. $E_{ave} = Q_0^2 k^2 \rho_0 c/8\pi, \; z = 1/(1+1/ikr)$

3.37. An infinite circular cylinder has a uniform membrane at its open end. The membrane vibrates with velocity $v = v_0 e^{i\omega t}$. Determine the reaction due to acoustic pressure on the membrane.
Ans. $p = \rho c v_0 e^{-ikx}$

SPECIFIC ACOUSTIC IMPEDANCE

3.38. If $kr = 100$, what is the ratio between the specific acoustic resistance and the specific acoustic reactance of a harmonic diverging spherical wave? *Ans.* 100

3.39. For harmonic diverging spherical waves, what is the maximum value of the specific acoustic resistance? *Ans.* $\frac{1}{2}\rho c$

RADIATION OF SOUND

3.40. Two simple sound sources of equal strength but pulsating with a phase difference of 180° are spaced a half wavelength apart. Determine the radiation pattern.
Ans. A figure eight with axis along the 0° line joining the sources

3.41. Determine the radiation pattern of two simple sound sources of equal strength but 90° out of phase with each other and separated by one quarter wavelength. *Ans.* Cardioid

3.42. Derive an expression for acoustic pressure at a point due to n equidistant simple sound sources all in a straight line and identical in strength, frequency, and phase angle.

Ans. $$p = \frac{\sin[(n\pi d/\lambda)\cos\theta]}{n\sin[(\pi d/\lambda)\cos\theta]}$$

3.43. Show that the directivity index for a nondirectional spherical source is equal to zero at all angles.

3.44. Six simple sound sources identical in strength, frequency, and phase angle are spaced a half wavelength apart in a straight line. Find the angles at which (*a*) maxima and (*b*) zero amplitudes occur. *Ans.* (*a*) 90°, 60°, 30°; (*b*) 71°, 49°, 0°

3.45. A piston source of radius 0.1 m radiates sound in still air at a frequency of 1000 cyc/sec. Find the beam width for down 6 db. *Ans.* 90°

3.46. The first lobe of acoustic pressure occurs at $ka \sin\theta_1 = 3.83$, while the second lobe occurs at $ka\sin\theta_2 = 7.02$ for pressure distribution by a piston source. Prove that $(p_{\theta_n})_{max} > (p_{\theta_{n+1}})_{max}$.

3.47. Derive an expression for acoustic pressure at a point a great distance r from a circular rigid piston source mounted flush in an infinite baffle.

Ans. $$p = \frac{i\rho ckQ}{2\pi t} e^{i(\omega t - kr)} \left[\frac{2J_1(ka\sin\theta)}{ka\sin\theta}\right]$$

Chapter 4

Transmission of Sound

NOMENCLATURE

a = radius, m
A = area, m^2
B = bulk modulus, nt/m^2
c = speed of sound in air, m/sec
f = frequency, cyc/sec
I_i = incident sound intensity, $watts/m^2$
I_t = transmitted sound intensity, $watts/m^2$
I_r = reflected sound intensity, $watts/m^2$
k = wave number
K = complex reflection coefficient
L = thickness, length, m
p = acoustic pressure, nt/m^2
R = characteristic impedance, rayls
r_n = normal specific acoustic resistance, rayls
s = condensation
SWR = standing wave ratio
TL = transmission loss, db
u = particle displacement, m
v = particle velocity, m/sec
W = acoustic power, watts
x_n = normal specific acoustic reactance, rayls
z = specific acoustic impedance, rayls
z_n = normal specific acoustic impedance, rayls
z_s = specific acoustic impedance, rayls
ω = circular frequency, rad/sec
λ = wavelength, m
α = absorbing coefficient; viscous attenuation constant, nepers/m
α_r = sound power reflection coefficient
α_t = sound power transmission coefficient
ρ = density, kg/m^3
η = viscosity coefficient, $nt\text{-}sec/m^2$
τ = viscous relaxation time, sec

INTRODUCTION

When sound waves are traveling through a medium, they may be reflected or refracted, diffracted or scattered, interferred or absorbed. The transmission of sound involves the transfer of acoustic energy through the medium in which sound waves travel.

TRANSMISSION THROUGH TWO MEDIA

For the transmission of sinusoidal plane acoustic waves from one fluid medium to another at normal incidence along the plane interface of the two media, *sound power reflection coefficient* α_r is defined as the ratio of the reflected flow of sound energy to the incident flow of sound energy:

$$\alpha_r = \left[\frac{\rho_2 c_2 - \rho_1 c_1}{\rho_2 c_2 + \rho_1 c_1}\right]^2 = \left[\frac{R_2 - R_1}{R_2 + R_1}\right]^2$$

Sound power transmission coefficient α_t is similarly defined as the ratio of the transmitted sound power to the incident sound power:

$$\alpha_t = \frac{4\rho_1 c_1 \rho_2 c_2}{(\rho_1 c_1 + \rho_2 c_2)^2} = \frac{4R_1 R_2}{(R_1 + R_2)^2}$$

where the ρ's are the densities and the c's are the speeds of sound. (See Problems 4.1-4.4.)

For normal incidence at surfaces of solids, the reflected and transmitted sound power coefficients can be expressed in terms of the *normal specific impedance* $z_n = r_n + ix_n$ which characterizes the behavior of solids with sound waves:

$$\alpha_r = \frac{(r_n - \rho_1 c_1)^2 + x_n^2}{(r_n + \rho_1 c_1)^2 + x_n^2}, \qquad \alpha_t = \frac{4\rho_1 c_1 r_n}{(r_n + \rho_1 c_1)^2 + x_n^2}$$

where r_n is the *resistive* component and x_n is the *reactive* component. (See Problems 4.5, 4.6.)

For the transmission of sound waves from one fluid medium to another at *oblique incidence*, the sound power reflection and transmission coefficients are given by

$$\alpha_r = \left[\frac{R_2 \cos\theta_i - R_1 \cos\theta_t}{R_2 \cos\theta_i + R_1 \cos\theta_t}\right]^2, \qquad \alpha_t = \frac{4R_1 R_2 \cos\theta_i \cos\theta_t}{(R_2 \cos\theta_i + R_1 \cos\theta_t)^2}$$

where θ_i is the angle of incidence and θ_t is the angle of refraction. (See Problems 4.5, 4.20.)

Sound power reflection and transmission coefficients for sound waves in air impinging at oblique incidence on the surface of a normally reacting solid are

$$\alpha_r = \frac{(r_n \cos\theta_i - R_1)^2 + x_n^2 \cos^2\theta_i}{(r_n \cos\theta_i + R_1)^2 + x_n^2 \cos^2\theta_i}, \qquad \alpha_t = \frac{4R_1 r_n \cos\theta_i}{(r_n \cos\theta_i + R_1)^2 + x_n^2 \cos^2\theta_i}$$

(See Problems 4.21, 4.22.)

TRANSMISSION THROUGH THREE MEDIA

The transmission of sinusoidal plane acoustic waves from one fluid medium through a second and into a third fluid medium is similar to transmission through two media. Reflected waves will be generated at the plane interfaces of the fluid media, and part of the incident wave will be transmitted through the boundaries. The sound power transmission coefficient from medium 1 through 2 into medium 3 is given by

$$\alpha_t = \frac{4R_1 R_3}{(R_1 + R_3)^2 \cos^2 k_2 L + (R_2 + R_1 R_3/R_2)^2 \sin^2 k_2 L}$$

where the R's are the characteristic impedances of the media, $k_2 = \omega/c$ is the wave number of medium 2, and L is the thickness of medium 2. (See Problems 4.10-4.14.)

Transmission loss is the difference in decibels between the sound energy striking the surface separating two spaces and the energy transmitted. It cannot be measured directly but is computed from sound pressure level measurements on both sides of the partition. Transmission loss TL can be expressed as

$$\text{TL} = 10 \log (I_i/I_t) \text{ db}$$

where I_i is the incident sound intensity and I_t is the transmitted sound intensity. (See Problems 7.12-7.14.)

REFLECTION OF SOUND

In general, a sound wave will be reflected whenever there is a discontinuity and interface of two media in which it is propagated. The reflected wave depends on the incident wave, the angle of incidence, the reflecting surface, and the characteristic impedances of the media. The reflected flow of sound energy is proportional to the square of the amplitude of the reflected sound wave.

Standing wave ratio SWR is defined as the ratio of acoustic pressure at an antinode to acoustic pressure at a node or as the ratio of maximum to minimum amplitudes in a standing wave. It serves as an indication of the amount of sound energy reflected at the boundary.

$$\text{SWR} = \frac{p_{\max}}{p_{\min}} = \frac{A_{\max}}{A_{\min}} = \frac{p_i + p_r}{p_i - p_r} \quad \text{or} \quad \frac{p_r}{p_i} = \frac{\text{SWR} - 1}{\text{SWR} + 1}$$

For total reflection of sound waves, $\text{SWR} = \infty$, or $p_r/p_i = 1$. For zero reflection of sound waves, $\text{SWR} = 1$ or $p_r/p_i = 0$.

Law of reflection: The angle of incidence equals the angle of reflection.

Snell's law:
$$\frac{c_{\text{before}}}{(\sin \theta)_{\text{before}}} = \frac{c_{\text{after}}}{(\sin \theta)_{\text{after}}}$$

Echo is a definite or distinct, separate or delayed sound heard by an observer as the result of reflection of sound. A reflected sound produced within 1/10 second interval of the original sound will not be detected by the human ear and thus merges with the original sound to give rise to *reverberation* or *overlapping echo*. A *musical echo* is the rapid and successive reflection of a sound, and *flutter echoes* are pulses reflecting back and forth from one end to the other end of an enclosure with diminishing amplitude.

The phenomenon of echo has many practical applications such as navigation and traveling, direction finding and ranging, detection of submerged vehicles and objects, and ultrasonic flaw detection. (See Problems 4.15-4.19.)

REFRACTION OF SOUND

When sound waves arrive at a discontinuity or boundary, some will be reflected and the rest cross the boundary to form transmitted waves. When the angle of incidence is greater than the *critical angle*, all the waves are reflected and none crosses the boundary. The direction of propagation of the transmitted waves is not the same as that of the incident waves. The transmitted waves are bent toward or away from the normal to the boundary in accordance with the speeds of sound in the media. This is *refraction* of sound. (See Chapter 8.)

Refraction of sound can take place in a single medium such as the earth's atmosphere or a large body of fluid such as the sea because of the effect of wind or temperature variations from place to place. (See Problems 4.20-4.22.)

DIFFRACTION OF SOUND

When sound waves meet an obstacle, they will spread around the edges of the obstacle to give rise to *diffraction* of sound. In other words, sound waves are bent or their directions of propagation are changed due to the obstacles placed in their paths. Also, sound waves are diffracted rather than reflected if their wavelengths are comparable with the dimensions of the reflecting objects. (See Problem 4.23.)

SCATTERING OF SOUND

Sound waves will be scattered in all directions when they strike obstacles of dimensions small compared with their wavelengths. This is in contrast with reflection or diffraction of sound.

The amplitude of the scattered waves at great distances from the obstacle is directly proportional to the volume of the obstacle and inversely proportional to the square of the wavelength. Hence sound of long wavelength will have little scattering effect whereas sound of short wavelength will have great scattering effect.

Diffuse echo is produced by the scattering of sound by a collection of small obstacles. A *harmonic echo* is the result of the differential scattering of a complex sound or noise of different frequencies.

INTERFERENCE

If sound waves of the same frequency and amplitude are superposed, they either neutralize or reinforce each other's effects. The phenomenon is described as *interference*, i.e. the resultant effect at each point in the medium is the algebraic sum of the effects of the two waves. Destructive interference occurs at points where they meet in opposite phase, and constructive interference occurs at points where they meet in phase.

Standing or *stationary waves* are formed from the interference of two sound waves of equal amplitude and frequency propagated through a medium along the same line in opposite directions. There will be fixed positions of zero amplitude (*nodes*) and fixed positions of maximum amplitude (*antinodes*), and the medium is set into steady state vibration.

Beats will be produced, as in the case of mechanical vibrations, from the interference of two sound waves of slightly different frequencies. (See Problems 4.15-4.19.)

FILTRATION OF SOUND

Filtration of sound, like any other forms of filtration, is a process employed to eliminate some portion of the sound waves of definite frequencies and wavelengths while letting the rest pass. In fact, this amounts to a selective passage of sound waves.

Acoustic filters, e.g. mufflers, plenum chambers, resonators, sound traps or silencers, and hydraulic filters, are devices used for separating components of a signal or sound on the basis of their frequency. They allow components of sound in one or more frequency bands to pass relatively unattenuated, but attenuate components of sound in other frequency bands. (Fig. 4-1 below.) (See Problems 4.24-4.26.)

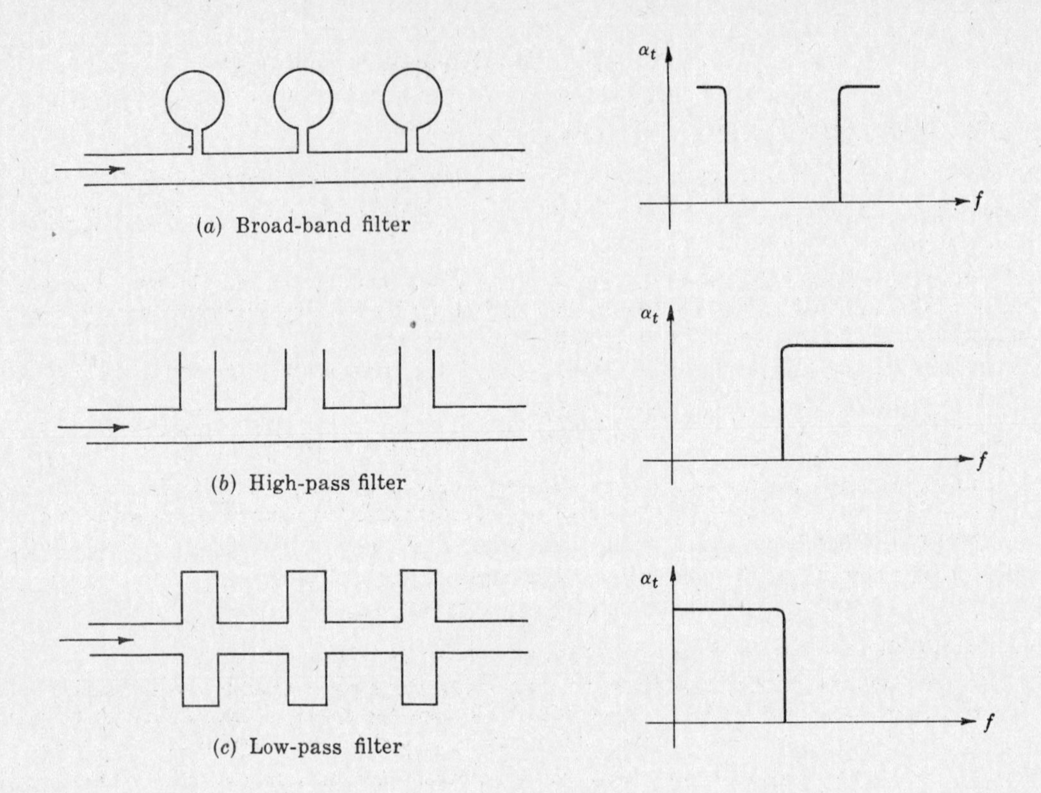

(a) Broad-band filter

(b) High-pass filter

(c) Low-pass filter

Fig. 4-1

ABSORPTION OF SOUND

A sound wave may lose some of its energy while propagating through a fluid or solid medium. This loss of acoustical energy is due to absorption.

Viscous losses of sound energy in fluid media arise from shear stresses set up in the media by the passage of compressive waves through the media. *Heat* or *conduction losses* are due to flow of heat from the slightly warmer compressed portion to the slightly cooler expanded portion of the fluid. *Molecular energy losses* result from thermal relaxation which causes exchanges of energy between different internal thermal states of the molecules. Absorption of sound energy occurs if the phase of these exchanges of energy differs from that of the sound waves. In air, for example, absorption of sound energy increases rapidly with increasing frequency; and in water, absorption of sound energy can be caused by scattering effect due to nonhomogeneities in the structure of water.

Absorption of sound in solids is caused by the interactions between sound waves and lattice vibrations, sound waves and electron motion, and ferromagnetic and ferroelectric effects. (See Problems 4.27-4.28.)

Solved Problems

TRANSMISSION THROUGH TWO MEDIA

4.1. For the transmission of sound waves from one fluid medium to another, derive an expression for (*a*) particle displacement, (*b*) particle velocity, (*c*) acoustic pressure, and (*d*) condensation.

When sound waves strike at right angles to a plane interface of two different fluid media, a wave will be reflected back along the original path in medium 1, and a second wave will be transmitted through the boundary into medium 2.

(*a*) Particle displacement.

The waves in medium 1 are given by

$$u_1 = A_i e^{i(\omega t - k_1 x)} + A_r e^{i(\omega t + k_1 x)} \tag{1}$$

where the first term represents an incident wave traveling in the positive x direction with speed $c_1 = \omega/k_1$, and the second term represents a reflected wave traveling in the negative x direction also with speed c_1.

The transmitted wave in medium 2 is given by

$$u_2 = A_t e^{i(\omega t - k_2 x)} \tag{2}$$

which travels in the positive x direction with speed $c_2 = \omega/k_2$.

We assume the transmitted wave always has the same frequency as the incident wave, and so we have ignored any Doppler effect. Because the speeds of sound are different in the two media, the magnitudes of the wave numbers k_1 and k_2 are different, i.e. $\omega = c_1 k_1 = c_2 k_2$.

At the plane interface of the two media, acoustic pressures on both sides of the boundary $(x = 0)$ are equal, and particle velocities normal to the interface are also equal, i.e. acoustic pressure must be continuous, and the two media must remain in contact at the boundary at all times.

Acoustic pressures in media 1 and 2 are

$$p_1 = -B_1 \frac{\partial u_1}{\partial x} = ik_1 B_1 e^{i\omega t}(A_i e^{-ik_1 x} - A_r e^{ik_1 x}) \tag{3}$$

$$p_2 = -B_2 \frac{\partial u_2}{\partial x} = ik_2 B_2 e^{i\omega t}(A_t e^{-ik_2 x}) \tag{4}$$

and particle velocities are

$$v_1 = \frac{\partial u_1}{\partial t} = i\omega e^{i\omega t}(A_i e^{-ik_1 x} + A_r e^{ik_1 x}) \tag{5}$$

$$v_2 = \frac{\partial u_2}{\partial t} = i\omega e^{i\omega t}(A_t e^{-ik_2 x}) \tag{6}$$

At the boundary, $x = 0$, equations (*3*), (*4*), (*5*), (*6*) become

$$p_1 = ik_1 B_1 e^{i\omega t}(A_i - A_r)$$
$$p_2 = ik_2 B_2 e^{i\omega t} A_t$$
$$v_1 = i\omega e^{i\omega t}(A_i + A_r)$$
$$v_2 = i\omega e^{i\omega t} A_t$$

Equating acoustic pressures and particle velocities at the boundary, we obtain

$$B_1 k_1 (A_i - A_r) = B_2 k_2 A_t \tag{7}$$

$$A_i + A_r = A_t \tag{8}$$

Eliminating A_t from equations (*7*) and (*8*),

$$\frac{A_r}{A_i} = \frac{\rho_1 c_1 - \rho_2 c_2}{\rho_1 c_1 + \rho_2 c_2} \tag{9}$$

and eliminating A_r from equations (*7*) and (*8*),

$$\frac{A_t}{A_i} = \frac{2\rho_1 c_1}{\rho_1 c_1 + \rho_2 c_2} \tag{10}$$

where the bulk modulus $B = \rho c^2$, $Bk = \rho c\omega$, ρ is the density of the medium, and c is the speed of sound in the medium. A_r/A_i is called the reflection coefficient for displacement while A_t/A_i is the transmission coefficient for displacement amplitude.

Using z_2, the specific acoustic impedance, instead of the purely resistive impedance $\rho_2 c_2$ at the boundary for the terminating medium, equations (9) and (10) become

$$\frac{A_r}{A_i} = \frac{\rho_1 c_1 - z_2}{\rho_1 c_1 + z_2} = \frac{R_1 - z_2}{R_1 + z_2}, \qquad \frac{A_t}{A_i} = \frac{2R_1}{R_1 + z_2} \qquad (9)', (10)'$$

which are known as the complex reflection and transmission coefficients for displacement amplitude. For the limiting case $R_1 = R_2$, we have $A_r = 0$ and $A_t = A_i$. This agrees with the physical situation of a continuous medium. If $R_2 \gg R_1$, $A_r = -A_i$ and $A_t = 0$. If $R_1 \gg R_2$, $A_r = A_i$ and $A_t = 2A_i$. (See Problem 4.2.)

(*b*) Particle velocity.

Particle velocities for incident, reflected and transmitted waves are

$$v_i = \frac{\partial u_i}{\partial t} = i\omega u_i, \qquad v_r = \frac{\partial u_r}{\partial t} = i\omega u_r, \qquad v_t = \frac{\partial u_t}{\partial t} = i\omega u_t$$

and hence the reflection coefficient for velocity amplitude from medium 1 to medium 2 is

$$\frac{v_r}{v_i} = \frac{i\omega u_r}{i\omega u_i} = \frac{u_r}{u_i} = \frac{A_r}{A_i} = \frac{\rho_1 c_1 - \rho_2 c_2}{\rho_1 c_1 + \rho_2 c_2} \qquad (11)$$

which is the same for particle displacement as given by (9). Similarly, the transmission coefficient for velocity amplitude is

$$\frac{v_t}{v_i} = \frac{2\rho_1 c_1}{\rho_1 c_1 + \rho_2 c_2} \qquad (12)$$

and if the terminating medium 2 is not infinite in extent, we have the corresponding complex reflection and transmission coefficients for velocity amplitude:

$$\frac{v_r}{v_i} = \frac{R_1 - z_2}{R_1 + z_2}, \qquad \frac{v_t}{v_i} = \frac{2R_1}{R_1 + z_2} \qquad (11)', (12)'$$

(*c*) Acoustic pressure.

The acoustic pressure in medium 1 consists of two parts:

$$p_1 = p_i + p_r = P_i e^{i(\omega t - k_1 x)} + P_r e^{i(\omega t + k_1 x)}$$

and the acoustic pressure in medium 2 is simply

$$p_2 = P_t e^{i(\omega t - k_2 x)}$$

Now the pressures at the boundary $x = 0$ are equal,

$$(p_1)_{x=0} = (p_2)_{x=0} \quad \text{or} \quad P_i + P_r = P_t \qquad (13)$$

We have defined the ratio of acoustic pressure in a medium to the associated particle velocity as the specific acoustic impedance, i.e. $z = p/v$; then

$$v_i = P_i/\rho_1 c_1, \quad v_r = -P_r/\rho_1 c_1, \quad v_t = P_t/\rho_2 c_2$$

Since the velocities at the boundary are also equal, we have

$$(v_i)_{x=0} + (v_r)_{x=0} = (v_t)_{x=0}$$

or

$$P_i/\rho_1 c_1 - P_r/\rho_1 c_1 = P_t/\rho_2 c_2 \qquad (14)$$

Eliminating P_t from equations (13) and (14),

$$\frac{P_r}{P_i} = \frac{\rho_2 c_2 - \rho_1 c_1}{\rho_2 c_2 + \rho_1 c_1} = \frac{R_2 - R_1}{R_2 + R_1} \qquad (15)$$

and eliminating P_r from equations (13) and (14),

$$\frac{P_t}{P_i} = \frac{2\rho_2 c_2}{\rho_1 c_1 + \rho_2 c_2} = \frac{2R_2}{R_1 + R_2} \qquad (16)$$

Equation (15) is the reflection coefficient for pressure amplitude while (16) is the transmission coefficient for pressure amplitude. If the terminating medium 2 is not infinite in extent, we obtain the corresponding complex reflection and transmission coefficients for pressure amplitude:

$$\frac{P_r}{P_i} = \frac{z_2 - R_1}{z_2 + R_1}, \qquad \frac{P_t}{P_i} = \frac{2R_2}{R_1 + z_2} \tag{15)', (16)'$$

Thus we may write the reflection coefficients for displacement, velocity, and pressure as

$$\frac{A_r}{A_i} = \frac{R_1 - R_2}{R_1 + R_2} = \frac{v_r}{v_i} = -\frac{P_r}{P_i}$$

We see that particle displacement and particle velocity in the reflected wave are in phase with each other, but 180° out of phase with acoustic pressure of the reflected wave.

The acoustic pressure of the reflected wave at the boundary is therefore either in phase or 180° out of phase with that of the incident wave at the boundary, depending on the values of the characteristic impedances of the media. If the second medium is very dense, R_2 is much greater than R_1, and $P_r = P_i$. The pressure amplitude at the boundary is an antinode, and no phase change takes place between the reflected and incident wave. If the second medium is a rarefied medium, R_2 is much less than R_1, and $P_r = -P_i$. The pressure amplitude at the boundary is a node, and a phase difference of 180° exists between the incident and reflected waves.

(*d*) Condensation.

The incident, reflected, and transmitted condensations are

$$s_i = -\frac{\partial u_i}{\partial x} = ik_1 u_1, \qquad s_r = -\frac{\partial u_r}{\partial x} = -ik_1 u_r, \qquad s_t = -\frac{\partial u_t}{\partial x} = ik_2 u_t$$

Therefore the reflection coefficient for condensation amplitude from medium 1 to medium 2 is given by

$$\frac{s_r}{s_i} = \frac{-ik_i u_r}{ik_1 u_i} = \frac{-u_r}{u_i} = \frac{-A_r}{A_i} = \frac{\rho_2 c_2 - \rho_1 c_1}{\rho_2 c_2 + \rho_1 c_1} \tag{17}$$

and the transmission coefficient for condensation amplitude from medium 1 to medium 2 is similarly given by

$$\frac{s_t}{s_i} = \frac{ik_2 u_t}{ik_1 u_i} = \frac{k_2 A_t}{k_1 A_i} = \frac{(\omega/c_2)A_t}{(\omega/c_1)A_i} = \frac{2\rho_1 c_1^2}{c_2(\rho_1 c_1 + \rho_2 c_2)} \tag{18}$$

If the terminating medium 2 is not infinite in extent, the complex reflection and transmission coefficients for condensation amplitude become

$$\frac{s_r}{s_i} = \frac{z_2 - R_1}{z_2 + R_1}, \qquad \frac{s_t}{s_i} = \frac{2R_1 c_1}{c_2(z_2 + R_1)} \tag{17)', (18)'$$

4.2. Derive expressions for the transmission of sound energy from one fluid medium to another.

The average power per unit area for the incident, reflected, and transmitted waves is respectively

$$W_i = \tfrac{1}{2}\rho_1 c_1 A_i^2 \omega^2, \qquad W_r = \tfrac{1}{2}\rho_1 c_1 A_r^2 \omega^2, \qquad W_t = \tfrac{1}{2}\rho_2 c_2 A_t^2 \omega^2 \tag{1}$$

where the A's are displacement amplitudes, ω is the angular frequency of the sound wave, $\rho_1 c_1 = R_1$ and $\rho_2 c_2 = R_2$ are the characteristic impedances of the media.

Therefore the ratio of the reflected flow of energy to the incident flow of energy is

$$\alpha_r = \frac{W_r}{W_i} = \frac{A_r^2}{A_i^2} = \left[\frac{\rho_1 c_1 - \rho_2 c_2}{\rho_1 c_1 + \rho_2 c_2}\right]^2 = \left[\frac{R_1 - R_2}{R_1 + R_2}\right]^2 \tag{2}$$

Since intensity $I = p^2/2\rho c$, and sound energy is proportional to its intensity, we can express equation (2) as

$$\alpha_r = \frac{I_r}{I_i} = \frac{p_r^2/2\rho_1 c_1}{p_i^2/2\rho_2 c_2} = \frac{p_r^2}{p_i^2} = \left[\frac{R_1 - R_2}{R_1 + R_2}\right]^2 \tag{3}$$

where α_r is the sound power reflection coefficient or simply reflection coefficient.

Similarly, the sound power transmission coefficient is the ratio of the transmitted flow of sound energy to the incident flow of sound energy:

$$\alpha_t = \frac{W_t}{W_i} = \frac{R_2 A_t^2}{R_1 A_i^2} = \frac{4R_1^2 R_2}{R_1(R_1+R_2)^2} = \frac{4R_1 R_2}{(R_1+R_2)^2} \tag{4}$$

or

$$\alpha_t = \frac{I_t}{I_i} = \frac{P_t^2/2R_2}{P_i^2/2R_1} = \frac{R_1}{R_2}\left[\frac{2R_2}{(R_1+R_2)}\right]^2 = \frac{4R_1 R_2}{(R_1+R_2)^2} \tag{5}$$

4.3. A plane sinusoidal longitudinal wave in water is incident normally on a boundary between water and ice. If each medium can be assumed to be infinite in extent, compute the following amplitude ratios: (*a*) u_r/u_i, (*b*) u_t/u_i, (*c*) v_r/v_i, (*d*) v_t/v_i, (*e*) p_r/p_i, (*f*) p_t/p_i, (*g*) s_r/s_i, (*h*) s_t/s_i, (*i*) I_r/I_i, (*j*) I_t/I_i where u, v, p, s and I are respectively the particle displacement, particle velocity, acoustic pressure, condensation and intensity amplitudes, and where the subscripts r, i and t indicate whether these terms are reflected, incident or transmitted. Find also their respective phase angles.

At standard temperature and atmospheric pressure, we have

$$(\rho c)_{\text{water}} = (\rho c)_1 = (998)(1480) \doteq 1{,}480{,}000 \text{ rayls}$$

$$(\rho c)_{\text{ice}} = (\rho c)_2 = (920)(3200) \doteq 2{,}940{,}000 \text{ rayls}$$

where ρ is the density in kg/m³ and c is the speed of sound in m/sec.

(*a*) $$\frac{u_r}{u_i} = \frac{(\rho c)_1 - (\rho c)_2}{(\rho c)_1 + (\rho c)_2} = \frac{(1.48 - 2.94)10^6}{(1.48+2.94)10^6} = -0.33$$

u_r is 180° out of phase with u_i.

(*b*) $$\frac{u_t}{u_i} = \frac{2(\rho c)_1}{(\rho c)_1 + (\rho c)_2} = \frac{2(1.48)10^6}{(1.48+2.94)10^6} = 0.67$$

u_t is in phase with u_i.

(*c*) $$\frac{v_r}{v_i} = \frac{(\rho c)_1 - (\rho c)_2}{(\rho c)_1 + (\rho c)_2} = -0.33 \text{ as in } (a)$$

v_r is 180° out of phase with v_i.

(*d*) $$\frac{v_t}{v_i} = \frac{2(\rho c)_1}{(\rho c)_1 + (\rho c)_2} = 0.67 \text{ as in } (b).$$

v_t is in phase with v_i.

(*e*) $$\frac{p_r}{p_i} = \frac{(\rho c)_2 - (\rho c)_1}{(\rho c)_2 + (\rho c)_1} = \frac{(2.94 - 1.48)10^6}{(2.94+1.48)10^6} = 0.33$$

p_r is in phase with p_i.

(*f*) $$\frac{p_t}{p_i} = \frac{2(\rho c)_2}{(\rho c)_2 + (\rho c)_1} = \frac{2(2.94)10^6}{(2.94+1.48)10^6} = 1.33$$

p_t is in phase with p_i.

(*g*) $$\frac{s_r}{s_i} = \frac{(\rho c)_2 - (\rho c)_1}{(\rho c)_2 + (\rho c)_1} = 0.33 \text{ as in } (e)$$

s_r is in phase with s_i.

(*h*) $$\frac{s_t}{s_i} = \frac{2(\rho_1 c_1^2)}{c_2(\rho_1 c_1 + \rho_2 c_2)} = \frac{2(1.48)(1480)10^6}{3200(1.48+2.94)10^6} = 0.3$$

s_t is in phase with s_i.

(*i*) $I_r/I_i = (u_r/u_i)^2 = (-0.33)^2 = 0.109$, see (*a*)

I_r is in phase with I_i.

(*j*) $I_t/I_i = 1 - I_r/I_i = 1 - 0.109 = 0.891$

I_t is in phase with I_i.

4.4. A beam of sound waves is incident normally on a plane interface of air and an infinite body of fluid of unknown impedance. If half of the sound energy is reflected, find the unknown impedance.

Sound energy reflected is described by the sound power reflection coefficient given by equation (2) of Problem 4.2,

$$\alpha_r = \left[\frac{R_1 - z_2}{R_1 + z_2}\right]^2 = \left[\frac{415 - z_2}{415 + z_2}\right]^2 = 0.5 \quad \text{or} \quad z_2 = 72 \text{ rayls}$$

where $R_1 = \rho_1 c_1 = 1.21(343) = 415$ rayls is the characteristic impedance of air, and z_2 is the characteristic impedance of the fluid.

4.5. Derive an expression for the sound power reflection coefficient for plane acoustic waves in air impinging at oblique incidence on the surface of a normally reacting solid.

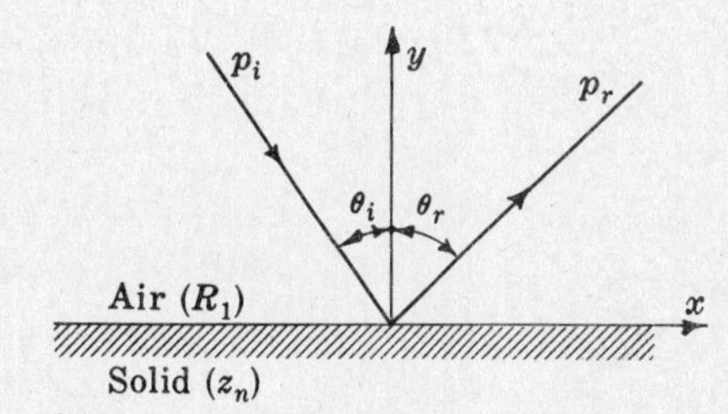

Fig. 4-2

The normal specific acoustic impedance z_n is defined as the ratio of the acoustic pressure to the particle velocity at the surface of the solid. For oblique incidence and at $x = 0$,

$$z_n = r_n + ix_n = \frac{p}{v}$$

or

$$z_n = \frac{p_i + p_r}{v_i \cos\theta_i + v_r \cos(180° - \theta_r)}$$

where $p_i = A_i e^{i(\omega t - k_1 x \cos\theta_i - k_1 y \sin\theta_i)}$ and $p_r = A_r e^{i(\omega t - k_2 x \cos\theta_r - k_2 y \cos\theta_r)}$.

Now $v_i = \dfrac{p_i}{\rho_1 c_1}$ and $v_r = \dfrac{-p_r}{\rho_1 c_1}$; then $\dfrac{\rho_1 c_1 (A_i + A_r)}{A_i - A_r} = z_n \cos\theta_i$ or

$$\frac{A_r}{A_i} = \frac{(r_n \cos\theta_i - \rho_1 c_1) + ix_n \cos\theta_i}{(r_n \cos\theta_i + \rho_1 c_1) + ix_n \cos\theta_i}$$

Hence the sound power reflection coefficient is given by

$$\alpha_r = \frac{A_r^2}{A_i^2} = \frac{(r_n \cos\theta_i - \rho_1 c_1)^2 + x_n^2 \cos^2\theta_i}{(r_n \cos\theta_i + \rho_1 c_1)^2 + x_n^2 \cos^2\theta_i}$$

Similarly, the sound power transmission coefficient is

$$\alpha_t = \frac{4\rho_1 c_1 r_n \cos\theta_i}{(r_n \cos\theta_i + \rho_1 c_1)^2 + x_n^2 \cos^2\theta_i}$$

Since for most solids $r_n > \rho_1 c_1$, the magnitude of the sound power transmission coefficient will reach a minimum when $r_n \cos\theta_i = \rho_1 c_1$.

4.6. An acoustic tile panel has a normal specific acoustic impedance of $1000 - 1300i$ rayls. Compute the sound power reflection and transmission coefficients for plane acoustic waves in air incident normally on the surface of the panel.

The normal specific acoustic impedance of a solid is the ratio of the acoustic pressure acting on the surface of the solid to the particle velocity of the fluid normal to the surface of the solid, i.e. $z_n = r_n + ix_n$, where r_n is the resistive component and x_n is the reactive component.

For normal incidence $\theta_i = 90°$, the sound power reflection coefficient is (see Problem 4.5)

$$\alpha_r = \frac{(r_n - \rho_1 c_1)^2 + x_n^2}{(r_n + \rho_1 c_1)^2 + x_n^2} = \frac{(1000 - 415)^2 + 1300^2}{(1000 + 415)^2 + 1300^2} = 0.55$$

where $\rho_1 c_1 = 1.21(343) = 415$ rayls is the characteristic impedance of air.

Similarly, the sound power transmission coefficient is

$$\alpha_t = \frac{4\rho_1 c_1 r_n}{(r_n + \rho_1 c_1)^2 + x_n^2} = \frac{4(415)1000}{(1000+415)^2 + 1300^2} = 0.45$$

or simply $\alpha_t = 1 - \alpha_r = 1 - 0.55 = 0.45$.

4.7. Plane acoustic waves in air strike the surface of an acoustic tile panel having a normal specific acoustic impedance of $1000 - 1300i$ rayls. Find the angle of incidence so that the sound power reflection coefficient will be a minimum. Find also the reflection coefficient for an angle of incidence of 80°.

The sound power reflection coefficient will be a minimum when (see Problem 4.5)

$$r_n \cos\theta_i = \rho_1 c_1$$

where $z_n = r_n + ix_n = 1000 - 1300i$ rayls,
$r_n = 1000$ rayls is the normal specific acoustic resistance,
$\theta_i = \cos^{-1}(\rho_1 c_1 / r_n)$ is the angle of incidence,
$\rho_1 = 1.21$ kg/m^3 is the density of air,
$c_1 = 343$ m/sec is the speed of sound in air. Hence $\theta_i = \cos^{-1}(415/1000) = 65.5°$.

The sound power reflection coefficient for $\theta_i = 80°$ is

$$\alpha_r = \frac{(r_n \cos\theta_i - R_1)^2 + x_n^2 \cos^2\theta_i}{(r_n \cos\theta_i + R_1)^2 + x_n^2 \cos^2\theta_i} = \frac{[(1000)0.174 - 415]^2 + 1300^2(0.174)^2}{[(1000)0.174 + 415]^2 + 1300^2(0.174)^2} = 0.27$$

4.8. Derive a general expression for the specific acoustic impedance for propagation of plane acoustic waves in a homogeneous and isotropic fluid medium where reflection is present.

The total acoustic pressure and total particle velocity at a point in terms of incident and reflected waves are

$$p = p_i e^{i(\omega t - kx)} + p_r e^{i(\omega t + kx)}$$
$$v = v_i e^{i(\omega t - kx)} + v_r e^{i(\omega t + kx)}$$

Now the complex reflection coefficients for acoustic pressure and particle velocity are given by

$$K_p = \frac{p_r}{p_i} = \frac{z_2 - R_1}{z_2 + R_1}, \qquad K_v = \frac{v_r}{v_i} = \frac{R_1 - z_2}{R_1 + z_2} = -K_p$$

The total acoustic pressure and total particle velocity can now be expressed as

$$p = p_i(e^{i(\omega t - kx)} + K_p e^{i(\omega t + kx)})$$
$$v = v_i(e^{i(\omega t - kx)} - K_p e^{i(\omega t + kx)})$$

Since specific acoustic impedance is defined as the ratio of the total acoustic pressure to the total particle velocity at a point, we have

$$z_s = \frac{p}{v} = \frac{p_i}{v_i}\frac{e^{-ikx} + Ke^{ikx}}{e^{-ikx} - Ke^{ikx}} = \rho c\left(\frac{e^{-ikx} + Ke^{ikx}}{e^{-ikx} - Ke^{ikx}}\right)$$

This ratio gives the specific acoustic impedance at any point of the medium as a function of the characteristic impedance ρc of the medium, the reflection coefficient at the boundary, and the distance x from the point in question to the boundary. In short, it controls the transmission of sound energy from one medium to another.

If there is no reflection at the boundary, $K = 0$, and the specific acoustic impedance z_s will be reduced to the characteristic impedance ρc of the medium.

It is interesting to note that at a distance L from the boundary in the first medium, $x = -L$; and if $K = 1$,

$$z_s = \rho c\left(\frac{e^{ikL} + e^{-ikL}}{e^{ikL} - e^{-ikL}}\right) = -i\rho c \cot kL$$

which corresponds to the driving impedance of a flexible string.

4.9. Determine the impedance of (*a*) a quarter-wavelength fluid column, and (*b*) a half-wavelength fluid column for the propagation of plane acoustic waves.

(*a*) Quarter-wavelength.

The specific acoustic impedance of a fluid column of finite length is given by

$$z_s = \rho c \frac{e^{-ikx} + Ke^{ikx}}{e^{-ikx} - Ke^{ikx}}$$

where ρc is the characteristic impedance of the fluid medium, $K = K_p$ is the complex reflection coefficient for acoustic pressure, and $k = \omega/c$ is the wave number. (See Problem 4.8.)

Now $k = \omega/c = 2\pi/\lambda$ and $kx = (2\pi/\lambda)(-\lambda/4) = -\pi/2$ (the minus sign is needed because $x = 0$ at the boundary). Hence

$$z_s = \rho c \frac{e^{i(\pi/2)} + Ke^{-i(\pi/2)}}{e^{i(\pi/2)} - Ke^{-i(\pi/2)}} = \rho c\left(\frac{1-K}{1+K}\right)$$

If $K = 1$, (e.g. $R_2 \gg R_1$, or $z_2 \to \infty$ for rigid terminating boundary) $z_s = 0$; i.e. the input impedance is zero.

If $K = -1$, (e.g. $R_1 \gg R_2$, or with a rarefied terminating medium) $z_s \to \infty$; i.e. the input impedance is very large.

If $K = 0$, (e.g. continuous medium, or matching impedances) $z_s = \rho c$; i.e. the input impedance is equal to the characteristic impedance of the medium.

If the column is terminated by a medium infinite in extent, $K = (R_2 - R_1)/(R_2 + R_1)$ and $z_s = \rho c R_1/R_2 = R_1^2/R_2$. The input impedance is therefore inversely proportional to the terminating impedance R_2.

(*b*) Half-wavelength.

$$z_s = \rho c\left(\frac{e^i + Ke^{-i}}{e^i - Ke^{-i}}\right) = \rho c\left(\frac{1+K}{1-K}\right)$$

If $K = 1$, $z_s = \infty$. If $K = -1$, $z_s = 0$. If $K = 0$, $z_s = \rho c$.

If the column is terminated by a medium infinite in extent, $K = (R_2 - R_1)/(R_2 + R_1)$, $z_s = \rho c R_2/R_1 = R_2$. This means that the input impedance of a half-wavelength fluid column equals its terminating impedance.

TRANSMISSION THROUGH THREE MEDIA

4.10. Derive an expression for the transmission coefficient of plane acoustic waves through three homogeneous and isotropic media.

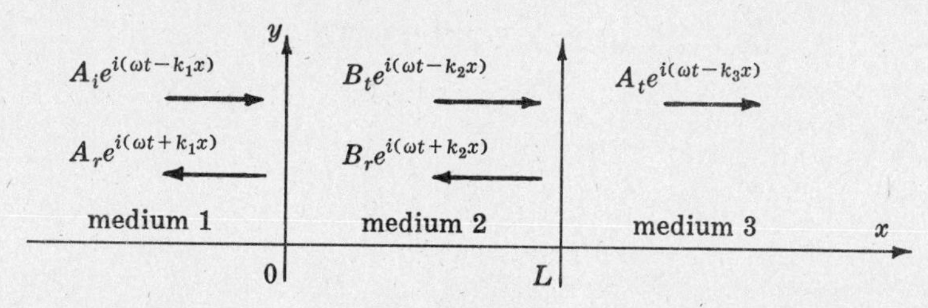

Fig. 4-3

As shown in Fig. 4-3, the incident wave $A_i e^{i(\omega t - k_1 x)}$ is traveling in the positive x direction and the reflected wave at boundary *I* ($x = 0$) is $A_r e^{i(\omega t + k_1 x)}$. Then the wave in medium 1 is represented by

$$u_1 = A_i e^{i(\omega t - k_1 x)} + A_r e^{i(\omega t + k_1 x)} \tag{1}$$

Now the transmitted wave at boundary *I* is $B_t e^{i(\omega t - k_2 x)}$, and the reflected wave at boundary *II* ($x = L$) is $B_r e^{i(\omega t + k_2 x)}$. Then the wave in medium 2 is given by

$$u_2 = B_t e^{i(\omega t - k_2 x)} + B_r e^{i(\omega t + k_2 x)} \tag{2}$$

Part of the wave incident normally on boundary *II* will be transmitted into medium 3 as

$$u_3 = A_t e^{i[\omega t - k_3(x-L)]} \tag{3}$$

where the A's and B's are amplitudes of sound waves, and $k_1 = \omega/c_1$, $k_2 = \omega/c_2$ and $k_3 = \omega/c_3$ are the wave numbers.

Under steady state condition, we have the following two boundary conditions at boundary *I* and *II*:

(1) the acoustic pressures at both sides of the boundary are equal,

(2) the particle velocities normal to the boundary are equal.

Thus at boundary *I* ($x = 0$)

$$p_1 = p_2 \qquad \text{or} \qquad -B_1 \frac{\partial u_1}{\partial x} = -B_2 \frac{\partial u_2}{\partial x}$$

where B is the bulk modulus of the medium and $k = \omega/c$ the wave number. Substituting (*1*) and (*2*) into the above conditions, we obtain

$$-B_1(-ik_1 A_i e^{i\omega t} + ik_1 A_r e^{i\omega t}) = -B_2(-ik_2 B_t e^{i\omega t} + ik_2 B_r e^{i\omega t})$$

or

$$\rho_1 c_1 (A_i - A_r) = \rho_2 c_2 (B_t - B_r) \tag{4}$$

where $c = \sqrt{B/\rho}$ is the speed of sound and $k = \omega/c$.

At $x = 0$,

$$\frac{\partial u_1}{\partial t} = \frac{\partial u_2}{\partial t} \qquad \text{or} \qquad A_i + A_r = B_t + B_r \tag{5}$$

Similarly at boundary *II* the acoustic pressures are equal, i.e., at $x = L$,

$$-B_2 \frac{\partial u_2}{\partial x} = -B_3 \frac{\partial u_3}{\partial x}$$

or

$$-B_2(-ik_2 B_t e^{i(\omega t - k_2 L)} + ik_2 B_r e^{i(\omega t + k_2 L)}) = -B_3(-ik_3 A_t e^{i\omega t})$$

which reduces to

$$\rho_2 c_2 (B_t e^{-ik_2 L} - B_r e^{ik_2 L}) = \rho_3 c_3 A_t \tag{6}$$

and the particle velocities normal to boundary *II* are also equal, i.e., at $x = L$,

$$\frac{\partial u_2}{\partial t} = \frac{\partial u_3}{\partial t} \qquad \text{or} \qquad B_t e^{-ik_2 L} + B_r e^{ik_2 L} = A_t \tag{7}$$

Solving equations (*4*) to (*7*) simultaneously, we obtain

$$\frac{A_t}{A_i} = \frac{2\rho_1 c_1 \rho_2 c_2}{\rho_2 c_2 (\rho_3 c_3 + \rho_1 c_1) \cos k_2 L + i(\rho_2^2 c_2^2 + \rho_3 c_3 \rho_1 c_1) \sin k_2 L} \tag{8}$$

We assume medium 3 extends to infinity, and write $z_3 = \rho_3 c_3 = R_3$, $\rho_2 c_2 = R_2$, $\rho_1 c_1 = R_1$; then equation (*8*) becomes

$$\frac{A_t}{A_i} = \frac{2R_2 R_1}{R_2(R_3 + R_1) \cos k_2 L + i(R_2^2 + R_3 R_1) \sin k_2 L} \tag{9}$$

Now the sound energy transmission coefficient is

$$\alpha_t = \frac{I_3}{I_1} = \frac{(p_t)_3^2/2R_3}{(p_i)_1^2/2R_1} = \frac{R_1}{R_3}\frac{A_t^2}{A_i^2} = \frac{R_1}{R_3}(A_t/A_i)^2$$

and from equation (*9*),

$$\alpha_t = \frac{4R_3 R_1}{(R_3 + R_1)^2 \cos^2 k_2 L + (R_2 + R_3 R_1/R_2)^2 \sin^2 k^2 L} \tag{10}$$

In the following cases we can further simplify the sound energy transmission coefficient given in equation (*10*):

(*a*) When medium 3 is the same as medium 1, we have $R_3 = R_1$ and

$$\alpha_t = \frac{4}{4 \cos^2 k_2 L + (R_2/R_1 + R_1/R_2)^2 \sin^2 k_2 L} \tag{11}$$

(*b*) For sound transmission from a rarefied medium through a dense medium into the same rarefied medium, such as sound waves from air in one room through a solid wall into air in an adjacent room, we have $R_2 \gg R_1$, and so *(11)* yields

$$\alpha_t = \frac{4}{4\cos^2 k_2L + (R_2/R_1)^2 \sin^2 k_2L} \tag{12}$$

(c) When the rarefied medium is air, we have $\dfrac{R_2 \sin k_2L}{R_1} \gg 2\cos k_2L$. Except for a very thick medium 2 (i.e. large L) and high frequency sound, we have $k_2L \ll 1$ and $\sin k_2L \doteq k_2L$. We obtain the simplest expression for sound energy transmission coefficient from equation *(10)*,

$$\alpha_t = \frac{4R_1^2}{R_2^2 k_2^2 L^2} \tag{13}$$

where L is the thickness of medium 2, $R_1 = R_3$ are the characteristic impedances of the media, and $k_2 = \omega/c_2$ is the wave number.

4.11. A plane sinusoidal acoustic wave in water is incident normally on the surface of a large steel plate of thickness 0.02 m. If the frequency of the wave is 3000 cyc/sec, find the transmission loss through the steel plate into water on the opposite side.

The sound energy transmission coefficient is defined as

$$\alpha_t = \frac{4}{4\cos^2 k_2L + (R_2/R_1)^2 \sin^2 k_2L}$$

where $k_2 = \omega/c = 3000(6.28)/5050 = 3.74$ is the wave number for steel, $k_2L = 3.74(0.02) = 0.075$, $R_2 = 39 \times 10^6$ rayls is the characteristic impedance of steel, and $R_1 = 1.48 \times 10^6$ rayls is the characteristic impedance of water.

Now $k_2L = 0.075 = 0.075(180°)/3.14 = 4.3°$, $\cos k_2L = \cos 4.3° \doteq 1.0$, $\sin k_2L = \sin 4.3° = 0.075$, and the transmission coefficient is

$$\alpha_t = \frac{4}{4 + (39/1.48)^2(0.075)^2} = 0.52$$

The transmission loss is $\text{TL} = 10 \log(1/\alpha_t) = 3.02$ db.

4.12. Maximum transmission of plane acoustic waves from water into steel is required. What should be the optimum characteristic impedance of the material to be placed between the water and the steel? If the thickness of the layer of material to be used is 0.02 m and the frequency of sound transmitted is 1000 cyc/sec, find the speed of sound in the material and the density of the material.

The sound energy transmission coefficient for transmission through three media at normal incidence is given by

$$\alpha_t = \frac{4R_1R_3}{(R_1 + R_3)^2 \cos^2 k_2L + (R_2 + R_1R_3/R_2)^2 \sin^2 k_2L}$$

where R_1, R_2, R_3 are the characteristic impedances of the media, $k_2 = \omega/c_2$ is the wave number of medium 2, and L the thickness of medium 2.

If $k_2L = (2n-1)\pi/2$ where $n = 1, 2, \ldots$, then $\sin k_2L \doteq 1$ and $\cos k_2L \doteq 0$, and the transmission coefficient becomes

$$\alpha_t = \frac{4R_1R_3}{(R_2 + R_1R_3/R_2)^2}$$

For maximum transmission of acoustic power, $R_2 = \sqrt{R_1R_3}$ where $R_1 = 1{,}480{,}000$ and $R_3 = 47{,}000{,}000$ rayls at standard temperature and atmospheric pressure; hence $R_2 = 8{,}350{,}000$ rayls. Therefore 100% transmission of sound occurs only for bands of frequencies centered about the particular frequencies for which

$$f = (2n-1)c_2/4L$$

or $c_2 = 4Lf = 4(0.02)1000 = 800$ m/sec, and $\rho = R_2/c_2 = 8{,}350{,}000/800 = 10{,}500$ kg/m³.

4.13. Show that a very thin layer of solid material of appropriate characteristic impedance may be employed to prevent two fluid media from mixing with each other and yet not interfere with the transmission of sound of low frequencies between them.

The sound power transmission coefficient from medium 1 through the thin layer into medium 3 is (see Problem 4.10, equation (*10*))

$$\alpha_t = \frac{4R_1R_3}{(R_3+R_1)^2 \cos^2 k_2L + (R_2+R_3R_1/R_2)^2 \sin^2 k_2L}$$

where the R's are the characteristic impedances of the three media, $k_2 = \omega/c_2$ is the wave number for medium 2, and L is the thickness of medium 2.

If ω is small, i.e. low frequency sound, k_2 is small. We have $k_2L \to 0$, $\cos k_2L \doteq 1$, and $\sin k_2L \doteq 0$. Hence $\cos^2 k_2L \doteq 1$, $\sin^2 k_2L \doteq 0$, and

$$\alpha_t = \frac{4R_1R_3}{(R_3+R_1)^2}$$

which is the same sound power transmission coefficient as for sound waves moving directly from medium 1 into medium 3. See Problem 4.2, equation (*4*).

4.14. A beam of plane sinusoidal acoustic waves in water is normally incident on a steel plate of thickness 0.04 m and emerges into water on the opposite side. If the frequency of the wave is 5000 cyc/sec, find the phase angle between the incident and transmitted waves.

The amplitude ratio of the incident and transmitted waves for sound transmission through three media is

$$\frac{A_i}{A_t} = \frac{(R_3+R_1)\cos k_2L}{2R_3} + i\frac{(R_2^2+R_3R_1)\sin k_2L}{2R_3R_2}$$

where the R's are the characteristic impedances of the media, L is the thickness of medium 2 and $k_2 = \omega/c_2$ is the wave number of medium 2. The phase angle between the incident and transmitted waves is therefore

$$\theta = \tan^{-1}\left[\frac{(R_2^2+R_1R_3)}{R_2(R_1+R_3)}\tan k_2L\right] = 17.6°$$

where $k_2 = \omega/c_2 = 5000(6.28)/6100 = 5.15$, $k_2L = 5.15(0.04) = 0.206$, $\tan k_2L = \tan 0.206 = \tan 10.5° = 0.181$, $R_1 = R_3 = 1{,}480{,}000$ rayls, $R_2 = 47{,}000{,}000$ rayls. The incident wave at $x = 0$ therefore leads the transmitted wave at $x = 0.04$ m by 17.6°.

REFLECTION OF SOUND

4.15. Plane sinusoidal acoustic waves in air are incident normally on a plane surface of characteristic impedance 785 rayls. Find the standing wave ratio.

At standard atmospheric pressure and temperature, the reflection coefficient for acoustic pressure amplitude is

$$\alpha_r = P_r/P_i = (R_2-R_1)/(R_2+R_1) = 0.31$$

where $R_1 = \rho_1c_1 = 415$ rayls is the characteristic impedance of air.

Standing wave ratio SWR $= (1+P_r/P_i)/(1-P_r/P_i) = 1.9$.

4.16. A ship is steaming toward a cliff with constant speed in the fog and the siren on the ship is sounded every minute. The echo of the first whistle is heard after 20 seconds and that of the second after 16.5 seconds. Compute the original distance of the ship from the cliff and her speed. What is the minimum distance for the observance of an echo?

Let s in meters and v in m/sec be the distance and speed of the ship. Also let the speed of sound be 343 m/sec. The first echo is heard after the ship has advanced $20v$ m:

$$343(20) = 2s - 20v \tag{1}$$

The second echo is heard after the ship has moved ahead $v(16.5 + 60)$ m:

$$343(16.5) = 2s - 76.5v \qquad (2)$$

Solving, $s = 3640$ m, $v = 21$ m/sec.

Sound waves emitted from the source will take a definite length of time to reach the surface and reflect back. Since the human ear is unable to distinguish separate sounds unless the time interval between the two sounds is more than 1/10 second, an echo will be observed when the time interval from emission to the arrival of sound is equal to or greater than 1/10 second. Then

$$2s = vt \quad \text{or} \quad s = vt/2 = 343(1/10)/2 = 17.2 \text{ m}$$

and so the minimum distance is 17.2 m.

4.17. A plane sinusoidal acoustic wave of effective pressure 100 nt/m^2 and frequency 1000 cyc/sec is incident normally on the plane surface of the water. Calculate (*a*) the acoustic pressure of the wave transmitted from water into air, (*b*) the intensity of the incident wave in water and of the transmitted wave in air, and (*c*) the ratio of the intensity of the transmitted wave in air to that of the incident wave in water.

(*a*) The transmission coefficient for acoustic pressure amplitude is

$$\frac{p_t}{p_i} = \frac{2\rho_2 c_2}{\rho_2 c_2 + \rho_1 c_1} = \frac{2(415)}{415 + 1{,}480{,}000} = 5.63 \times 10^{-4}$$

where $\rho_1 c_1 = 1{,}480{,}000$ and $\rho_2 c_2 = 415$ rayls are the characteristic impedances of water and air respectively at standard temperature and atmospheric pressure. Thus the pressure of the transmitted wave in air is

$$p_t = 5.63 \times 10^{-4}(100) = 5.63 \times 10^{-2} \text{ nt/m}^2$$

(*b*) The intensity of the incident wave in water is

$$I_i = \frac{p_i^2}{\rho_1 c_1} = \frac{100^2}{1{,}480{,}000} = 6.78 \times 10^{-3} \text{ watt/m}^2$$

and the intensity of the transmitted wave in air is

$$I_t = \frac{p_t^2}{\rho_2 c_2} = \frac{[5.63(10)^{-2}]^2}{415} = 7.6 \times 10^{-6} \text{ watt/m}^2$$

(*c*)
$$\frac{I_t}{I_i} = \frac{7.6(10)^{-6}}{6.8(10)^{-3}} = 1.13 \times 10^3 \quad \text{or} \quad 10 \log (1.13 \times 10^{-3}) = -29.5 \text{ db}$$

4.18. For normal incidence of plane sinusoidal acoustic waves from hydrogen to oxygen, find the ratio of the reflected sound energy to the incident sound energy.

The ratio of the reflected sound energy to the incident sound energy is

$$\alpha_r = \left[\frac{\rho_1 c_1 - \rho_2 c_2}{\rho_1 c_1 + \rho_2 c_2}\right]^2 = 0.36$$

where $\rho_1 = \rho_{\text{hydrogen}} = 0.09$ kg/m^3, $\rho_2 = \rho_{\text{oxygen}} = 1.43$ kg/m^3, $c_1 = c_{\text{hydrogen}} = 1269$ m/sec and $c_2 = c_{\text{oxygen}} = 317$ m/sec.

Since $c_1/c_2 = \sqrt{\rho_2/\rho_1}$, we can express the sound power reflection coefficient as

$$\alpha_r = \left[\frac{\sqrt{\rho_2} - \sqrt{\rho_1}}{\sqrt{\rho_2} + \sqrt{\rho_1}}\right]^2 = 0.36 \quad \text{or} \quad \alpha_r = \left[\frac{c_1 - c_2}{c_1 + c_2}\right]^2 = 0.36$$

4.19. Derive general expressions for the reflection and transmission coefficients of plane acoustic waves incident normally on the plane interface of two absorbing media.

Assuming linear absorbing media, the particle displacements of the incident, reflected, and transmitted waves can be expressed as

$$u_i = A_i e^{-\alpha_1 x} e^{i(\omega t - k_1 x)}$$
$$u_r = A_r e^{-\alpha_1 x} e^{i(\omega t + k_1 x)}$$
$$u_t = A_t e^{-\alpha_2 x} e^{i(\omega t - k_2 x)}$$

where the A's are the amplitudes of the waves, α_1 and α_2 are the linear absorbing coefficients of the two media, $k_1 = \omega/c_1$ and $k_2 = \omega/c_2$ are the wave numbers.

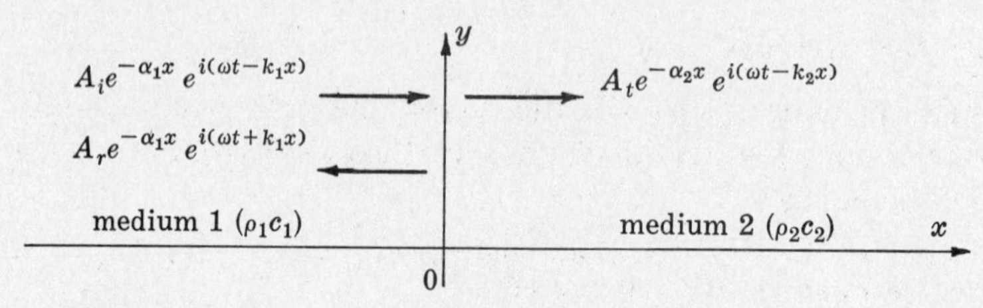

Fig. 4-4

The boundary conditions are:

(1) The particle velocities normal to the interface are always equal, i.e., at $x = 0$,

$$\frac{\partial u_i}{\partial t} + \frac{\partial u_r}{\partial t} = \frac{\partial u_t}{\partial t}$$

or

$$A_i + A_r = A_t \tag{1}$$

(2) Acoustic pressure on both sides of the interface is the same, i.e., at $x = 0$,

$$-B_1 \frac{\partial u_i}{\partial x} - B_1 \frac{\partial u_r}{\partial x} = -B_2 \frac{\partial u_t}{\partial x}$$

or

$$(\alpha_1 + ik_1)B_1 A_i + (\alpha_1 - ik_1)B_1 A_r = (\alpha_2 + ik_2)B_2 A_t \tag{2}$$

where B is the bulk modulus.

Elimination of A_t from equations (*1*) and (*2*) gives

$$\frac{A_r}{A_i} = -\frac{B_2(\alpha_2 + ik_2) - B_1(\alpha_1 + ik_1)}{B_2(\alpha_2 + ik_2) - B_1(\alpha_1 - ik_1)} \tag{3}$$

and elimination of A_r from equations (*1*) and (*2*) yields

$$\frac{A_t}{A_i} = \frac{2ik_1 B_1}{B_2(\alpha_2 + ik_2) - B_1(\alpha_1 - ik_1)} \tag{4}$$

If we write $B_1 = \rho_1 c_1^2$, $B_2 = \rho_2 c_2^2$, $B_1 k_1 = \rho_1 c_1 \omega$ and $B_2 k_2 = \rho_2 c_2 \omega$, (*3*) and (*4*) become

$$\frac{A_r}{A_i} = -\frac{(\rho_2 c_2^2 \alpha_2 + i\rho_2 c_2 \omega) - (\rho_1 c_1^2 \alpha_1 + i\rho_1 c_1 \omega)}{(\rho_2 c_2^2 \alpha_2 + i\rho_2 c_2 \omega) - (\rho_1 c_1^2 \alpha_1 - i\rho_1 c_1 \omega)} \tag{5}$$

$$\frac{A_t}{A_i} = \frac{2i\rho_1 c_1 \omega}{(\rho_2 c_2^2 \alpha_2 + i\rho_2 c_2 \omega) - (\rho_1 c_1^2 \alpha_1 - i\rho_1 c_1 \omega)} \tag{6}$$

If the absorbing coefficients α_1 and α_2 are equal to zero, equations (*5*) and (*6*) reduce to (*9*) and (*10*) of Problem 4.1 for the nondissipative case.

Now the average incident, reflected, and transmitted acoustic powers per unit area are respectively

$$W_i = \tfrac{1}{2}\rho_1 c_1 \omega^2 A_i^2, \quad W_r = \tfrac{1}{2}\rho_1 c_1 \omega^2 A_r^2, \quad W_t = \tfrac{1}{2}\rho_2 c_2 \omega^2 A_t^2$$

Hence the ratio of the reflected to incident sound power is

$$\alpha_r = \frac{W_r}{W_i} = \frac{(\rho_1 c_1^2 \alpha_1 - \rho_2 c_2^2 \alpha_2)^2 + (\rho_1 c_1 \omega - \rho_2 c_2 \omega)^2}{(\rho_2 c_2^2 \alpha_2 - \rho_1 c_1^2 \alpha_1)^2 + (\rho_1 c_1 \omega + \rho_2 c_2 \omega)^2} \tag{7}$$

and the ratio of the transmitted to the incident sound power is

$$\alpha_t = \frac{W_t}{W_i} = \frac{4\rho_2 c_2 \rho_1 c_1}{(\rho_1 c_1 + \rho_2 c_2)^2 + [(\rho_2 c_2^2 \alpha_2/\omega) - (\rho_1 c_1^2 \alpha_1/\omega)]^2} \tag{8}$$

If $\alpha_1 = \alpha_2 = 0$, equations (7) and (8) reduce to the reflection and transmission coefficients for the nondissipative case of Problem 4.1.

If $\alpha_1/\alpha_2 = \rho_2 c_2^2/\rho_1 c_1^2$, the sound power reflection and transmission coefficients of equations (7) and (8) also reduce to those for the nondissipative media.

REFRACTION OF SOUND

4.20. Derive general expressions for the sound power reflection and transmission coefficients for the transmission of plane acoustic waves from one fluid medium to another at oblique incidence.

The acoustic pressures for plane sinusoidal longitudinal waves at normal incidence are

$$p_i = P_i e^{i(\omega t - k_1 x)}$$

$$p_r = P_r e^{i(\omega t + k_1 x)}$$

$$p_t = P_t e^{i(\omega t - k_2 x)}$$

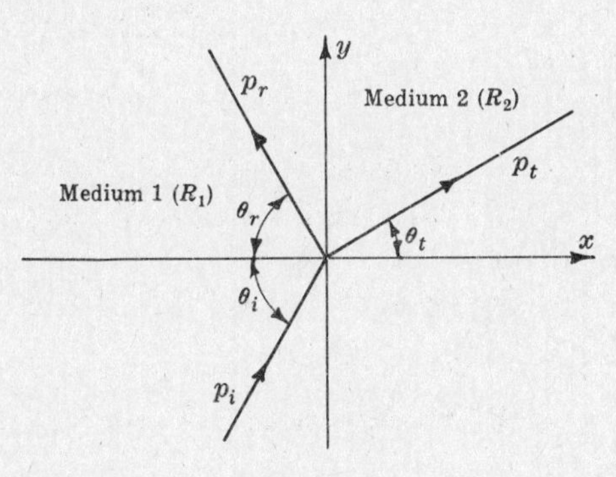

Fig. 4-5

and at oblique incidence,

$$p_i = P_i e^{i(\omega t - k_1 x \cos\theta_i - k_1 y \sin\theta_i)}$$

$$p_r = P_r e^{i(\omega t + k_1 x \cos\theta_r - k_1 y \sin\theta_r)}$$

$$p_t = P_t e^{i(\omega t - k_2 x \cos\theta_t - k_2 y \sin\theta_t)}$$

where θ_i is the angle of incidence, θ_r the angle of reflection, and θ_t the angle of refraction as shown in Fig. 4-5.

At the plane interface of the two media ($x = 0$), the acoustic pressure must be continuous, i.e.

$$p_i + p_r = p_t$$

or

$$P_i e^{-ik_1 y \sin\theta_i} + P_r e^{-ik_1 y \sin\theta_r} = P_t e^{-ik_2 y \sin\theta_t}$$

From the laws of reflection and refraction of plane waves, we have the angle of incidence θ_i is equal to the angle of reflection θ_r. And from Snell's law, we have $(\sin\theta_i)/(\sin\theta_t) = c_1/c_2 = k_2/k_1$. The previous boundary condition of continuity of acoustic pressure becomes

$$P_i + P_r = P_t$$

The second boundary condition states that the particle velocities normal to the interface must be equal, i.e.

$$v_i \cos\theta_i + v_r \cos(180° - \theta_r) = v_t \cos\theta_t$$

or in terms of acoustic pressure and characteristic impedance,

$$(P_i/R_1)\cos\theta_i - (P_r/R_1)\cos\theta_r = (P_t/R_2)\cos\theta_t$$

where $R_1 = \rho_1 c_1$, $R_2 = \rho_2 c_2$ are the characteristic impedances of the two media.

Solving for the ratios of P_r/P_i and P_t/P_i from the two boundary conditions,

$$\frac{P_r}{P_i} = \frac{R_2\cos\theta_i - R_1\cos\theta_t}{R_2\cos\theta_i + R_1\cos\theta_t}$$

and so the sound power reflection coefficient is

$$\alpha_r = \left[\frac{P_r}{P_i}\right]^2 = \left[\frac{R_2\cos\theta_i - R_1\cos\theta_t}{R_2\cos\theta_i + R_1\cos\theta_t}\right]^2$$

The corresponding sound power transmission coefficient is

$$\alpha_t = 1 - \alpha_r = \frac{4R_1R_2\cos\theta_i\cos\theta_t}{(R_2\cos\theta_i + R_1\cos\theta_t)^2}$$

If the angle of refraction θ_t is 90°, we have from Snell's law,

$$\frac{\sin\theta_i}{\sin\theta_t} = \frac{c_1}{c_2} \quad \text{or} \quad \sin\theta_i = \sin\theta_c = \frac{c_1}{c_2}$$

Since no acoustic energy is transmitted for angle of incidence greater than θ_c, we call θ_c the *critical angle of incidence.*

If the angle of incidence approaches 90°, $\cos\theta_i \to 0$ and

$$\alpha_r = \left[-\frac{R_1\cos\theta_t}{R_1\cos\theta_t}\right]^2 = 1$$

Again, no acoustic energy is transmitted. This is known as the condition of *grazing incidence.*

4.21. The density of a given solution is 800 kg/m³ and the speed of sound is 1300 m/sec. (*a*) Find the critical angle of incidence for plane acoustic waves traveling from the given solution into water. (*b*) If the angle of incidence in the given solution is 40°, what is the sound transmission coefficient into water?

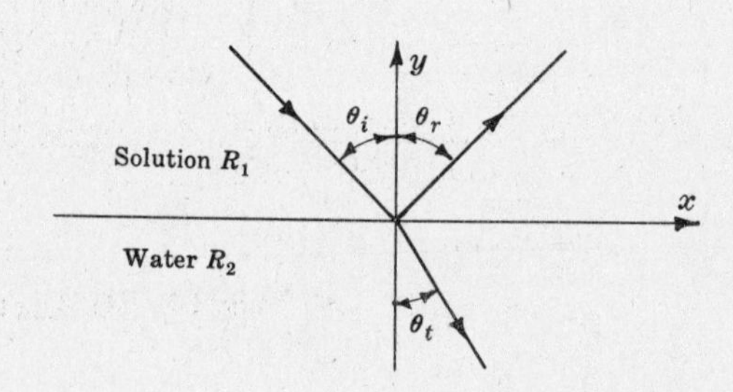

Fig. 4-6

(*a*) The critical angle of incidence is given by

$$\sin\theta_c = \frac{c_1}{c_2} = \frac{c_{\text{solution}}}{c_{\text{water}}} = \frac{1300}{1480} = 0.879 \quad \text{or} \quad \theta_c = 61.5°$$

(*b*) The sound transmission coefficient is

$$\alpha_t = \frac{4R_1R_2\cos\theta_i\cos\theta_t}{(R_2\cos\theta_i + R_1\cos\theta_t)^2} = 0.96$$

where $R_1 = \rho_1 c_1 = 800(1300)$ is the characteristic impedance of the given solution, $R_2 = \rho_2 c_2 = 998(1480)$ is the characteristic impedance of water, $\theta_i = 40°$ is the angle of incidence, $\theta_t = 47°$ is the angle of refraction, obtained from Snell's law of refraction.

4.22. If the velocity of sound in oil changes suddenly from 1350 m/sec to 1340 m/sec along a horizontal plane at a certain depth while the density of oil remains constant at 850 kg/m³, calculate the sound reflection coefficient for plane acoustic waves incident from above the plane interface where velocity changes take place at angles of incidence of (*a*) 88°, (*b*) 80°, and (*c*) at normal incidence.

The sound reflection coefficient for plane acoustic waves is given by

$$\alpha_r = \left[\frac{R_2\cos\theta_i - R_1\cos\theta_t}{R_2\cos\theta_i + R_1\cos\theta_t}\right]^2$$

where $R_1 = \rho_1 c_1 = 850(1350)$ rayls and $R_2 = \rho_2 c_2 = 850(1340)$ rayls are the characteristic impedances of oil above and below the plane interface where velocity changes take place, θ_i is the angle of incidence, and θ_t is the angle the transmitted wave makes with the normal.

(*a*) $\sin\theta_t = \dfrac{c_2\sin\theta_i}{c_1} = \dfrac{1340\sin 88°}{1350} = 0.988$ or $\theta_t = 82°$, and $\alpha_r = 0.36$.

(*b*) $\sin\theta_t = \dfrac{c_2\sin\theta_i}{c_1} = \dfrac{1340\sin 80°}{1350} = 0.98$ or $\theta_t = 78°$, and $\alpha_r = 8.7\times 10^{-3}$.

(*c*) Since $\theta_i = 0$ for normal incidence, $\theta_t = 0$ and $\alpha_r = 1.9\times 10^{-5}$.

DIFFRACTION OF SOUND

4.23. Sketch the diffraction of high frequency and low frequency sound waves around bends.

Figure 4-7 shows the diffraction of sound waves around bends. It is clear that low frequency sound waves readily diffract around bends where bend openings are small compared to wavelength. High frequency sound waves, as shown in Fig. 4-7(*b*) do not easily diffract around bends where bend openings are large compared to wavelength. Moreover, multiple reflections occur at the bend resulting in scattering and cancellation of high frequency sound waves.

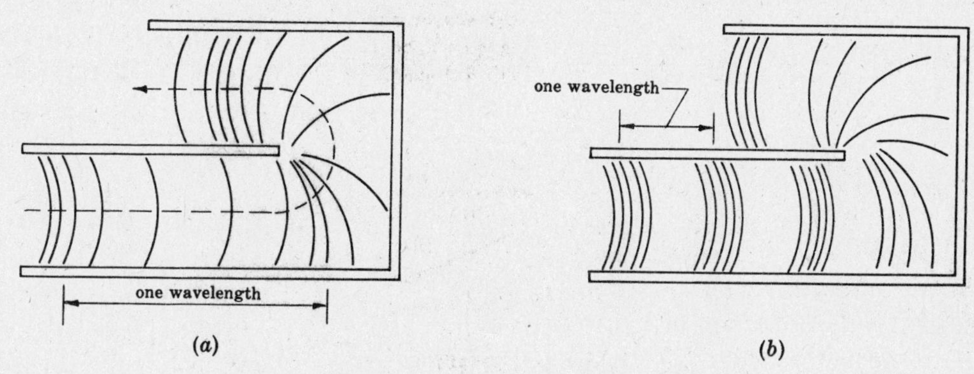

Fig. 4-7. Diffraction of sound waves around bends

FILTRATION OF SOUND

4.24. A rigid smooth pipe of radius 0.04 m has a hole of radius 0.02 m in its thin wall. Find the sound power transmission coefficient for plane acoustic waves along the pipe at the following frequencies: 100, 400, 800 cyc/sec. If a similar hole is drilled directly across the first hole, what will be the sound power transmission coefficient at a frequency of 400 cyc/sec?

The sound power transmission coefficient for a hole drilled in the thin wall of the main pipe is given by

$$\alpha_t = \frac{1}{1 + (\pi a^2/2ALk)^2}$$

where a is the radius of the hole, A is the area of the cross section of the pipe, $L = 1.7a$, $k = 2\pi f/c$ is the wave number, and $c = 343$ m/sec is the speed of sound in air. Now $a^2 = (0.02)^2 = 0.0004\ \text{m}^2$, $L = 1.7a = 0.034$ m, $A = 3.14(0.04)^2 = 0.0051\ \text{m}^2$; and for $f = 100$ cyc/sec, $k = 6.28(100)/343 = 1.83$. Substituting values, we find $\alpha_t = 0.21$.

For $f = 400$ cyc/sec, $k = 6.28(400)/343 = 7.31$, and $\alpha_t = 0.81$. For $f = 800$ cyc/sec, $k = 6.28(800)/343 = 14.7$, and $\alpha_t = 0.94$. In other words, sound power transmits better at higher frequencies for plane acoustic waves in air along rigid smooth pipes with holes.

When a similar hole is drilled directly across the first hole, the result will be equivalent to two identical impedances in parallel. The sound power transmission coefficient for $f = 400$ cyc/sec therefore becomes $\alpha_t = \dfrac{1}{1 + (\pi a^2/ALk)^2} = 0.51$.

The plot of the transmission coefficient versus frequency for the transmission of plane acoustic waves through an acoustic line with an orifice as a branch is shown in Fig. 4-8. α_t is zero for $f = 0$, and increases to unity as f approaches infinity.

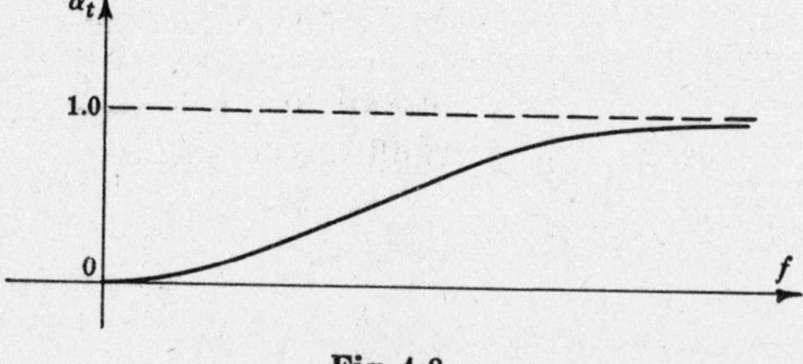

Fig. 4-8

4.25. A section of pipe of length 1 m and cross-sectional area 0.8 m² is inserted into a main transmitting pipe of cross-sectional area 0.2 m² as shown in Fig. 4-9. Compute the sound power transmission coefficient at (*a*) 0 cyc/sec, (*b*) 100 cyc/sec, (*c*) 200 cyc/sec, (*d*) 512 cyc/sec.

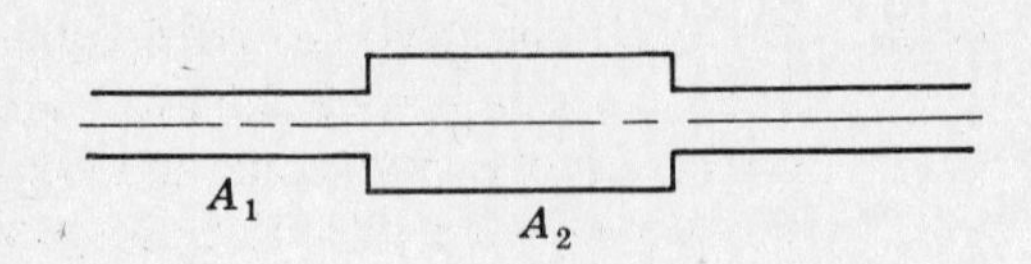

Fig. 4-9

The sound power transmission coefficient for pipes with expanded sections is

$$\alpha_t = \frac{4}{4 + [(A_2/A_1)^2 - 2] \sin^2 kL}$$

where $A_2 = 0.8\text{ m}^2$ is the cross-sectional area of the expanded pipe, $A_1 = 0.2\text{ m}^2$ is the cross-sectional area of the main pipe, $k = \omega/c$ is the wave number, $c = 343$ m/sec is the speed of sound, and $L = 1$ m is the length of the expanded pipe.

(*a*) $f = 0$, $k = 0$, $\sin kL = 0$, and $\alpha_t = 1.0$.

(*b*) $f = 100$, $k = 100(6.28)/343 = 1.83$, $\sin^2 kL = \sin^2 1.83 = 0.95$, and $\alpha_t = 0.23$.

(*c*) $f = 200$, $k = 200(6.28)/343 = 3.68$, $\sin^2 kL = \sin^2 3.68 = 0.28$, and $\alpha_t = 0.49$.

(*d*) $f = 512$, $k = 512(6.28)/343 = 9.4$, $\sin^2 kL = \sin^2 9.4 = 0$, and $\alpha_t = 1.0$.

Thus a plot of transmission coefficients versus frequencies has the following general form of selective transmission or filtration of sound. Note that the result for a constriction is theoretically identical with that for an expansion.

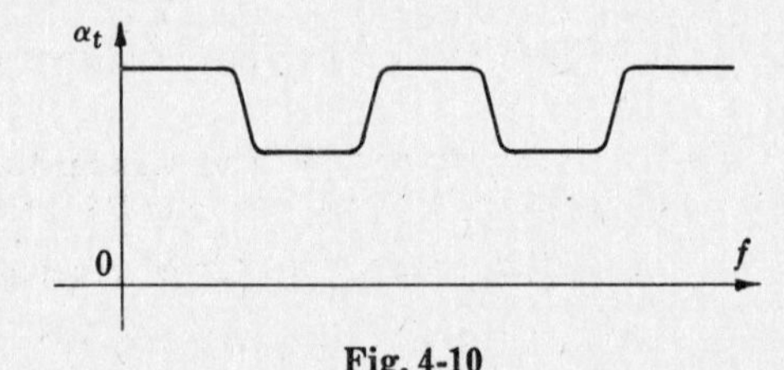

Fig. 4-10

For $A_1 < A_2$, the incident and reflected waves are in opposite phase. This corresponds to the passage of sound from a dense to a rare medium.

For $A_1 > A_2$, the incident and reflected waves are in phase with each other. This corresponds to the passage of sound from a rare to dense medium.

For $A_1 = A_2$, there is no reflected wave, and the transmitted wave is always in phase with the incident wave.

4.26. A plenum chamber designed to trap and absorb sound is installed in a ventilating system of radius 0.2 m. (*a*) Find the minimum length of the chamber that will most effectively filter out fan-induced sound of frequency 10,000 rpm. (*b*) What will be the sound transmission coefficient if the radius of the chamber is 0.5 m? (*c*) What will be the reduction in sound level? (*d*) If a 30 db sound reduction is desired, how many chambers are required?

The sound power transmission coefficient for pipes with expansion type of acoustic filters is

$$\alpha_t = \frac{4}{4 \cos^2 kL + (A_2/A_1 + A_1/A_2)^2 \sin^2 kL}$$

where the A's are the cross-sectional areas of the pipes, $k = \omega/c$ is the wave number, and L is the length of the expanded pipe.

(*a*) When sound is effectively filtered, there is a minimum transmission of sound through the plenum chamber. This occurs at $kL = \pi/2$. Now $k = \omega/c = 2\pi f/c$. Hence the minimum length of the chamber is $L_{min} = c/4f = 0.52$ m, where $c = 343$ m/sec is the speed of sound in air and $f = 10{,}000/60$ cyc/sec is the frequency of sound.

(*b*) $\alpha_t = \dfrac{(2A_1A_2)^2}{A_1^4 + A_2^4 + 2A_1^2A_2^2} = 0.1$ where $A_1 = 0.25\pi$ and $A_2 = 0.04\pi$.

(*c*) The reduction in sound level is $10 \log (1/\alpha_t) = 10$ db.

(*d*) Three chambers are required.

ABSORPTION OF SOUND

4.27. Plane acoustic waves of frequency 10,000 cyc/sec are being propagated in a water-filled pipe of radius 0.01 m. Determine the attenuation constant α in nepers/m due to the effects of viscous and heat conduction losses at the walls of the pipe. What is the attenuation in a 10 m length of this pipe?

The attenuation constant due to the effects of viscous and heat conduction is

$$\alpha = (1/ac)\sqrt{\eta\omega/2\rho} = 0.012 \text{ nepers/m}$$

where $a = 0.01$ m is the radius of the pipe, $c = 1480$ m/sec is the speed of sound in water, $\eta = 0.001$ nt-sec/m^2 is the coefficient of viscosity for water, $\rho = 998$ kg/m^3 is the density of water, and $\omega = 62{,}800$ rad/sec is the frequency of sound.

The attenuation in a 10 m length of this pipe is $8.7(10)0.012 = 1.05$ db.

4.28. Compute the viscous relaxation time and the viscous attenuation constant in air at 20°C and standard atmospheric pressure.

The relaxation time is defined as the time required for a process to proceed to within $1/e$ of its equilibrium value. For viscous relaxation time,

$$\tau = 4\eta/3\rho c^2 \text{ sec}$$

where η is the coefficient of viscosity in nt-sec/m^2, ρ is density in kg/m^3, and c is the speed of sound in m/sec.

For air at 20°C and standard atmospheric pressure, $\eta = 1.8 \times 10^{-5}$ nt-sec/m^2, $\rho = 1.21$ kg/m^3, $c = 343$ m/sec, and so $\tau = 1.7 \times 10^{-10}$ sec.

The viscous attenuation constant is given by

$$\alpha = 2\omega^2\eta/3\rho c^3 = 9.85 \times 10^4 \text{ nepers/m}$$

where $f = 100$ megacycles/sec.

MISCELLANEOUS PROBLEMS

4.29. Plane acoustic waves are propagated in a pipe in the longitudinal direction as shown in Fig. 4-11. The pipe is frictionless and its cross section changes abruptly from $A_1 = 1.00$ m^2 to $A_2 = 0.80$ m^2. Find the sound power transmission and reflection coefficients.

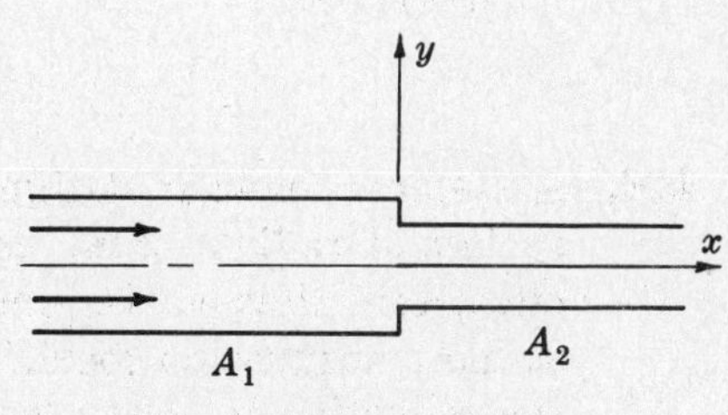

Fig. 4-11

At the junction $x = 0$, the acoustic pressure and the volume velocity must be continuous:

$$p_i + p_r = p_t, \qquad A_1(v_i + v_r) = A_2 v_t$$

where the subscript i refers to incident waves to the left of the junction, r refers to reflected waves to the left of the junction, and t refers to transmitted waves to the right of the junction. These boundary conditions yield

$$\frac{p_i + p_r}{v_i + v_r} = \frac{A_1}{A_2}\left(\frac{p_t}{v_t}\right)$$

Substituting $v_i = p_i/\rho c$, $v_r = -p_r/\rho c$, $v_t = p_t/\rho c$,

$$\frac{\rho c(p_i + p_r)}{p_i - p_r} = \frac{A_1 p_t}{A_2(p_t/\rho c)} = \frac{\rho c A_1}{A_2} \quad \text{or} \quad \frac{p_r}{p_i} = \frac{A_1 - A_2}{A_1 + A_2}$$

Then the sound power reflection coefficient is

$$\alpha_r = \left[\frac{A_1 - A_2}{A_1 + A_2}\right]^2 = \left[\frac{1.0 - 0.8}{1.0 + 0.8}\right]^2 = 0.012$$

Similarly, the sound power transmission coefficient is

$$\alpha_t = 1 - \left[\frac{A_1 - A_2}{A_1 + A_2}\right]^2 = 0.988$$

Note that the magnitudes of these two coefficients remain the same whether A_1 is greater than A_2 or A_2 is greater than A_1. No sound waves are reflected when A_1 equals A_2.

4.30. Two pipes of cross-sectional areas A_1 and A_2 contain fluid media of characteristic impedances $\rho_1 c_1$ and $\rho_2 c_2$ respectively. The pipes are connected as shown in Fig. 4-12, and the two fluid media are separated by means of a thin diaphragm. Determine the sound power transmission coefficient for plane acoustic waves traveling from pipe A_1 to pipe A_2, and the condition for 100% sound power transmission.

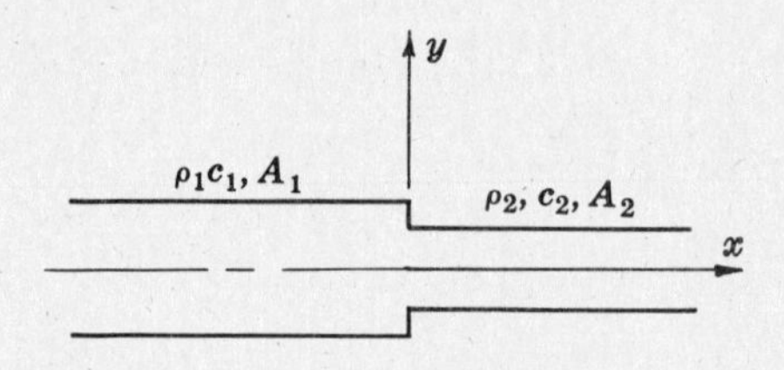

Fig. 4-12

At the junction $x = 0$, the acoustic pressure and the volume velocity must be continuous:

$$p_i + p_r = p_t, \quad A_1(v_i + v_r) = A_2 v_t \quad \text{or} \quad \frac{p_i + p_r}{v_i + v_r} = \frac{A_1 p_t}{A_2 v_t}$$

where the subscript i refers to incident waves to the left of the junction, r refers to reflected waves also to the left of the junction, and t refers to transmitted waves to the right of the junction.

Now $v_i = p_i/R_1$, $v_r = -p_r/R_1$, $v_t = p_t/R_2$ where $R_1 = \rho_1 c_1$, $R_2 = \rho_2 c_2$. Then

$$\frac{p_i + p_r}{v_i + v_r} = \frac{R_1(p_i + p_r)}{p_i - p_r} = \frac{A_1 p_t}{A_2 v_t} = \frac{A_1 R_2}{A_2}$$

from which

$$\frac{p_r}{p_i} = \frac{A_1 R_2 - A_2 R_1}{A_1 R_2 + A_2 R_1}$$

and hence the sound power reflection coefficient is

$$\alpha_r = (p_r/p_i)^2 = \left[\frac{A_1 R_2 - A_2 R_1}{A_1 R_2 + A_2 R_1}\right]^2$$

The sound power transmission coefficient is

$$\alpha_t = 1 - \alpha_r = \frac{4 A_1 A_2 R_1 R_2}{(A_1 R_2 + A_2 R_1)^2}$$

The condition for 100% power transmission is obtained by having zero power reflection or $\alpha_r = 0$, i.e. $A_1 R_2 = A_2 R_1$; and if $R_1 = R_2$, this condition becomes $A_1 = A_2$.

4.31. Harmonic plane acoustic waves of pressure amplitude P_0 are propagated into a pipe of constant cross-sectional areas $A_1 = 2A_2$ as shown in Fig. 4-13. Determine the pressure amplitude acting on the closed end of the pipe.

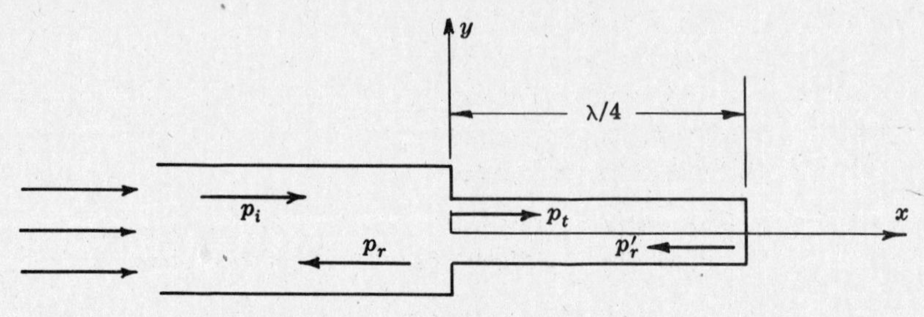

Fig. 4-13

For harmonic progressive plane acoustic waves, the incident, reflected, transmitted, and secondary reflected waves are given respectively by

$$p_i = P_0 e^{i(\omega t - kx)}, \quad p_r = P_r e^{i(\omega t + kx)}, \quad p_t = P_t e^{i(\omega t - kx)}, \quad p'_r = P'_r e^{i(\omega t + kx)}$$

where P_0, P_r, P_t, P'_r are the pressure amplitudes of the waves, ω is the frequency of the sound waves in rad/sec, and $k = \omega/c = 2\pi/\lambda$ is the wave number.

At $x = 0$ (at the junction of the pipes) the boundary condition of continuity of acoustic pressure yields

$$p_i + p_r = p_t + p'_r$$

Substituting the expressions for acoustic pressures into the above boundary condition, we obtain

$$P_0 + P_r = P_t + P'_r \tag{1}$$

At $x = 0$, the forces acting are

$$A_1(p_i - p_r) = A_2(p_t - p'_r)$$

and from expressions for acoustic pressure, we obtain

$$A_1(P_0 - P_r) = A_2(P_t - P'_r) \tag{2}$$

At the closed end, $p_t = p'_r$ or

$$P_t e^{i(\omega t - k\lambda/4)} - P'_r e^{i(\omega t + k\lambda/4)} = 0$$

Since $k = 2\pi/\lambda$, we can reduce the above expression to

$$P_t e^{-i\pi/2} - P'_r e^{i\pi/2} = 0 \tag{3}$$

Using $e^{-i\pi/2} = \cos \pi/2 - i \sin \pi/2 = -i$, and $e^{i\pi/2} = \cos \pi/2 + i \sin \pi/2 = i$, equation (3) becomes

$$-P_t = P'_r \tag{4}$$

Substituting (4) into (1) and (2), we have

$$P_0 + P_r = 0, \quad A_1(P_0 - P_r) = 2A_2 P_t$$

from which

$$P_t = (A_1/A_2)P_0 \tag{5}$$

Now the sound wave in the small pipe is

$$p(x, t) = P_t e^{i(\omega t - kx)} - P'_r e^{i(\omega t + kx)} \tag{6}$$

Since $x = \lambda/4$ at the closed end and $P_t = -P'_r$ by equation (4), we may rewrite (6) as

$$p(\lambda/4, t) = -2iP_t e^{i\omega t}$$

The pressure amplitude at the closed end is therefore equal to

$$2P_t = 2(A_1/A_2)P_0 = 2(2A_2/A_2)P_0 = 4P_0$$

i.e. four times that of the incident wave.

4.32. An infinitely long rigid smooth pipe of cross-sectional area $A_b = 1.0\ \text{m}^2$ is connected with another infinitely long rigid smooth pipe as shown in Fig. 4-14. If the cross-sectional area of the main pipe is $A = 4.00\ \text{m}^2$, find the sound power transmission coefficient in the main pipe and the branch pipe.

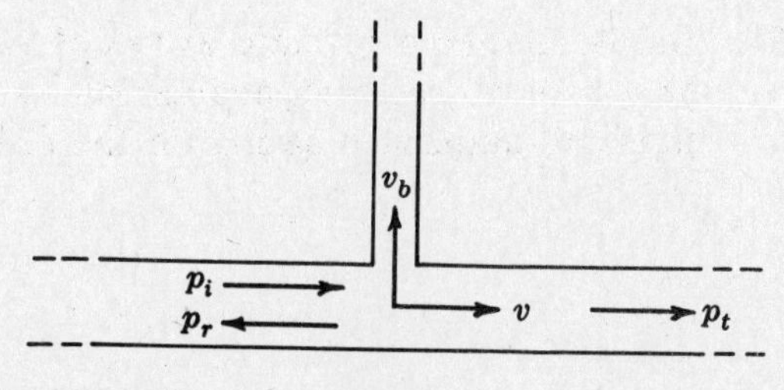

Fig. 4-14

When plane acoustic waves approach the junction, there is a change in acoustic impedance because of the branch pipe. The equivalent acoustic impedance at the junction is

$$z_{\text{eq}} = \frac{p}{v_b + v} = \frac{p}{p/z_b + p/z} = \frac{1}{1/z_b + 1/z} = \frac{z_b z}{z + z_b}$$

where p is the common acoustic pressure at the junction because of continuity of pressure, v is the particle velocity in the main pipe, v_b is the particle velocity in the branch pipe, z and z_b are the acoustic impedances to the right of the junction into the main pipe and the branch pipe respectively.

Since $z = \rho c/A$ for an infinitely long main pipe, we obtain

$$\frac{p_r}{p_i} = \frac{z_{eq} - z}{z_{eq} + z} = \frac{z_b z/(z+z_b) - z}{z_b z/(z+z_b) + z} = \frac{-\rho c/2A}{\rho c/2A + z_b} \qquad \text{(see Problem 4.1)}$$

and so the sound power reflection coefficient is

$$\alpha_r = (p_r/p_i)^2 = \frac{(\rho c/2A)^2}{(\rho c/2A + z_b)^2} = \frac{(\rho c/2A)^2}{(\rho c/2A + R_b)^2 + X_b^2}$$

For a branch pipe of infinite length, $z_b = R_b = \rho c/A_b$ and

$$\alpha_r = \frac{(\rho c/2A)^2}{(\rho c/2A + \rho c/A_b)^2} = \frac{A_b^2}{(2A + A_b)^2} = \frac{1}{(8+1)^2} = 0.012$$

The sound power transmission coefficient for sound waves going into the main pipe from the junction is similarly given by

$$\alpha_t = \frac{R_b^2 + X_b^2}{(\rho c/2A + R_b)^2 + X_b^2} = \frac{(\rho c)^2/A_b^2}{(\rho c/2A + \rho c/A_b)^2} = \frac{4A^2}{(2A + A_b)^2} = 0.79$$

and finally the sound power transmission coefficient for sound waves going into the branch pipe from the junction is

$$(\alpha_t)_b = \frac{\rho c R_b/A}{(\rho c/2A + R_b)^2 + X_b^2} = \frac{(\rho c)^2/AA_b}{(\rho c/2A + \rho c/A_b)^2} = \frac{4A_bA}{(2A + A_b)^2} = 0.198$$

i.e. $(\alpha_t)_b = 1 - \alpha_t - \alpha_r = 1 - 0.79 - 0.012 = 0.198$.

Supplementary Problems

4.33. Find the sound power transmission and reflection coefficients for sinusoidal plane acoustic waves traveling from steel into air. *Ans.* $\alpha_t = 10^{-6}$, $\alpha_r = 1$

4.34. For normal incidence of plane acoustic waves, determine the percent of sound energy passing into steel from water, and into water from air. *Ans.* 14%, 12%

4.35. Show that a 2 to 1 mismatch of characteristic impedances between two media in contact will cause a sound transmission loss of 2.5 db.

4.36. A compound rod is formed by joining the ends of two rods of constant cross-sectional areas A_1 and A_2. For normal incidence of plane acoustic waves at the junction of the compound rod, find the condition for no reflection of sound waves. *Ans.* $A_1/A_2 = (\rho_2 E_2/\rho_1 E_1)^{1/2}$

4.37. For maximum transmission of sound energy, two media should have matching impedances. Why?

4.38. The free vibrations of a steel bar decay much more rapidly when immersed in water than in air. Explain.

4.39. A pipe of cross-sectional area A_1 is connected to a second pipe of cross-sectional area A_2. For propagation of plane acoustic waves from pipes A_1 to A_2, find the standing wave ratio in pipe A_1 if A_1 is smaller than A_2. *Ans.* $\text{SWR} = A_2/A_1$

4.40. Plane acoustic waves travel from the open end of a pipe to the other end of the pipe where a piston of mass M is free to move within the pipe. Determine the sound power transmission and reflection coefficients for the pipe.

Ans. $\alpha_t = \dfrac{1}{1 + i^2M^2/4\rho^2}, \quad \alpha_r = \dfrac{1}{1 + 4\rho^2/i^2M^2}, \quad i = 1, 2, 3, \ldots$

4.41. Determine the specific acoustic impedance at a distance L to the left of the interface of two media of characteristic impedances z_1 and z_2.

Ans. $z = z_1 \dfrac{(z_1 + z_2)e^{ikL} - (z_1 - z_2)e^{-ikL}}{(z_1 + z_2)e^{ikL} + (z_1 - z_2)e^{-ikL}}$

4.42. Determine the input specific acoustic impedance of a fluid column of length L if the absorption factor of the fluid is γ. *Ans.* $z = \rho c(e^{\gamma L} + Ke^{-\gamma L})/(e^{\gamma L} - Ke^{-\gamma L})$

4.43. The sound from an aircraft flying at a great altitude from an observer on the ground is found to be limited to the lowest frequencies in the emitted complex noise. Explain.

4.44. For transmission of plane acoustic waves from one medium to another at oblique incidence, find the angle of incidence (known as the angle of intromission) for 100% transmission.

Ans. $\theta_i = \cot^{-1} \dfrac{c_1^2 - c_2^2}{(\rho_2^2 c_2^2 - \rho_1^2 c_1^2)/\rho_1^2}$

4.45. A pipe of length 1 m and cross-sectional area A_2 is inserted into a main pipe of cross-sectional area A_1. If $A_2/A_1 = 10$ and $f = 100$ cyc/sec, find the sound power transmission coefficient.
Ans. $\alpha_t = 0.185$

4.46. A hole of radius $3.4a_0^2\omega/c$ is drilled into the wall of a pipe of radius a_0. Find the sound power transmission coefficient for plane acoustic waves of frequency $2\pi f$ in the pipe.
Ans. $\alpha_t = 0.5$

4.47. An infinitely long pipe of radius 0.5 m is submerged in water. It has a hole of radius 0.1 m in its wall. Plane acoustic waves of frequency 1000 cyc/sec and 1.0 watt power are being propagated through the pipe. Determine the sound power transmitted through the pipe.
Ans. $W = 0.93$ watt

4.48. Plane acoustic waves are being propagated in a pipe closed at one end. The measured standing wave ratios of pressure at the open end and at a point 1.0 m from the open end are 10 and 9.6 respectively. Determine the attenuation constant in nepers/m. *Ans.* $\alpha = 0.0045$ neper/m

4.49. Derive an expression for the sound power transmission coefficient for plane acoustic waves through a pipe of cross-sectional areas A_1, A_2 and A_3. See Fig. 4-15.

Ans. $\alpha_t = \dfrac{4}{(A_3/A_1 + 1)^2 \cos^2 kL + (A_2/A_1 + A_3/A_2)^2 \sin^2 kL}$

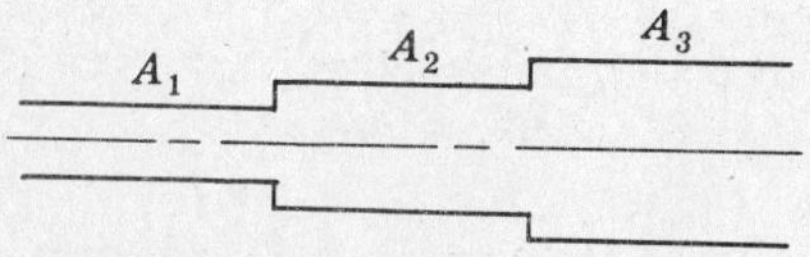

Fig. 4-15

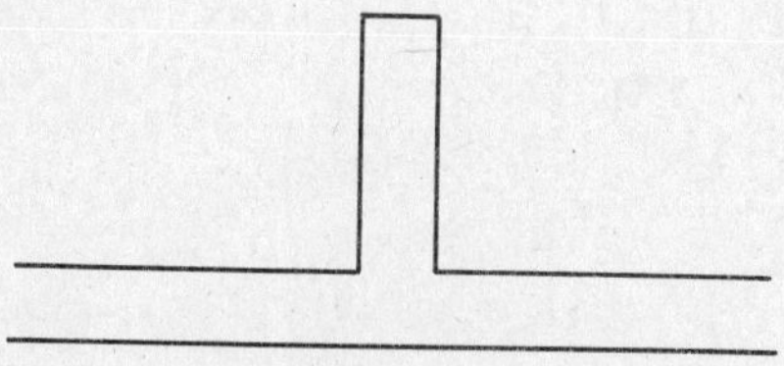

Fig. 4-16

4.50. A closed pipe is attached as a branch to the main transmitting pipe as shown in Fig. 4-16. If both pipes are made of the same material and have the same cross section, find an expression for the sound power transmission coefficient for plane acoustic waves through the main pipe.

Ans. $\alpha_t = \dfrac{4}{\sec^2 kx + 3}$

Chapter 5

Loudspeaker and Microphone

NOMENCLATURE

a = radius, m
A = area, m^2
B = magnetic flux density, webers/m^2; bulk modulus, nt/m^2
c = speed of sound in air, m/sec
C = capacitance, farads
C_a = acoustical compliance, m^5/nt
d = spacing, m
E = voltage, volts
E_L = voltage generated in the load resistor, volts
f = frequency, cyc/sec; force, nt
f_c = cut-off frequency, cyc/sec
f_r = resonant frequency, cyc/sec
h = resistance constant, ohms/m
i, I = current, amperes
k = wave number; spring constant, nt/m
k_a = acoustical stiffness, kg-m^2/sec^2
L = length, m; inductance, henrys
m = flare constant of horns; mass, kg
M = sensitivity, volts/nt/m^2
M_a = acoustical inertance or mass, kg/m^4
n_0 = sound pressure level gain, db
p = acoustic pressure, nt/m^2
Q = quality factor
R_a = acoustical resistance, nt-sec/m^5
R_E = resistance of voice coil, ohms
R_m = mechanical resistance, kg/sec
R_0 = internal impedance of microphone, ohms
R_r = radiation resistance, kg/sec
s = stiffness of the suspension, nt/m
t = thickness, m
T = tension, nt
u = displacement along the x axis, m
v = voltage, volts
V = volume, m^3
W = power, watts
X = volume displacement, m^3

$\dot{X}$ = volume velocity, m^3/sec

$\ddot{X}$ = volume acceleration, m^3/sec^2

X_r = radiation reactance, kg/sec

Z_E = total electrical impedance, ohms

Z_I = input electrical impedance, ohms

Z_m = total mechanical resistance, kg/sec

ω = circular frequency, rad/sec

λ = wavelength, m

ρ = density, kg/m^3

γ = $m/2$; ratio of the specific heat of gas at constant pressure to that at constant volume

τ = transmission coefficient

η = electroacoustic efficiency

INTRODUCTION

A *loudspeaker* is an electroacoustic transducer which converts electrical energy to acoustical energy. A *microphone* is also an electroacoustic transducer, but it converts acoustical energy to electrical energy. In general, loudspeakers are used to reproduce and amplify sound while microphones are used to record sound and to make acoustical measurements.

ELECTROACOUSTICAL ANALOGY

Like mechanical systems, acoustical systems are represented and analyzed by their equivalent electroacoustical analogues which are easier to construct than models of the corresponding acoustical systems and from which experimental results are more conveniently taken than from the acoustical models.

The equivalent electrical analogues are obtained by comparing the differential equations of motion for both systems. The acoustical and electrical systems are analogous if their differential equations of motion are mathematically the same. When this happens, the corresponding terms in the differential equations of motion are analogous to one another. The equivalent electrical circuits can then be constructed using Kirchhoff's laws of voltage and current.

There are two electrical analogies for mechanical systems: the *voltage-force* or mass-inductance analogy and the *current-force* or mass-capacitance analogy, as given in Table 5-1 below. Similarly, there are two electrical analogies for acoustical systems: the *voltage-pressure* analogy and the *current-pressure* analogy, as given in Table 5-2 below.

Acoustical inertance M_a is defined as

$$M_a = \frac{\text{acoustic pressure}}{\text{rate of change of volume velocity}} = \frac{p}{d\dot{X}/dt} = \text{kg/m}^4$$

Acoustical resistance R_a is defined as

$$R_a = \frac{\text{acoustic pressure}}{\text{volume velocity}} = \frac{p}{dX/dt} = \text{nt-sec/m}^5$$

Acoustical compliance C_a is defined as

$$C_a = \frac{\text{volume displacement}}{\text{acoustic pressure}} = \frac{X}{p} = \text{m}^5\text{/nt}$$

(See Problems 5.1-5.7.)

Table 5-1

Mechanical System		Electrical System			
		Voltage-force Analogy		Current-force Analogy	
	D'Alembert's principle		Kirchhoff's voltage law		Kirchhoff's current law
	Degree of freedom		Loop		Node
	Force applied		Switch closed		Switch closed
F	Force (lb)	v	Voltage (volt)	i	Current (ampere)
m	Mass (lb-sec^2/in)	L	Inductance (henry)	C	Capacitance (farad)
x	Displacement (in)	q	Charge (coulomb)		$\phi = \int v\,dt$
$\dot{x}$	Velocity (in/sec)	i	Loop current (ampere)	v	Node voltage (volt)
c	Damping (lb-sec/in)	R	Resistance (ohm)	$1/R$	Conductance (mho)
k	Spring (lb/in)	$1/C$	1/Capacitance	$1/L$	1/Inductance
	Coupling element		Element common to two loops		Element between nodes

Table 5-2

Acoustical System		Electrical System			
		Voltage-pressure Analogy		Current-pressure Analogy	
p	Pressure (nt/m^2)	v	Voltage (volt)	i	Current (ampere)
M_a	Inertance (kg/m^4)	L	Inductance (henry)	C	Capacitance (farad)
X	Volume displacement (m^3)	q	Charge (coulomb)	$\int v\,dt$	impulse (volt-sec)
$\dot{X}$	Volume velocity (m^3/sec)	i	Current (ampere)	v	Voltage (volt)
R_a	Resistance (nt-sec/m^5)	R	Resistance (ohm)	$1/R$	Conductance (mho)
C_a	Compliance (m^5/nt)	C	Capacitance (farad)	L	Inductance (henry)
Z_a	Impedance (ohm)	Z	Impedance (ohm)	$1/Z$	Admittance (mho)

LOUDSPEAKERS

The loudspeaker is the prime source of sound in the sound reproduction system. It provides mechanical vibrations of its own as it is energized, and vibrates the air in contact with it. As an important source of sound, loudspeakers must have high efficiency, good power-handling capacity, uniform frequency response, and minimum distortion.

The most widely used *dynamic loudspeaker* has the voice coil immersed in a fixed magnetic field generated by a powerful permanent magnet. Current flowing through the voice coil reacts with the magnetic field to produce motion which in turn actuates the diaphragm into vibration to produce sound. This type of speaker has low impedance and offers little resistance to the flow of current through it.

The *electrodynamic loudspeaker* operates like the dynamic loudspeaker and is thus *current sensitive*. Unlike the dynamic loudspeaker, the magnetic field of an electrodynamic speaker is electrically energized from an external power source.

The *condenser* or *electrostatic loudspeaker* is a *voltage sensitive* device and has high impedance. It transfers electrical signals into mechanical motion of the diaphragm through electrostatic attraction or repulsion force at the electrodes energized by voltage to produce variation in capacitance. Hence this type of loudspeaker is not suitable for low frequency operation because of the close spacing of the electrodes.

The *crystal* or *piezoelectric loudspeaker* has limited application because of its restricted low frequency response and low power output. It operates on the theory that crystal material will expand or contract when alternating electric current is applied to the surfaces of the crystal.

Acoustic power output of loudspeakers is given by

$$W = \frac{\phi^2 R_r I^2}{Z_m^2} \quad \text{or} \quad \frac{\phi^2 R_r E^2}{Z_m^2 Z_I^2} \text{ watts}$$

where $\phi = BL$, B is the magnetic flux density in webers/m^2, L the length of voice coil in meters, R_r the radiation resistance in kg/sec, Z_m the total mechanical resistance in kg/sec, Z_I the total input electrical impedance in ohms, I the current in amperes, and E the applied voltage in volts.

For *multispeaker system,* the speakers must be matched in efficiency to produce smooth overall response, and their ranges must also overlap to ensure no holes in the response curve. (See Problems 5.8-5.13.)

LOUDSPEAKER ENCLOSURES

In general the shape, size, and construction of the loudspeaker enclosure affect its overall performance. The loudspeaker enclosure generally directs the sound waves, determines the frequency response of the system, and controls sound intensity. Closed enclosure also stops front-to-end cancellation of sound waves and at the same time raises the response frequency of the system. A back-enclosed cabinet will increase the stiffness of the suspension system of the speaker cone by

$$s = \frac{\rho c^2 A^2}{V} \text{ nt/m}$$

where ρ is the density of air in kg/m^3, c the speed of sound in m/sec, A the area of the piston in m^2, and V the volume of the cabinet in m^3. (See Problems 5.10-5.12.)

HORNS

Loudspeaker horns, like loudspeaker enclosures, are designed to achieve various patterns of sound distribution and to act as acoustic transformers to couple high impedance at the throat to low impedance at the mouth of the horn. Moreover, horns usually increase the electroacoustic efficiency of the speakers and provide better reproduction of sound.

Basically there are three types of horn: (1) the *conical horn,* (2) the *exponential horn,* and (3) the *hyperbolic horn.* The cross-sectional area of the conical horn expands the most rapidly while that of the hyperbolic horn expands the slowest, as shown in Fig. 5-1.

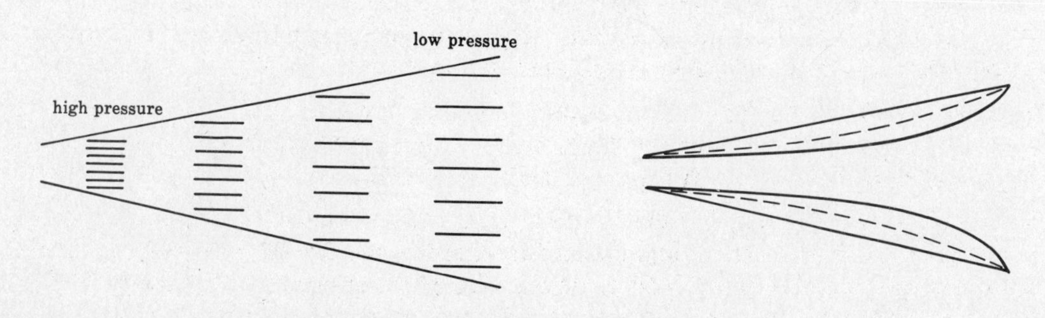

Fig. 5-1

The equation of motion for plane acoustic waves in horns is

$$\frac{\partial^2 u}{\partial t^2} = c^2 \frac{\partial^2 u}{\partial x^2} + \frac{c^2}{A}\frac{\partial A}{\partial x}\frac{\partial u}{\partial x}$$

with solution

$$u(x,t) = e^{-\gamma x}[Ce^{i(\omega t - \beta x)} + De^{i(\omega t + \beta x)}]$$

where u = displacement along the x axis, $\gamma = m/2$, m = flare constant of the horn, c = speed of sound in air, $k = \omega/c$ is the wave number, $\beta = \sqrt{k^2 - m^2/4}$. (See Problems 5.14-5.17.)

Transmission coefficient or *radiating efficiency* of a horn is the ratio of the actual acoustic power radiated out of a given horn to the acoustic power radiated by the same diaphragm which moves at the same velocity into a cylindrical tube of infinite length and having the same cross-sectional area as the throat of the given horn. For the exponential horn, the transmission coefficient is

$$\tau = \frac{1}{\sqrt{1 - (f_c/f)^2}}$$

where f is the frequency of sound and f_c is the cutoff frequency.

Cutoff frequency of horns is the minimum frequency below which propagation of sound waves inside the horn is not possible. For the exponential horn, the cutoff frequency is

$$f_c = mc/2\pi \text{ cyc/sec}$$

where m is the flare constant of the horn and c is the speed of sound in air.

A *multicellular* is a group of horns; each radiates sound as a separate and distinct horn but they are driven by a common source. To achieve wide distribution of sound waves, different arrays of obstacles are built into the *acoustic lens,* a horn designed to control the directional spread of sound. A *diffraction horn* is a narrow horn that expands uniformly in the vertical direction but is unflared in the horizontal direction. Thus a diffraction horn approximates a point source.

MICROPHONES

As dynamic air pressure transducers, microphones can be classified into two appropriate groups: (1) the *constant-velocity,* e.g. moving-coil, velocity-ribbon, and magnetostriction; (2) the *constant-amplitude,* e.g. carbon, condenser, and crystal. Depending on the nature of the operational force obtained from sound pressure to drive the diaphragm, microphones are either pressure-operated, pressure-gradient operated, or phase-shift operated. This determines whether the microphone will accept or discriminate against sounds from a particular direction.

PRESSURE-OPERATED MICROPHONES

Basically pressure-operated microphones utilize the cyclic variation in air pressure resulting from the vibration of an elastic body. The pressure inside the casting is maintained at atmospheric level, hence the force acting on the diaphragm is proportional to sound pressure and is independent of frequency, as shown in Fig. 5-2.

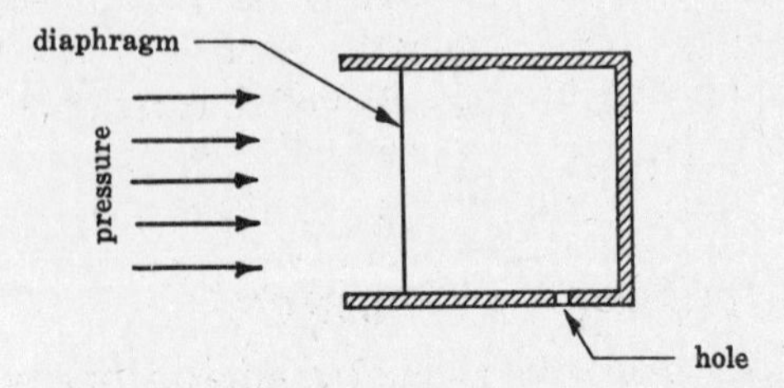

Fig. 5-2

PRESSURE GRADIENT MICROPHONES

Because both front and back faces of the diaphragm are exposed to sound pressure as shown in Fig. 5-3, a pressure gradient microphone experiences a phase difference in sound pressure. This pressure difference or gradient causes the diaphragm to move and produce a force that is proportional to frequency and path length d. A pressure gradient microphone thus discriminates against sounds arriving at an angle to the axis of the microphone.

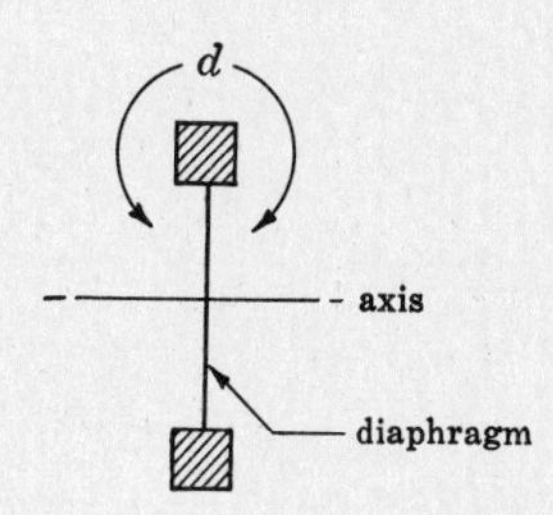

Fig. 5-3

SENSITIVITY

Sensitivity or *open-circuit voltage response* of microphones is the voltage output for a sound pressure input of one microbar, i.e. 74 db re 0.0002 microbar. For carbon microphones, for example, the sensitivity is expressed as

$$M_c = \frac{E_0 h A}{R_0 s} \text{ volts per nt/m}^2 \qquad \text{or} \qquad 20 \log (M_c/10) \text{ db}$$

where E_0 is the voltage of the battery in volts, h the resistance constant in ohms/m, A the area of the diaphragm in m², R_0 the internal impedance of the microphone in ohms and s the effective stiffness in nt/m. (See Problems 5.18-5.23.)

DIRECTIVITY

Directivity or *directional response characteristics* of microphones is the variation of microphone output with different angles of incidence, and is usually represented by a polar graph or directivity characteristics as shown in Fig. 5-4.

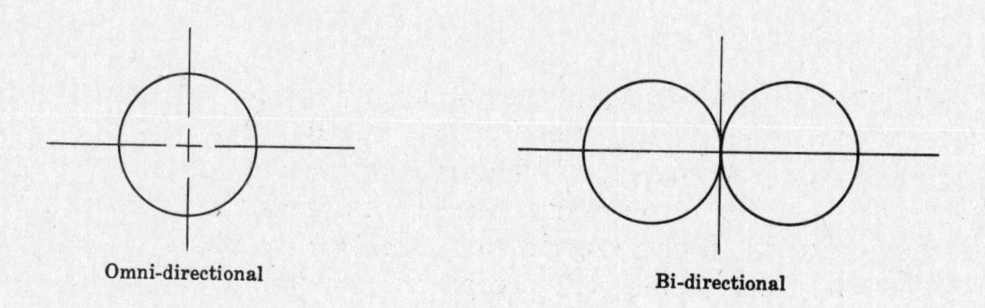

Fig. 5-4

The directional response characteristics of an uni-directional or *cardioid microphone*, for instance, is the combination of the response characteristics of an omni-directional and a bi-directional microphones. It discriminates against sounds from its sides and back, but will receive sounds from its front. Other uni-directional response characteristics may be obtained by the combination of different sizes of omni-directional and bi-directional response characteristics.

DIRECTIONAL EFFICIENCY

Directional efficiency of a microphone is the ratio of energy output due to simultaneous sounds at all angles to energy output which would be obtained from an omni-directional microphone of the same axial sensitivity. (See Problems 5.27-5.28.)

RESONANCE

The effects of resonance on microphone performances may be controlled and made negligible by: (1) *resistance control*: heavy damping is built-in to reduce the amplitude of vibration of the diaphragm; (2) *mass control*: the resonant frequency is made much lower than the working frequency; (3) *compliance control*: the resonant frequency is made much higher than the working frequency.

CALIBRATION

Microphones can be calibrated by one of the following methods: direct known sound source, comparison, Rayleigh disc, radiometer, hot-wire microphone, motion of suspended particles, and the reciprocity technique. Calibration can be carried out either in a free field with purely progressive waves as in an anechoic chamber or in a closed chamber such as a reverberation chamber where acoustic intensity and energy are constant throughout. (See Problems 5.24-5.25.)

The choice of microphone is therefore determined by the environmental conditions such as temperature, humidity, range of pressure level, and frequency response. Microphones should have high sensitivity, favorable directivity, uniform frequency response, minimum phase distortion, and very little inherent or external noise.

Solved Problems

ELECTRO-MECHANICAL ANALOGY

5.1. Investigate the electrical analogues of the single-degree-of-freedom vibratory system as shown in Fig. 5-5(*a*).

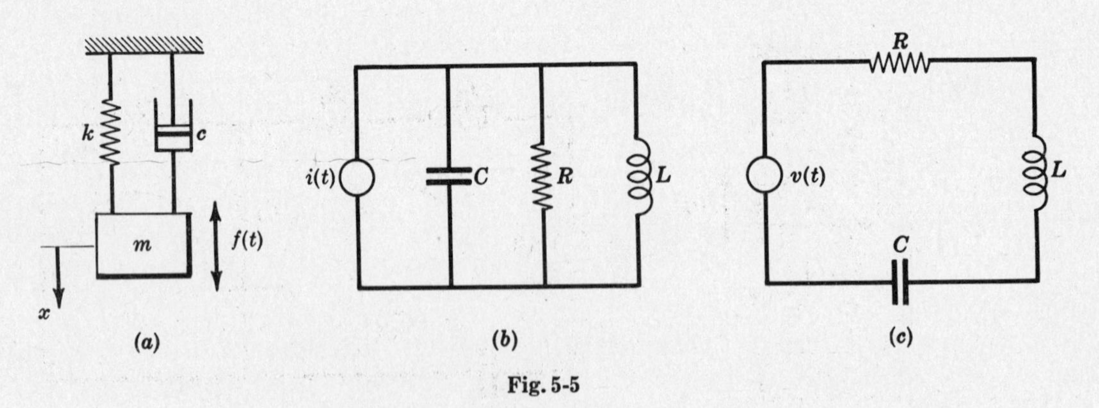

Fig. 5-5

Employing Newton's law of motion, the differential equation of motion is given by

$$m\frac{d^2x}{dt^2} + c\frac{dx}{dt} + kx = f(t) \tag{1}$$

For an electrical network as shown in Fig. 5-5(*b*), an equation of the following form can be written:

$$C\frac{d^2v}{dt^2} + \frac{1}{R}\frac{dv}{dt} + \frac{1}{L}v = \frac{di(t)}{dt} \tag{2}$$

where C = capacitance; $\left(i = C\dfrac{dv}{dt}\right)$,

R = resistance; $(i = v/R)$,

L = inductance; $\left[i = \frac{1}{L}\int v\,dt + i(0)\right]$,

$i(t)$ = current source,

v = voltage.

Since equations (1) and (2) are of the same form, i.e. they are identical mathematically, the two systems represented by these two equations are analogous.

Using *Kirchhoff's voltage law,* the voltage equation for the electrical network as shown in Fig. 5-5(*c*) is given by

$$L\frac{di}{dt} + Ri + \frac{1}{C}\int i\,dt = v(t) \tag{3}$$

Rewrite equation (1) as

$$m\frac{d\dot{x}}{dt} + c\dot{x} + k\int \dot{x}\,dt = f(t) \tag{4}$$

where dx/dt is replaced by $\dot{x}$, and x by $\int \dot{x}\,dt$. Now equations (3) and (4) are of the same form, which means that the two systems represented by these two equations are analogous. In other words, the excitation voltage $v(t)$ is analogous to the excitation force $f(t)$, the loop current i is analogous to the mass velocity $\dot{x}$, and so on. This is known as the *mass-inductance* or *voltage-force analogy.*

Integrating equation (2) once with respect to time, we obtain the current equation for the network shown in Fig. 5-5(*b*):

$$C\frac{dv}{dt} + \frac{v}{R} + \frac{1}{L}\int v\,dt = i(t) \tag{5}$$

(Equation (5) can also be obtained by *Kirchhoff's current law.*)

Now equations (4) and (5) are of the same form; which means that the two systems represented by these two equations are analogous. Hence the excitation current $i(t)$ is analogous to the excitation force $f(t)$, the network voltage v is analogous to the mass velocity $\dot{x}$, and so on. This is known as the *mass-capacitance* or *current-force analogy.*

5.2. A two-degrees-of-freedom spring-mass system is shown in Fig. 5-6(*a*). Use both the voltage-force and current-force analogy to set up the equivalent electrical circuits for the system.

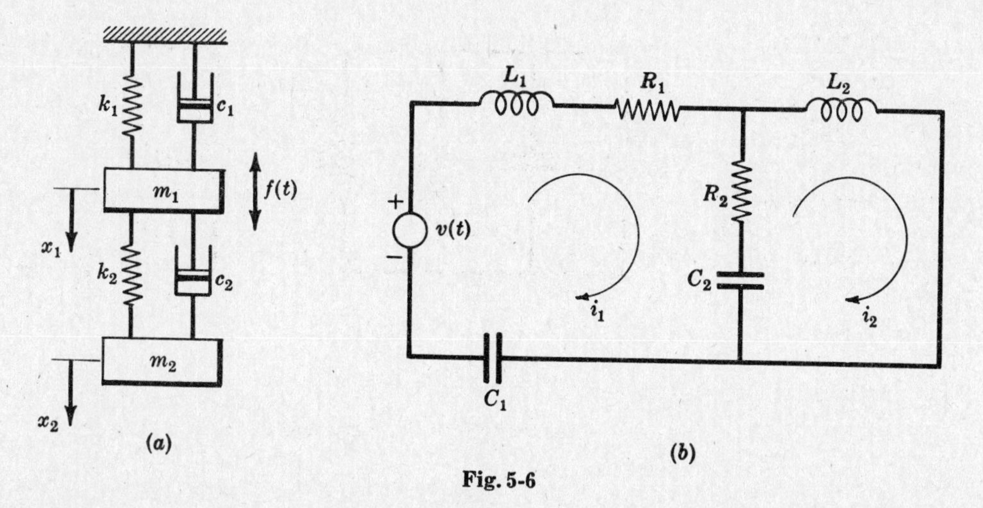

Fig. 5-6

The equations of motion given by $\Sigma F = ma$ are

$$m_1\frac{d^2x_1}{dt^2} + (c_1 + c_2)\frac{dx_1}{dt} + (k_1 + k_2)x_1 - c_2\frac{dx_2}{dt} - k_2x_2 = f(t)$$

$$m_2\frac{d^2x_2}{dt^2} + c_2\frac{dx_2}{dt} + k_2x_2 - c_2\frac{dx_1}{dt} - k_2x_1 = 0$$

Using the voltage-force analogy given in Table 5.1, the analogous electrical equations are

$$L_1\frac{di_1}{dt} + (R_1+R_2)i_1 + \left[\frac{1}{C_1}+\frac{1}{C_2}\right]\int i_1\,dt - R_2 i_2 - \frac{1}{C_2}\int i_2\,dt = v(t)$$

$$L_2\frac{di_2}{dt} + R_2 i_2 + \frac{1}{C_2}\int i_2\,dt - R_2 i_1 - \frac{1}{C_2}\int i_1\,dt = 0$$

and the analogous electrical circuit is shown in Fig. 5-6(*b*).

Using the current-force analogy as shown in Table 5.1, the analogous electrical equations are

$$C_1\frac{dv_1}{dt} + \left[\frac{1}{R_1}+\frac{1}{R_2}\right]v_1 + \left[\frac{1}{L_1}+\frac{1}{L_2}\right]\int v_1\,dt - \frac{v_2}{R_2} - \frac{1}{L_2}\int v_2\,dt = i(t)$$

$$C_2\frac{dv_2}{dt} + \frac{v_2}{R_2} + \frac{1}{L_2}\int v_2\,dt - \frac{v_1}{R_2} - \frac{1}{L_2}\int v_1\,dt = 0$$

and the analogous electrical circuit is shown in Fig. 5-6(*c*).

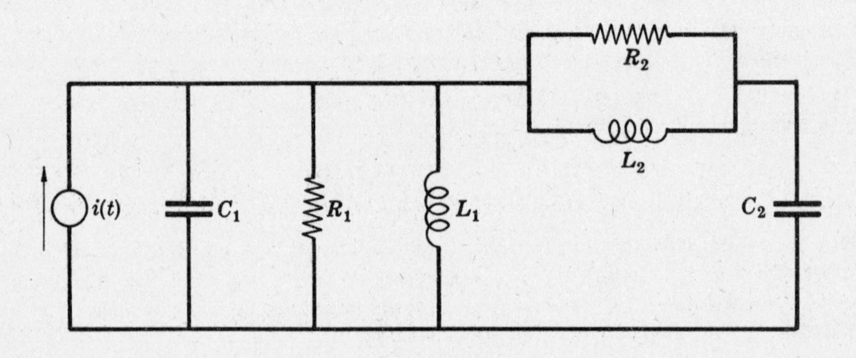

Fig. 5-6(*c*)

ELECTRO-ACOUSTICAL ANALOGY

5.3. A rigid enclosure of volume V with a small opening of radius a and length L is subjected to harmonic plane acoustic waves as shown in Fig. 5-7. Investigate the motion of the air in the enclosure.

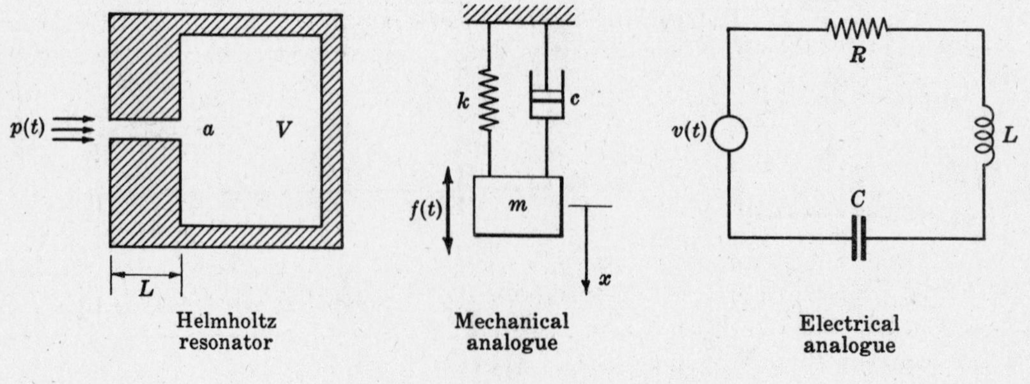

Fig. 5-7

The mass of air in the neck of the enclosure is ρAL, where ρ is the density of air and A is the cross-sectional area of the neck. This volume of air can be considered to move as a unit and thus provides the mass element of the system. The force required to move this mass is $\rho AL\ddot{x}$.

Neglecting viscous forces, the resistance element of the system is due to the radiation of sound at the opening in the form of acoustic energy dissipation, i.e. resistance resulting from radiation of sound from a simple source. This is $\rho c^2k^2A^2/2\pi$, where c is the speed of sound in air and $k = c/\omega$ is the wave number.

The volume of air inside the enclosure acts like a spring to provide the stiffness element of the system. When it is compressed the pressure increases, and when it is expanded the pressure decreases.

Now acoustic pressure $p = \rho c^2 s$, and $s = dV/V$ is the condensation. Then the force acting on area A of the opening due to acoustic pressure is

$$f = pA = \rho c^2 A\, dV/V = \rho c^2 A^2 x/V$$

and thus the effective spring stiffness is

$$k = f/x = \rho c^2 A^2/V$$

The driving force of the system is due to harmonic acoustic pressure, i.e.

$$f(t) = AP_0 \sin \omega t$$

where P_0 is the amplitude of the pressure.

Summing all the forces,

$$A\rho L\ddot{x} + \frac{\rho c^2k^2A^2}{2\pi}\dot{x} + \frac{\rho c^2A^2}{V}x = AP_0 \sin \omega t$$

and dividing by A,

$$\frac{\rho L}{A}(A\ddot{x}) + \frac{\rho c^2k^2}{2\pi}(A\dot{x}) + \frac{\rho c^2}{V}(Ax) = P_0 \sin \omega t$$

or

$$M_a\ddot{X} + R_a\dot{X} + k_aX = P_0 \sin \omega t$$

where $M_a = \rho L/A$ is the acoustical mass, $R_a = \rho c^2k^2/2\pi$ is the acoustical resistance, $k_a = \rho c^2/V$ is the acoustical stiffness ($C_a = 1/k_a = V/\rho c^2$ is the acoustical compliance), and $\dot{X} = A\dot{x}$ is the volume velocity.

Thus we have reduced a simple acoustic system to an analogous simple oscillator, i.e. a mechanical system having lumped mechanical elements of mass, resistance and stiffness. The final equation of motion corresponds to the equation of motion for a forced oscillation of a mechanical system with damping.

The steady state acoustical oscillation is therefore given by

$$X(t) = \frac{P_0}{R_a + i(\omega M_a - 1/\omega C_a)}$$

where the denominator represents the acoustic impedance.

Resonance or maximum volume velocity (air flow) in the neck occurs at a frequency which makes the total reactance zero, i.e.

$$\omega M_a - \frac{1}{\omega C_a} = 0 \qquad \text{or} \qquad \omega_n = \sqrt{\frac{1}{M_aC_a}} = c\sqrt{\frac{A}{LV}} \text{ rad/sec}$$

This basic acoustic system is represented by the *Helmholtz resonator* and its mechanical and electrical analogues as shown in Fig. 5-7. Because of the restoring force due to the volume of air inside the resonator opposite to the displacement of the volume of air in the neck, the air in the neck has harmonic motion. The Helmholtz resonator plays an important role in musical acoustics.

The resonant frequency f_r and the quality factor Q of the three systems are

Acoustical: $$f_r = \frac{1}{2\pi\sqrt{M_aC_a}} \text{ cyc/sec}, \qquad Q = \omega M_a/R_a$$

Mechanical: $$f_r = \frac{1}{2\pi\sqrt{m/k}} \text{ cyc/sec}, \qquad Q = \omega m/c$$

Electrical: $$f_r = \frac{1}{2\pi\sqrt{LC}} \text{ cyc/sec}, \qquad Q = \omega L/R$$

5.4. An air column of length 0.2 m and diameter 0.02 m is exposed to standard atmospheric pressure. For small adiabatic changes in length, find its spring constant.

For adiabatic changes, the relation between the absolute pressure and the volume is $pV^{\gamma} = \text{constant}$, where γ is the ratio of the specific heat of the gas at constant pressure to the specific heat at constant volume.

Using $d(u^n) = nu^{n-1}\,du$, we have

$$V^{1.4}\,dp + 1.4pV^{0.4}\,dV = 0 \quad \text{or} \quad dp/dV = -1.4p/V$$

But the bulk modulus for fluid is defined as $B = -V\,dp/dV = -V(-1.4p/V) = 1.4p$ and so the spring constant is

$$k = AB/L = \pi r^2(1.4p)/L = 3.14(0.01)^2(1.4)(1.01)(10)^5/0.2 = 223 \text{ nt/m}$$

5.5. Using both the voltage-pressure and the current-pressure analogy, set up the electrical analogue circuits for the low-pass acoustic filter as shown in Fig. 5-8(a).

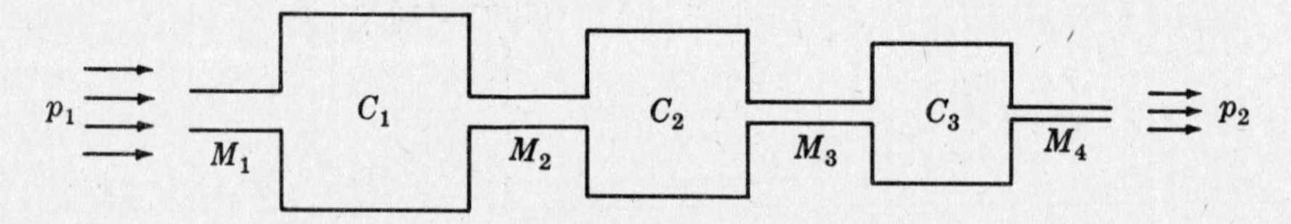

Fig. 5-8(a)

The differential equations of motion of air inside the low-pass filter are given by

$$M_1\ddot{X}_1 + \frac{X_1 - X_2}{C_1} = p_1 \cos \omega t \qquad M_3\ddot{X}_3 + \frac{X_3 - X_2}{C_2} + \frac{X_3 - X_4}{C_3} = 0$$

$$M_2\ddot{X}_2 + \frac{X_2 - X_1}{C_1} + \frac{X_2 - X_3}{C_2} = 0 \qquad M_4\ddot{X}_4 + \frac{X_4 - X_3}{C_3} = p_2 \cos \omega t$$

where the M's are the inertances in kg/m^4, X's are the volume displacements, and C's are the compliances in m^5/nt.

Using the voltage-pressure analogy, the electrical analogue equations are

$$L_1\frac{di_1}{dt} + \frac{1}{C_1}\int (i_1 - i_2)\,dt = v_1 \cos \omega t$$

$$L_2\frac{di_2}{dt} + \frac{1}{C_1}\int (i_2 - i_1)\,dt + \frac{1}{C_2}\int (i_2 - i_3)\,dt = 0$$

$$L_3\frac{di_3}{dt} + \frac{1}{C_2}\int (i_3 - i_2)\,dt + \frac{1}{C_3}\int (i_3 - i_4)\,dt = 0$$

$$L_4\frac{di_4}{dt} + \frac{1}{C_3}\int (i_4 - i_3)\,dt = v_2 \cos \omega t$$

where i's are the currents in amperes, L's are the inductances in henrys, C's are the capacitances in farads, and v's are the voltages in volts. The corresponding electrical analogue circuit is shown in Fig. 5-8(b).

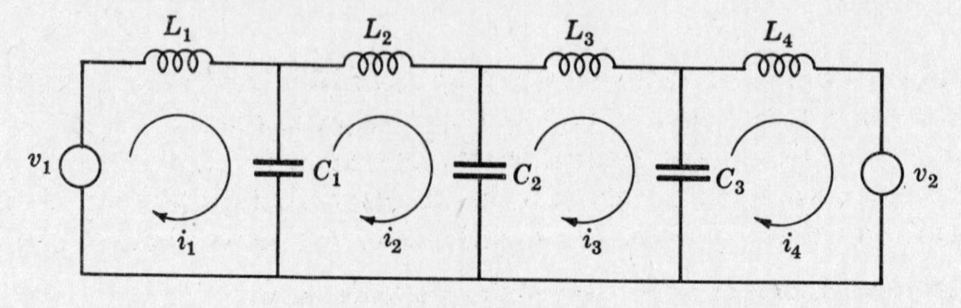

Fig. 5-8(b)

Using the current-pressure analogy, the electrical analogue equations are

$$C_1 \frac{dv_1}{dt} + L_1 \int (v_1 - v_2)\, dt \;=\; i_1 \cos \omega t$$

$$C_2 \frac{dv_2}{dt} + L_1 \int (v_2 - v_1)\, dt + L_2 \int (v_2 - v_3)\, dt \;=\; 0$$

$$C_3 \frac{dv_3}{dt} + L_2 \int (v_3 - v_2)\, dt + L_3 \int (v_3 - v_4)\, dt \;=\; 0$$

$$C_4 \frac{dv_4}{dt} + L_3 \int (v_4 - v_3)\, dt \;=\; i_2 \cos \omega t$$

where v's are the voltages in volts, C's are the capacitances in farads, L's are the inductances in henrys, and i's are the currents in amperes. The corresponding electrical analogue circuit is shown in Fig. 5-8(*c*).

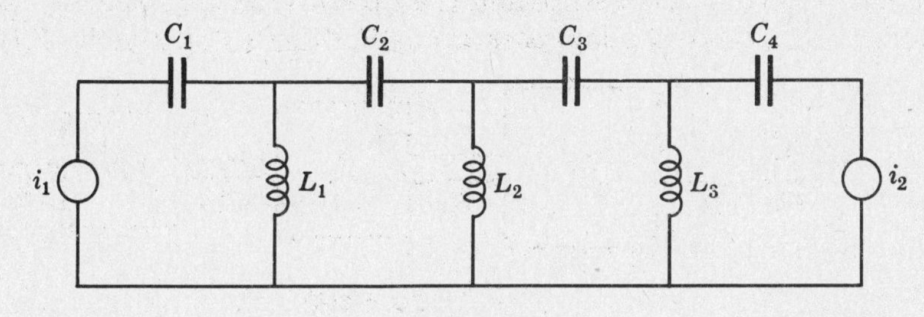

Fig. 5-8(*c*)

5.6. A Helmholtz resonator has a volume of $0.001\ \text{m}^3$ and a neck of radius 0.01 m and length 0.002 m. Find (*a*) the frequency at resonance, (*b*) the quality factor, and (*c*) the sound pressure level gain.

(*a*) The resonant frequency of the Helmholtz resonator is

$$\omega_0 \;=\; c\sqrt{\frac{A}{LV}} \;=\; 343\sqrt{\frac{0.000314}{0.002(0.001)}} \;=\; 4300 \text{ rad/sec}$$

where $c = 343$ m/sec is the speed of sound in air, $A = 3.14(0.01)^2\ \text{m}^2$ is the cross-sectional area of the neck, $L = 0.002$ m is the effective length of the neck, and $V = 0.001\ \text{m}^3$ is the volume of the resonator.

When used as a band filter, e.g. a Helmholtz resonator constructed around a ventilating duct, this resonator will most effectively filter sound at a frequency of 4300 rad/sec or 685 cyc/sec.

(*b*) The quality factor is an indication of the sharpness of resonance of a Helmholtz resonator and can be obtained by

$$Q \;=\; 2\pi\sqrt{\frac{L^3V}{A^3}} \;=\; 6.28\sqrt{\frac{8(10)^{-9}(0.001)}{3.14^3(10)^{-12}}} \;=\; 10$$

(*c*) The sound pressure level gain is acoustic pressure amplification at resonance in decibels, i.e.

$$n_0 \;=\; 20 \log (P/P_0) \;=\; 20 \log Q \;=\; 20 \log 10 \;=\; 20 \text{ db}$$

5.7. A small hole is drilled in the sphere of a Helmholtz resonator of radius 0.05 m. (*a*) If the frequency of resonance is 300 cyc/sec, what is the radius of the hole? (*b*) If the internal pressure of the resonator at resonance is 30 microbars, find the pressure amplitude of an incident plane acoustic wave that produced it. (*c*) Find also the resonant frequency if two additional holes of the same size are drilled in the sphere.

(*a*) The resonant frequency of a Helmholtz resonator is

$$\omega_0 \;=\; c\sqrt{A/LV} \text{ rad/sec}$$

where $c = 343$ m/sec is the speed of sound in air, $A = \pi r^2$ m^2 is the area of the hole, $L = 16r/3\pi$ m is the effective length of the hole, $V = 4\pi r^3/3$ m^3 is the volume of the resonator, and $\omega_0 = 2\pi(300)$ rad/sec. Substituting values into the above expression, we find $r = 0.0093$ m.

(b) The acoustic pressure amplification of the resonator at resonance is

$$P/P_0 = 2\pi\sqrt{L^3V/A^3} = 67.1 \quad \text{or} \quad P_0 = 0.446 \text{ microbar}$$

where $L = 16(0.0093)/3\pi$ m, $V = 4\pi(0.0093)^3/3$ m^3, $A = \pi(0.0093)^2$ m^2, and $P = 30$ microbars.

(c) For a total of three holes of the same size, the area is three times the original area while the effective length and volume of the resonator remain the same.

$$\omega_0 = c\sqrt{3A/LV} = 343(9.76) \text{ rad/sec} \quad \text{or} \quad f_0 = 535 \text{ cyc/sec}$$

LOUDSPEAKERS

5.8. A direct-radiator dynamic loudspeaker has a total mass of 0.01 kg (the cone and voice coil) and operates in a magnetic field of flux density 1 weber/m^2. The radius of the speaker is 0.1 m, its mechanical resistance is 1 kg/sec, its radiation resistance is 2 kg/sec, its radiation reactance is 2 kg/sec, and the stiffness of the cone system is 2000 nt/m. The length of the voice coil is 7.5 m, its inductance is 0.0005 henry, and its resistance is 10 ohms. Compute the following quantities at a frequency of 200 cyc/sec: (a) the frequency of mechanical resonance, (b) the electroacoustic efficiency, and (c) the acoustic power output W for an input current of 2 amperes.

(a) The frequency f_0 of mechanical resonance is determined by

$$(X_r + \omega_0 m - s/\omega_0) = 0$$

where $X_r = 2$ kg/sec is the radiation reactance, $m = 0.01$ kg is the mass, $s = 2000$ nt/m is the stiffness. Substitute values and solve for $\omega_0 = 360$ rad/sec or $f_0 = 57.3$ cyc/sec.

(b) The electroacoustic efficiency is

$$\eta = \frac{\phi^2 R_r}{\phi^2(R_r + R_m) + R_E Z_m^2} = 0.058 \text{ or } 5.8\%$$

where $\phi^2 = (BL)^2 = (1.0)^2(7.5)^2 = 56.1$,

$B = 1.0$ weber/m^2 is the magnetic flux density,

$L = 7.5$ m is the length of voice coil,

$R_r = 2$ kg/sec is the radiation resistance,

$R_m = 1$ kg/sec is the mechanical resistance,

$R_E = 10$ ohms is the resistance of voice coil,

$Z_m = \sqrt{(R_r + R_m)^2 + (X_r + \omega m - s/\omega)^2} = \sqrt{177}$ kg/sec is the total mechanical impedance,

$\omega = 200(6.28)$ rad/sec.

(c)

$$W = \frac{\phi^2 R_r I^2}{Z_m^2} = \frac{56.1(2)4}{177} = 2.6 \text{ watts}$$

5.9. For the direct-radiator dynamic loudspeaker of Problem 5.8, compute the acoustic power output produced by a driving voltage of 20 volts and the rms displacement amplitude of the speaker cone at resonance.

The acoustic power output is

$$W = \phi^2 R_r E^2 / Z_m^2 Z_I^2 = 1.86 \text{ watts}$$

where $\phi^2 = (BL)^2 = 56.1$,

$R_r = 2$ kg/sec is the radiation resistance,

$E = 20$ volts is the driving voltage,

$Z_m = \sqrt{177}$ kg/sec is the total mechanical impedance,

$Z_I = \sqrt{(R_E + R_M)^2 + (\omega L_E + X_M)^2} = \sqrt{(10 + 0.95)^2 + [(57.3)(6.28)(0.0005) - 4.14]^2}$

$= 11.7$ ohms is the total input electrical impedance.

For an applied voltage E_0, the current in the voice coil is

$$i = \frac{E_0}{Z_E + (\phi^2/Z_m)} = 1.4 \text{ amperes}$$

where $Z_E = 10 + i(0.628)$ ohms is the total electrical impedance of the voice coil, or its magnitude is $\sqrt{10^2 + (0.628)^2} = 10.1$ ohms.

Now the velocity of the voice coil is $v_0 = BLi/Z_m$ and so the displacement amplitude $u_0 = v_0/\omega = BLi/\omega Z_m = 0.00215$ m. Thus the root mean square displacement amplitude of the speaker cone at resonance is 0.00215 m.

5.10. If the loudspeaker of Problem 5.8 is mounted in a back-enclosed cabinet of volume 0.1 m^3, compute the frequency of mechanical resonance and the acoustic power output.

The increase in the suspension stiffness due to the back-enclosed cabinet is

$$s = \frac{\rho c^2(\pi a^2)^2}{V} = \frac{1.21(343)^2(0.01\pi)^2}{0.1} = 3160 \text{ nt/m}$$

The total stiffness constant of the suspension system is $3160 + 2000 = 5160$ nt/m.

The frequency of mechanical resonance is obtained from

$$(X_r + \omega_0 m - s/\omega_0) = (0.01\omega_0^2 + 2\omega_0 - 5160) = 0$$

which gives $\omega_0 = 625$ rad/sec or $f_0 = 99.8$ cyc/sec.

The acoustic power output for an input current of 2 amperes is

$$W = \frac{\phi^2 R_r I^2}{Z_m^2} = \frac{56.1(2)4}{121} = 3.72 \text{ watts} \quad \text{or} \quad 42.5\% \text{ increase}$$

where

$$Z_m^2 = (R_r + R_m)^2 + (X_r + \omega m - s/\omega)^2 = (2+1)^2 + [2 + 200(6.28)0.01 - 5160/1256]^2 = 121 \text{ ohms}^2$$

By mounting the loudspeaker in a back-enclosed cabinet, an increase in power output is achieved. For loudspeakers operating in the low frequency ranges, this effect is much greater.

5.11. A direct-radiator dynamic loudspeaker of radius 0.1 m and mass of the cone 0.01 kg has a suspension system of stiffness 1500 nt/m. If the loudspeaker is mounted in a back-enclosed rigid-walled cabinet of inside dimensions $0.4 \times 0.5 \times 0.6$ m and wall thickness 0.02 m, find the resonant frequency of the cabinet which can be considered as a Helmholtz resonator. What is the resonant frequency of the loudspeaker cone?

The resonant frequency of the Helmholtz resonator is

$$\omega_0 = c\sqrt{A/LV} = 343\sqrt{0.0314/0.189(0.12)} = 406 \text{ rad/sec}$$

where $c = 343$ m/sec is the speed of sound in air,

$A = \pi r^2 = 0.0314 \text{ m}^2$ is the cross-sectional area of the opening,

$L = 0.02 + 16r/3\pi = 0.189$ m is the effective length of the opening,

$V = 0.4(0.5)(0.6) = 0.12 \text{ m}^3$ is the volume of the resonator.

By considering the loudspeaker and the cabinet as a system, the effective mass is the sum of the mass of the cone and the fluid in the opening, and so the acoustic inertance is

$$M_a = m/A^2 = (\rho LA + 0.01)/A^2 = 17.43 \text{ kg/m}^4 \quad \text{where } \rho = 1.2 \text{ kg/m}^3$$

The effective stiffness of the system is the sum of the stiffness of the cone and of the cabinet. Hence the acoustic compliance of the system is

$$C_a = VA^2/(\rho c^2 A^2 + sV) = 3.7 \times 10^{-7} \text{ sec}^2\text{m}^4/\text{kg} \quad \text{where } s = 1500 \text{ nt/m}$$

The resonant frequency of the loudspeaker cone is

$$f_0 = (1/6.28)\sqrt{1/M_a C_a} = 62.4 \text{ cyc/sec}$$

5.12. A direct-radiator dynamic loudspeaker, mounted in an infinite baffle, has a radius of 0.1 m and a frequency of mechanical resonance of 20 cyc/sec. When mounted in a back-enclosed cabinet of volume 0.1 m^3, the same loudspeaker has a frequency of mechanical resonance of 40 cyc/sec. Find the mass of the speaker cone and the stiffness constant of its suspension system.

A back-enclosed cabinet will increase the stiffness of the suspension system of the speaker cone by

$$s = \rho c^2 A^2/V = 1420 \text{ nt/m}$$

where $\rho = 1.21$ kg/m^3 is the density of air, $c = 343$ m/sec is the speed of sound in air, $A = 3.14(0.01)$ m^2 is the area of the piston, and $V = 0.1$ m^3 is the volume of the back-enclosed cabinet.

Now the frequency f_0 of mechanical resonance is determined from $(X_r + 2\pi f_0 m - s/2\pi f_0) = 0$, where $X_r = 1.0$ kg/sec is the radiation reactance acting on one side of the speaker. Two such equations can be written for two frequencies f_{01} and f_{02} of mechanical resonance, i.e.

$$1.0 + 126m - s/126 = 0$$

$$1.0 + 252m - (s + 1420)/252 = 0$$

Solving these two equations, we obtain $s = 555$ nt/m, $m = 0.027$ kg.

5.13. Two identical loudspeakers are radiating acoustic power of 0.1 watt separately at a frequency of 50 cyc/sec. If they are brought together to a distance of 0.5 m between their centers and if they radiate sound waves in opposite phase, find the total acoustic power output.

Assume the sound radiation coming from each loudspeaker possesses hemispherical symmetry. Apply the *acoustic doublet theory*

$$W_d/W_s = k^2L^2/3$$

where W_d is the acoustic power output of two identical sources radiating in opposite phase, W_s is the acoustic power output of one such source, $k = \omega/c = 50(6.28)/343 = 0.92$ is the wave number, $c = 343$ m/sec is the speed of sound in air, and $L = 0.5$ m is the distance between the centers of the two sound sources. Then

$$W_d = (0.92)^2(0.5)^2(0.1)/3 = 0.007 \text{ watt}$$

HORNS

5.14. Investigate the propagation of plane acoustic waves along the axis of an infinite exponential horn.

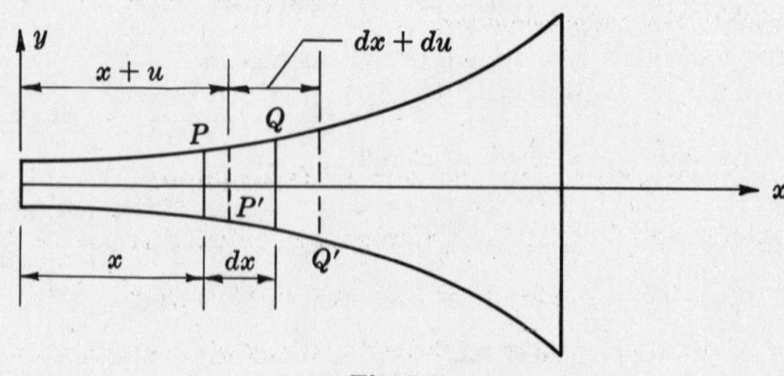

Fig. 5-9

An infinite exponential horn is a pipe whose cross-sectional area A increases exponentially with distance from its throat.

Consider an incremental section of air PQ of length dx, and its displaced position $P'Q'$. We have shown that

$$\rho_0 \frac{\partial^2 u}{\partial t^2} = -\frac{\partial p}{\partial u}$$

where ρ_0 is the density of air and p is the acoustic pressure.

But the mass of air in PQ is the same as that in $P'Q'$, i.e.

$$\rho_0 A(x)\,dx = \rho A(x+u)(dx+du)$$

or

$$\rho_0 A(x) = \rho\left[A(x) + u\frac{\partial A}{\partial x}\right]\left(1 + \frac{\partial u}{\partial x}\right)$$

Neglecting higher order terms, we have

$$\rho_0 = \rho\left(1 + \frac{\partial u}{\partial x} + \frac{u}{A}\frac{\partial A}{\partial x}\right)$$

We can express the density of air as

$$\rho = \rho_0\left(1 - \frac{\partial u}{\partial x} - \frac{u}{A}\frac{\partial A}{\partial x}\right) = \rho_0\left[1 - \frac{1}{A}\frac{\partial}{\partial x}(Au)\right]$$

The equation of motion for the incremental section becomes

$$\rho_0\frac{\partial^2 u}{\partial t^2} = -\frac{dp}{d\rho}\frac{\partial \rho}{\partial x} = c^2\rho_0\frac{\partial}{\partial x}\left[\frac{1}{A}\frac{\partial}{\partial x}(Au)\right]$$

where $c^2 = dp/d\rho$. Thus the equation of motion is

$$\frac{\partial^2 u}{\partial t^2} = c^2\frac{\partial^2 u}{\partial x^2} + \frac{c^2}{A}\frac{\partial A}{\partial x}\frac{\partial u}{\partial x}$$

which has the same form as the equation of motion for the free longitudinal vibration of a bar with variable cross section.

For an infinite exponential horn, the cross-sectional area varies with the distance according to

$$A(x) = A_0 e^{mx}$$

where A_0 is the throat area. Substituting this expression into the equation of motion, we obtain

$$\frac{\partial^2 u}{\partial t^2} = c^2\left(\frac{\partial^2 u}{\partial x^2} + m\frac{\partial u}{\partial x}\right)$$

and the solution is

$$u(x,t) = e^{-\gamma x}(Ae^{i(\omega t - \beta x)} + Be^{i(\omega t + \beta x)})$$

where $\gamma = m/2$, $\beta = \sqrt{k^2 - m^2/4}$, $k = \omega/c$. The first term on the right represents a wave going outwards and the second a wave coming inwards. The plane waves decrease in amplitude because of the attenuation factor $e^{-\gamma x}$ as a result of the spreading of waves over an increasing cross-sectional area within the horn. Since sound waves travel outward with a velocity c which is approximately independent of the frequency and with an attenuation factor which is also independent of the frequency, good reproduction of whatever waves are generated at the narrow end of the exponential horn is possible. Other forms of horn such as the conical, hyperbolic, etc., in general will not give rise to the same behavior.

5.15. Determine an expression for the cutoff frequency of an infinite exponential horn.

The motion of sound waves in an infinite exponential horn is (see Problem 5.14)

$$u(x,t) = e^{-\gamma x}(Ae^{i(\omega t - \beta x)} + Be^{i(\omega t + \beta x)})$$

which represents waves traveling in opposite directions with velocity $v = \omega/\beta$. Since $\beta = \sqrt{k^2 - m^2/4}$, the velocity of sound propagation can be expressed as

$$v = \frac{\omega}{\sqrt{k^2 - m^2/4}} = \frac{\omega/k}{\sqrt{1 - m^2/4k^2}}$$

where the quantity under the radical sign cannot be negative. Then

$$1 = m^2/4k^2 \quad \text{or} \quad k = \omega/c = m/2$$

and the *cutoff frequency* is $f_c = \omega_c/2\pi = mc/4\pi$. This is the minimum frequency, below which propagation of sound waves inside an infinite exponential horn is not possible.

5.16. An infinite exponential horn of length 0.75 m has a radius of 0.02 m at the throat and a radius of 0.2 m at its mouth. Find (*a*) the flare constant of this horn and its cutoff frequency, (*b*) the peak volume velocity at the throat for 0.5 watt acoustic output of the horn. (*c*) If the radius of the driving diaphragm is 0.03 m, find the peak displacement amplitude in order to produce the above volume velocity at the throat of the horn.

(*a*) For infinite exponential horns, the cross-sectional area at a distance L from the throat is

$$A_L = A_0 e^{mL}$$

where A_0 is the throat area and m is the flare constant. Thus

$$(0.2)^2 = (0.02)^2 e^{0.75m} \quad \text{or} \quad 100 = e^{0.75m}$$

Taking natural logarithms, we obtain the flare constant $m = 6.15$.

The cutoff frequency of infinite exponential horns is $f_c = mc/4\pi = 167$ cyc/sec.

(*b*) Acoustic output for an infinite exponential horn can be expressed as

$$W = R_r v^2 = (A_0^2 R_0)v^2 = A_0^2(\rho c/A_0)v^2 \text{ watts}$$

Since volume velocity at the throat is $V_0 = A_0 v$, we have $W = V_0^2(\rho c/A_0)$ or

$$V_0 = \sqrt{WA_0/\rho c} = \sqrt{0.5\pi(0.02)^2/415} = 0.00123 \text{ m}^3/\text{sec}$$

(*c*) The peak velocity at the throat is $v_0 = V_0/\pi A_0^2 = 0.979$ m/sec. Thus the peak displacement amplitude at the throat is

$$u_0 = v_0/\omega = 0.979/[167(6.28)] = 0.000928 \text{ m}$$

The volume displacement at the throat must equal the volume displacement at the driving diaphragm, i.e. $u_0 A_0 = u_d A_d$, and hence the peak displacement amplitude at the driving diaphragm is

$$u_d = u_0 A_0/A_d = [0.000464\pi(0.02)^2]/[\pi(0.03)^2] = 0.00021 \text{ m}$$

5.17. Investigate the propagation of sound waves along the axis of a conical horn as shown in Fig. 5-10.

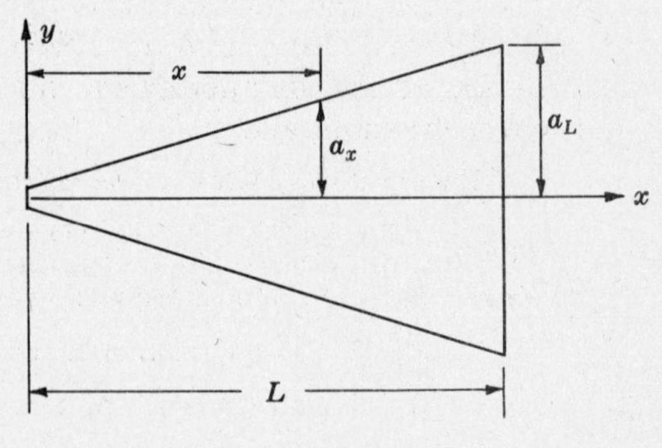

Fig. 5-10

The equation of motion for horns with variable cross-sectional area A is

$$\frac{\partial^2 u}{\partial t^2} = c^2 \frac{\partial^2 u}{\partial x^2} + \frac{c^2}{A}\frac{\partial A}{\partial x}\frac{\partial u}{\partial x}$$

where u is the displacement along the axis and c is the speed of sound in air. Rewriting,

$$\frac{\partial^2 u}{\partial t^2} = c^2 \frac{\partial^2 u}{\partial x^2} + c^2 \frac{\partial u}{\partial x}\frac{\partial \ln A}{\partial x}$$

From the geometry of the conical horn,

$$\frac{a_x}{a_L} = \frac{x}{L} \quad \text{or} \quad \frac{A_x}{A_L} = \frac{\pi a_x^2}{\pi a_L^2} = \frac{x^2}{L^2}$$

where A_L is the area of cross section at the mouth, A_x is the area of cross section at a length x from the throat, and the area at the throat is assumed to be negligible. Taking the natural logarithm of the last expression,

$$\ln A_x = 2 \ln x + \ln(A_L/L^2)$$

or

$$\frac{\partial \ln A_x}{\partial x} = \frac{2}{x}$$

Putting the above expression into the general equation of motion,

$$\frac{\partial^2 u}{\partial t^2} = c^2\left(\frac{\partial^2 u}{\partial x^2} + \frac{2}{x}\frac{\partial u}{\partial x}\right)$$

This can be written as
$$\frac{\partial^2(xu)}{\partial t^2} = c^2 \frac{\partial^2(xu)}{\partial x^2}$$

which is similar to the equation governing the propagation of spherical acoustic waves from a point source. Thus we conclude that spherical instead of plane acoustic waves will be propagated in a conical horn with a velocity c independent of frequency, and with attenuation of intensity in accordance with the inverse square law. (See Problem 3.8.)

MICROPHONES

5.18. A crystal microphone has a sensitivity of -50 db re 1 volt/microbar and an internal capacitive impedance of 150,000 ohms at 500 cyc/sec. Plane acoustic waves of frequency 500 cyc/sec and acoustic pressure 0.5 microbar are incident on the microphone. Determine the voltage generated in a load resistor of 400,000 ohms connected across the output terminals of the microphone. What power will be generated in this load resistor?

The sensitivity of the microphone is

$$M_c = 20 \log (E/p) = 20 \log (E/0.5) = -50$$

or the output voltage of the microphone is $E = 0.5 \text{ antilog}(-2.5) = 0.00161$ volt. Hence the voltage generated in the load resistor is

$$E_L = i_L R_L = \frac{E R_L}{R_L + R} = \frac{0.00161(400{,}000)}{400{,}000 + 150{,}000} = 0.00117 \text{ volt}$$

The power generated by the load resistor is $W_L = E_L^2/R_L = 3.42 \times 10^{-12}$ watt.

5.19. A carbon microphone diaphragm of radius 0.01 m and effective stiffness 10^6 nt/m is connected to a 12-volt battery. If the internal impedance of this microphone is 120 ohms and its resistance constant is 7.5×10^8 ohms/m, find the microphone sensitivity. Find also the ratio of the second harmonic to fundamental voltage developed in this microphone for an incident plane acoustic wave of 150 microbars pressure amplitude.

The sensitivity of the carbon microphone is

$$M_c = E_0 hA/R_0 s = 12(7.5 \times 10^8)(0.000314)/1.2(10)^8 = 2.35 \times 10^{-2} \text{ volt/nt-m}^2$$

where $E_0 = 12$ volts is the voltage of the battery,

$h = 7.5 \times 10^8$ ohms/m is the resistance constant,

$A = 0.000314 \text{ m}^2$ is the area of the diaphragm,

$R_0 = 120$ ohms is the internal impedance of the microphone,

$s = 10^6$ nt/m is the effective stiffness.

The response of the microphone can be expressed as a decibel level relative to one volt/microbar or one volt per 0.1 nt/m², i.e. $20 \log (M_c/10) = -52.6$ db re 1 volt/microbar.

The ratio of the second harmonic voltage to fundamental voltage is $hy_0/2R_0$, where y_0 is the displacement amplitude at the center of the diaphragm due to sound pressure and is given by

$$y_0 = P_0 A/s = 4.7 \times 10^{-9} \text{ m}$$

where $P_0 = 150$ microbars or 15 nt/m² is the pressure amplitude of the incident sound waves. Thus the ratio of the second harmonic to the fundamental is $hy_0/2R_0 = 0.015$.

It is interesting to note that both the microphone response and the ratio of second harmonic distortion depend on the factor h/R_0. By increasing h/R_0 either by increasing the value for h or decreasing the internal impedance R_0, we obtain better microphone sensitivity but greater distortion, and vice versa. For very intense sound waves, the output will have considerable harmonic distortion due to large y_0.

5.20. A moving-coil microphone has a moving element of radius 0.05 m, 0.002 kg mass, 50,000 nt/m stiffness, and 20 kg/sec mechanical resistance. The coil is 0.3 m long and moves in a magnetic field of 1.5 webers/m² flux density. What is the open-circuit response at 1000 cyc/sec frequency? Find the amplitudes of the velocity and displacement of the diaphragm when it is subjected to an acoustic pressure of 1.0 nt/m². What is the open-circuit voltage generated in the coil?

The open-circuit response is

$$M_m = BLA/Z_m = 1.75 \times 10^{-4} \text{ volt per nt/m}^2 = 1.75 \times 10^{-5} \text{ volt/microbar}$$

where $B = 1.5$ webers/m² is the flux density,

$L = 0.3$ m is the length of the coil,

$A = \pi a^2 = 3.14(0.05)^2 = 7.84 \times 10^{-3}$ m² is the cross section,

$Z_m = \sqrt{R_m^2 + (\omega m - 1/\omega C_m)^2} = 20.6$ ohms is the impedance,

$\omega = 1000(2\pi) = 6280, \quad \omega m = 6280(0.002) = 12.56,$

$C_m = 1/s = 1/50{,}000 = 2 \times 10^{-5}$ is the compliance in m/nt, $1/\omega C_m = 8.0,$

$R_m = 20$ kg/sec is the mechanical resistance.

Using one volt per microbar as reference, the open-circuit response in decibels is

$$M_m = 20 \log 1.75 \times 10^{-5} = -95 \text{ db}$$

The amplitude of velocity of the diaphragm is $v_0 = F/Z_m = 1/20.6 = 0.0485$ m/sec. Hence the amplitude of displacement of the diaphragm is $u_0 = v_0/\omega = 7.72 \times 10^{-6}$ m. The open-circuit voltage generated in the coil is $V = BLv_0 = 0.0218$ volt.

5.21. A condenser microphone diaphragm of radius 0.02 m is stretched to a tension of 20,000 nt/m. The spacing between diaphragm and the backing plate is 0.00001 m, and the polarizing voltage of the microphone is 400 volts. (*a*) What is the open-circuit voltage response of the microphone? (*b*) Find the amplitude of the average displacement of the diaphragm when it is acted upon by a sound wave of 15 nt/m² pressure amplitude. (*c*) Determine the voltage generated in a load resistor of 3 megohms if the frequency of the incident sound waves is 150 cyc/sec.

(*a*) The open-circuit voltage response of the condenser microphone is

$$M_c = \frac{E_0 a^2}{8dT} = \frac{400(0.0004)}{8(0.00001)(20{,}000)} = 0.1 \text{ volt per nt/m}^2 = 0.01 \text{ volt/microbar}$$

where $E_0 = 400$ volts is the polarizing voltage, a is the radius of the diaphragm in meters, $d = 0.00001$ m is the spacing between diaphragm and backing plate, and $T = 20{,}000$ nt/m is the tension.

The response in decibels is $M_c = 20 \log 0.01 = -40$ db re 1 volt/microbar.

(*b*) The amplitude of the average displacement of the diaphragm is

$$y_{av} = P_0 a^2/8T = 15(0.02)^2/8(20{,}000) = 3.75 \times 10^{-8} \text{ m}$$

(*c*) The voltage drop across the load resistor is

$$E_L = \frac{E_0 C_1 R_L}{C_0\sqrt{(1/\omega C_0)^2 + R_L^2}} = 1.48 \text{ volts}$$

where $C_0 = (27.8a^2/d) \times 10^{12} = [(27.8)(0.0004)/0.00001]10^{12} = 1120 \times 10^{12}$ farads,

$C_1 = C_0 P_0 a^2/8dT = \dfrac{1120 \times 10^{12}(15)(0.0004)}{8(0.00001)(20{,}000)} = 4.2 \times 10^{12}$ farads,

$E_0 = 400$ volts,

$R_L = 3$ megohms $= 3 \times 10^6$ ohms.

5.22. A velocity-ribbon microphone has an aluminum strip of width 0.004 m, length 0.03 m and mass 3×10^{-6} kg. The strip moves in a magnetic field of flux density 0.3 weber/m^2 inside a circular baffle of radius 0.05 m. If a plane acoustic wave of frequency 300 cyc/sec and pressure 2.5 nt/m^2 is incident normally on the face of the ribbon, find (*a*) the voltage generated in the ribbon, (*b*) the sensitivity of the microphone M_v at this frequency, and (*c*) the amplitude of the velocity and displacement of the ribbon.

(*a*) Voltage generated is

$$E = BL_cLAP_0/cm = 1.31 \times 10^{-4} \text{ volt}$$

where $B = 0.3$ weber/m^2 is the flux density of the magnetic field,

$L_c = 0.03$ m is the length of the ribbon,

$L = 0.05$ m is the radius of the circular baffle,

$A = 0.004(0.03) = 1.2 \times 10^{-4}$ m^2 is the area of the strip,

$P_0 = 2.5$ nt/m^2 is the acoustic pressure amplitude,

$c = 343$ m/sec is the velocity of sound,

$m = 3 \times 10^{-6}$ kg is the mass of the ribbon.

(*b*) $M_v = (2BL_cA/\omega m) \sin(\frac{1}{2}kL \cos\theta) = 5.35 \times 10^{-5}$ volt/(nt/m^2) $= -85.4$ db re 1 volt/microbar where $k = \omega/c = 300(6.28)/343 = 5.49$, $kL = 5.49(0.05) = 0.274$, $\cos\theta = 1$ at normal incidence, $\sin(\frac{1}{2}kL) = \sin 7.85° = 0.14$.

(*c*) The amplitude of velocity of the ribbon is $v_0 = E/BL_c = 0.0146$ m/sec, and hence the amplitude of displacement of the ribbon is $u_0 = v_0/\omega = 7.76 \times 10^{-6}$ m.

5.23. If the diaphragm of the condenser microphone of Problem 5.21 is made of steel of thickness 0.00001 m, compute the fundamental frequency of the diaphragm. What is the internal impedance of the condenser microphone?

The fundamental frequency of a flexible circular diaphragm stretched to a high tension at the edges is given by

$$f_1 = (2.4/2\pi a)\sqrt{T/\rho_a} = 9780 \text{ cyc/sec}$$

where $a = 0.02$ m is the radius of the diaphragm, $T = 20{,}000$ nt/m is the tension, $\rho_a = \rho t = 7700(0.00001) = 0.077$ kg/m^2 is the density per unit area of the diaphragm, and $t = 0.00001$ m is the thickness of the diaphragm.

The internal impedance $= 1/\omega C_0 = t/150(6.28)27.8a^2 = 0.95 \times 10^{-6}$ ohm.

5.24. In a reciprocity type of calibration of two identical reversible microphones spaced 1.5 m from each other, the measured open-circuit voltage output of one microphone is 0.01 volt when a driving current of 0.15 ampere is supplied to the other microphone at a frequency of 1500 cyc/sec. Calculate the sensitivity of the microphones.

The open-circuit voltage response of the microphones calibrated by the reciprocity method is

$$M_a = M_b = \sqrt{2dE_a/\rho f I_b} = \sqrt{2(1.5)(0.01)/(1.21)(1500)(0.15)} = 0.0106 \text{ volt/(nt/m}^2)$$
$$= -59.94 \text{ db re 1 volt/microbar}$$

where $d = 1.5$ m is the spacing between the two identical microphones, $E_a = 0.01$ volt is the measured open-circuit voltage output of one of the microphones, $\rho = 1.21$ kg/m^3 is the density of air, $f = 1500$ cyc/sec is the frequency of the driving current, and $I_b = 0.15$ ampere is the driving current.

5.25. A reversible electroacoustic transducer and a loudspeaker are used in the reciprocity calibration of a microphone. The open-circuit voltages in the transducer and the microphone are 0.16 and 0.64 volts respectively when they are placed the same distances from the loudspeaker. When the microphone is 2.0 m from the transducer which acts as the source, an open-circuit voltage of 0.02 volt is generated in the microphone while the transducer is supplied with a driving current of 12 amperes at a frequency of 1500 cyc/sec. Determine the open-circuit response of the microphone and the acoustic pressure p acting on the microphone.

The open-circuit response of microphones calibrated by the reciprocity method is given by (see Problem 5.24)

$$M_a = \sqrt{2dE_aE_a'/\rho f I_b E_b} = \sqrt{2(2)(0.64)(0.02)/(1.21)(1500)(12)(0.16)} = 0.0038 \text{ volt/(nt/m}^2)$$
$$= -88 \text{ db re 1 volt/microbar}$$

and so $p = E_a'/M_a = 0.02/0.0038 = 5.3$ nt/m^2.

5.26. A microphone of impedance 100 ohms and frequency 1000 cyc/sec is connected to an amplifier by 25 m of coaxial cable having a capacitance of 0.01 microfarad per meter of cable. If the impedance of the microphone is entirely reactance, find the voltage loss in decibels due to the capacitance of the cable.

The capacitance of the microphone is $C_m = 1/\omega X_c = 1/6280(100) = 1.6$ microfarads, and the capacitance of the coaxial cable is $C_c = 25(0.01) = 0.25$ microfarad. Hence voltage loss is $20 \log (1.6 + 0.25)/1.6 = 1.24$ db.

Cables connecting microphones to amplifiers should be short in length, well screened, and of low capacitance. Otherwise the voltage output of the microphone will be affected.

5.27. Find an expression for the ratio of the pressure gradient in spherical acoustic waves and the pressure gradient in plane acoustic waves for a first-order pressure gradient microphone.

For harmonic spherical acoustic waves, the instantaneous pressure p at a distance r from the source is

$$p = \frac{P_0}{r} \cos(\omega t - kr)$$

where $k = \omega/c$ is the wave number and P_0 is the maximum pressure amplitude. The pressure gradient is therefore given by

$$\frac{dp}{dr} = \frac{P_0 k}{r} \sin(\omega t - kr) - \frac{P_0}{r^2} \cos(\omega t - kr)$$
$$= \frac{P_0}{r}\left[k \sin(\omega t - kr) - \frac{1}{r}\cos(\omega t - kr)\right]$$

and hence the rms value of the pressure gradient is

$$\left(\frac{dp}{dr}\right)_{\text{rms}} = \frac{P_0}{r}\sqrt{\tfrac{1}{2}(k^2 + 1/r^2)} = \frac{P_0 k}{r\sqrt{2}}\sqrt{1 + 1/r^2k^2}$$

Similarly, for harmonic plane acoustic waves of the same amplitude the instantaneous pressure p at a distance r from the source is

$$p = P_0' \cos(\omega t - kr)$$

where the amplitude $P_0' = P_0/r$ remains constant. For the pressure gradient we have

$$\frac{dp}{dr} = P_0' k \sin(\omega t - kr) = \frac{P_0 k}{r} \sin(\omega t - kr)$$

and the rms value is $P_0 k/r\sqrt{2}$.

Thus the ratio of the first-order pressure gradient is

$$P_n = \frac{(P_0 k/r\sqrt{2})\sqrt{(1+1/r^2k^2)}}{P_0 k/r\sqrt{2}} = \sqrt{1+1/r^2k^2}$$

This ratio indicates that pressure gradient microphones favor spherical acoustic waves (e.g. close sound sources) but discriminate against plane acoustic waves (e.g. distant ambient noises). This is based on the assumption that the path length of the microphone is very small compared with the wavelength (this is true for low frequencies but inaccurate at high frequencies).

It can be shown in a similar manner that the ratio of pressure gradient for second-order pressure gradient microphones is $\sqrt{1+4/k^4r^4}$. As in the previous case, plane acoustic waves are being discriminated against while spherical acoustic waves are being favored.

5.28. An array of n pressure-sensitive microphones are connected in series and equally spaced a distance d meters as shown in Fig. 5-11. If the microphones have identical response and sensitivity, determine an expression for the output of the array for plane acoustic waves with angle of incidence θ.

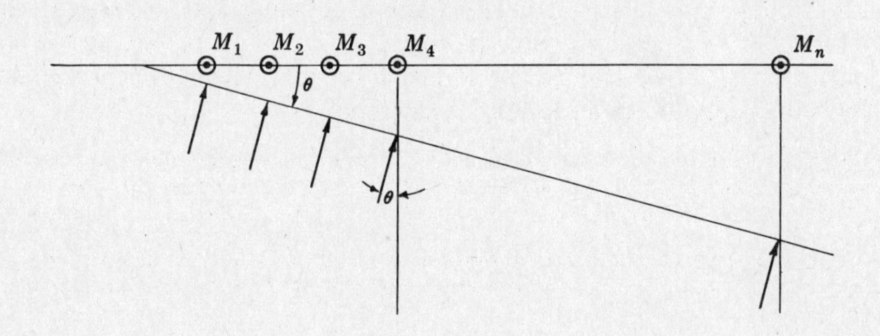

Fig. 5-11

Since sound waves arrive with angle of incidence θ, the wavefront reaches different microphones at different times and the output from each unit will vary in phase. Let AB and BC represent the outputs of microphones M_1 and M_2 respectively. If the angle of incidence $\theta = 0$, AC' would represent the total output of microphones M_1 and M_2. Now the output BC from microphone M_2 lags the output AB from microphone M_1 by an angle $\phi = kd \sin\theta$ as shown in Fig. 5-12, where $k = \omega/c = 2\pi/\lambda$ is the wave number.

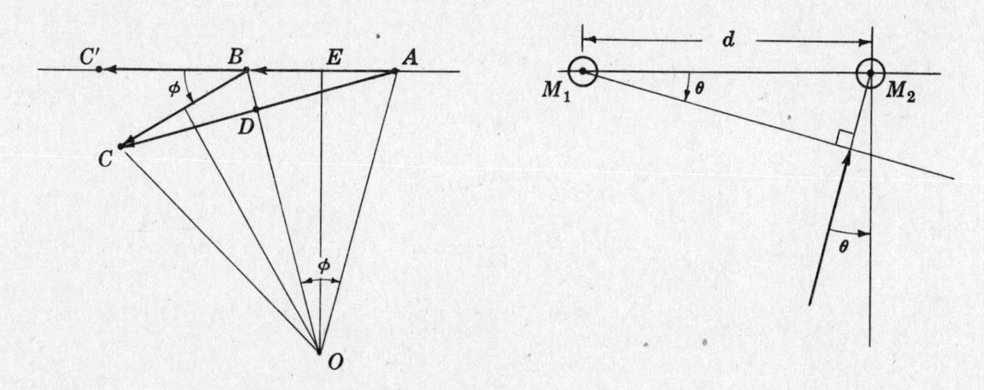

Fig. 5-12

Triangles OAB and OBC are similar isosceles triangles, so

$$\phi = 180° - \angle CBO - \angle OBA \quad \text{and} \quad \phi = \angle BOA = \angle BOC$$

The output for microphones M_1 and M_2 is therefore $AC = 2AD = 2AO \sin \phi$. Thus for an array of n identical units, the total output is

$$E_n = 2AO \sin (n\phi/2)$$

But $AO = AE/\sin (\phi/2)$ where $2AE = E_1$ is the output of one microphone. Then

$$E_n = \frac{2AE}{\sin (\phi/2)} \sin (n\phi/2) = \frac{\sin (n\phi/2)}{\sin (\phi/2)} E_1$$

At low frequencies the wavelength $\lambda = c/f$ is considerably larger than the spacing distance d; hence $\phi = kd \sin \theta = (2\pi/\lambda)d \sin \theta$ and so

$$\frac{\sin (n\phi/2)}{\sin (\phi/2)} \doteq \frac{n\phi/2}{\phi/2} = n \quad \text{for small values of } \phi$$

In other words, the total output of an array of n microphones at an angle of incidence θ is $E_n = nE_1$, which is the same for an array at an angle of incidence $\theta = 0$.

At high frequencies the values for ϕ are no longer small. Consequently the output depends on the angle of incidence θ, i.e. the array is highly directional.

For an array of 10 microphones spaced evenly at a distance 0.12 m apart, for example, the angles of incidence for zero output for sound waves at a frequency of 343 cyc/sec are given by

$$\sin (n\phi/2) = 0 \quad \text{or} \quad 10\phi/2 = \pi \quad \text{or} \quad \phi = \pi/5$$

Now $$\phi = (2\pi d/\lambda) \sin \theta \quad \text{or} \quad 2(3.14)0.12 \sin \theta = (3.14/5)\lambda$$

where $\omega/c = 2\pi/\lambda$ and $\lambda = 2\pi c/343(6.28) = 1.0$. Hence $\sin \theta = 0.83$ and $\theta = 56°, 124°$.

Supplementary Problems

5.29. Determine the equivalent electrical circuit for the acoustical system consisting of a series of Helmholtz resonators as shown in Fig. 5-13.

Ans.

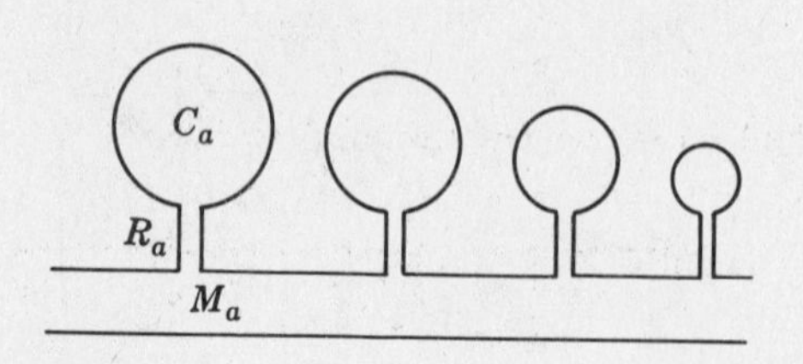

Fig. 5-13

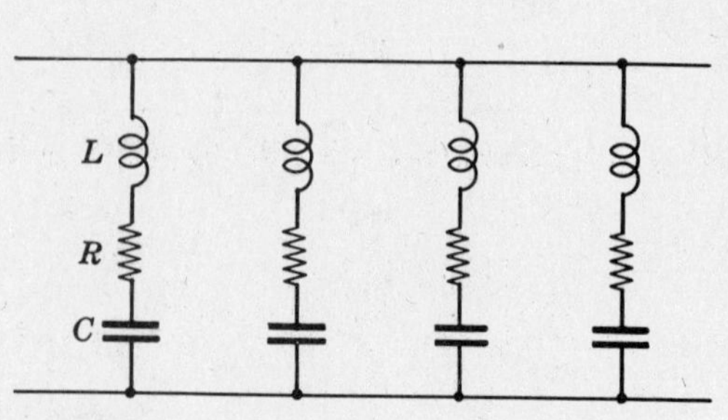

5.30. Find an equivalent electrical circuit for the acoustical system shown in Fig. 5-14.

Ans.

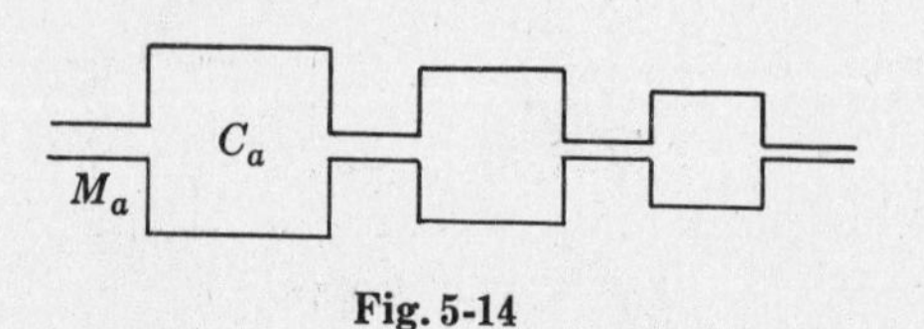

Fig. 5-14

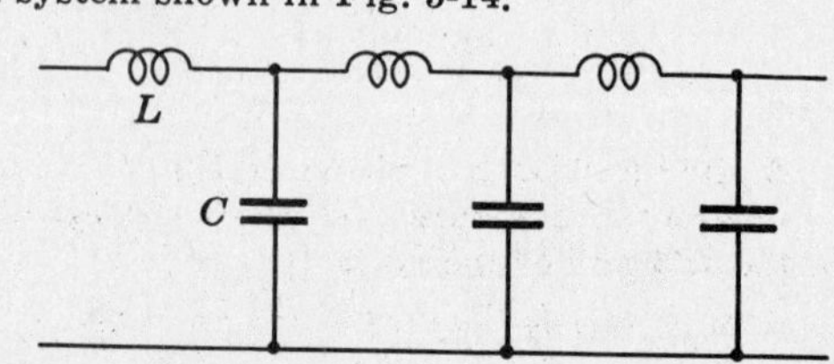

5.31. For the acoustical system shown in Fig. 5-15, find the equivalent electrical circuit.

Ans.

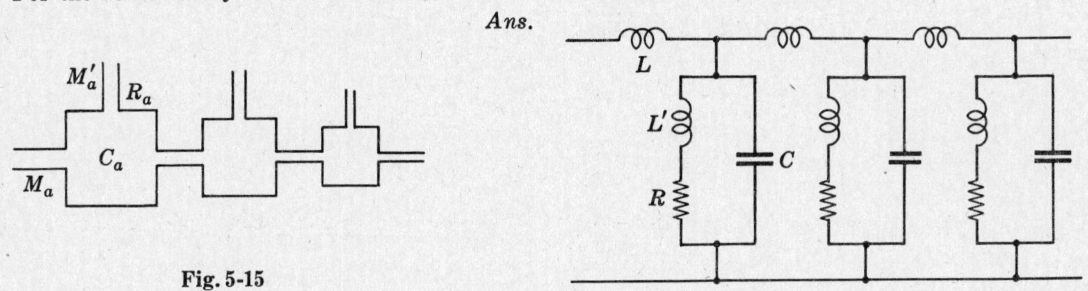

Fig. 5-15

5.32. An acoustic filter as shown in Fig. 5-16 is subjected to steady harmonic sound pressure $p_0 \sin \omega t$. Find the amplitude ratio of the steady state response. *Ans.* $p/p_0 = 1/(1 - \omega^2/\omega_n^2)$

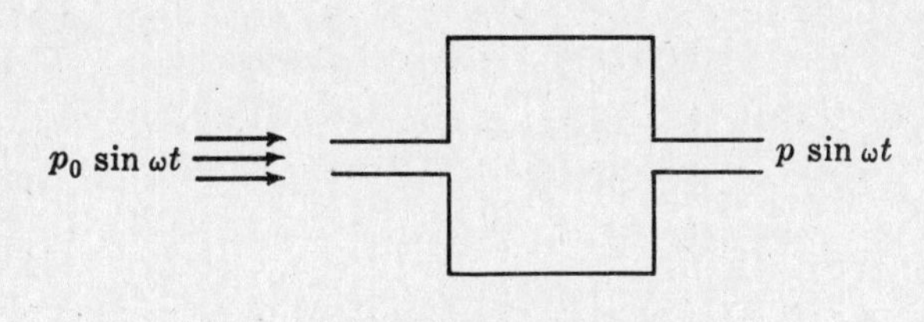

Fig. 5-16

5.33. Determine the equivalent electrical circuit for the mechanical-acoustical system shown in Fig. 5-17.

Ans.

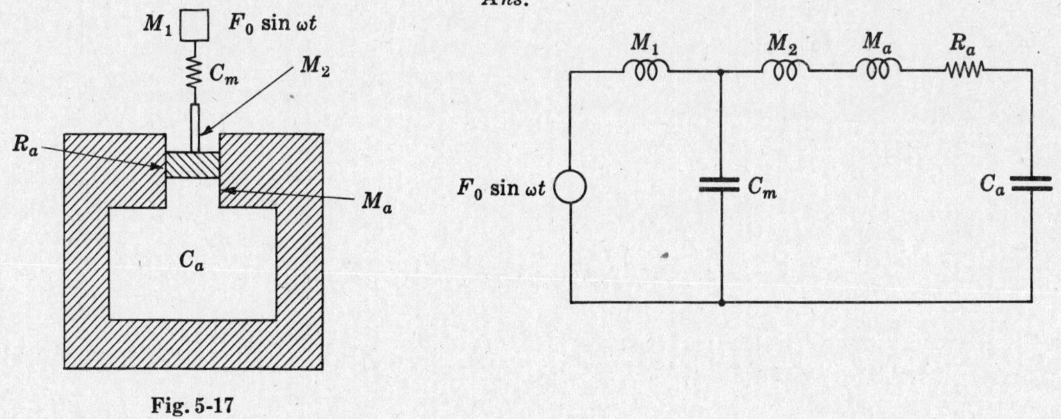

Fig. 5-17

5.34. A high-pass acoustic filter is shown in Fig. 5-18. Find its equivalent electrical circuit.

Ans.

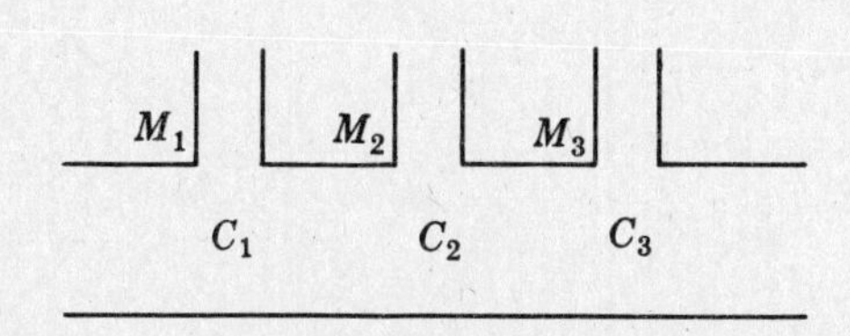

Fig. 5-18

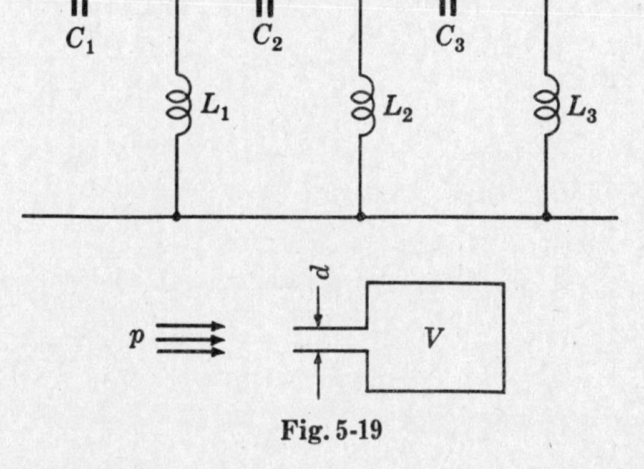

5.35. A rigid enclosure as shown in Fig. 5-19 is subjected to acoustic pressure p. Derive an expression for the stiffness of the system.

Ans. $k = (1.4p/V)(\pi d^2/4)^2$

Fig. 5-19

MICROPHONES

5.36. Calculate the lowest natural frequency of a conical horn of radius 1 meter open at its wide end. *Ans.* 166 cyc/sec

5.37. What is the sound loss in decibels for a bi-directional pressure gradient microphone if the sound on the axis is moved to an angle of 50°? *Ans.* 3.8 db

5.38. Obtain an expression for the force acting on the diaphragm of a pressure gradient microphone when it is exposed to an acoustic pressure $p_0 \sin \omega t$. *Ans.* $F = 2Ap_0 \sin kd$

5.39. For a second-order pressure gradient microphone, derive an expression for the ratio of pressure gradients for spherical and plane acoustic waves. *Ans.* $(1 + 4/k^4r^4)^{1/2}$

5.40. If the directional response characteristics of a second-order pressure gradient microphone is proportional to $\cos^2 \theta$, find an expression for the pressure gradient for spherical and plane acoustic waves. *Ans.* $(5 + 20/k^4r^4)^{1/2}$

5.41. Compute the directional efficiency of bi-directional and uni-directional microphones. *Ans.* 1/3, 1/3

5.42. An array of n identical microphones are spaced evenly in a distance L. For incident acoustic waves of wavelength $\lambda = nL$, show that the output of the array at $\theta = 90°$ is $1/n$ of the output at the axis of the array.

5.43. An array of 10 identical microphones are spaced equally at 1/9 meter apart. For incident sound waves of frequency 343 cyc/sec, determine the angles of incidence that will give zero output. *Ans.* $\theta = 64°, 116°$

5.44. Plane acoustic waves are incident at an angle θ to the axis of a multi-tube microphone as shown in Fig. 5-20. Find the phase angle ϕ between acoustic pressures for adjacent tubes, and the resultant pressure on the diaphragm.

Ans. $\phi = \dfrac{kL}{n-1}(1 - \cos \theta), \quad p_\theta = p_0 \left[\dfrac{\sin (n\phi/2)}{\sin (\phi/2)}\right]$

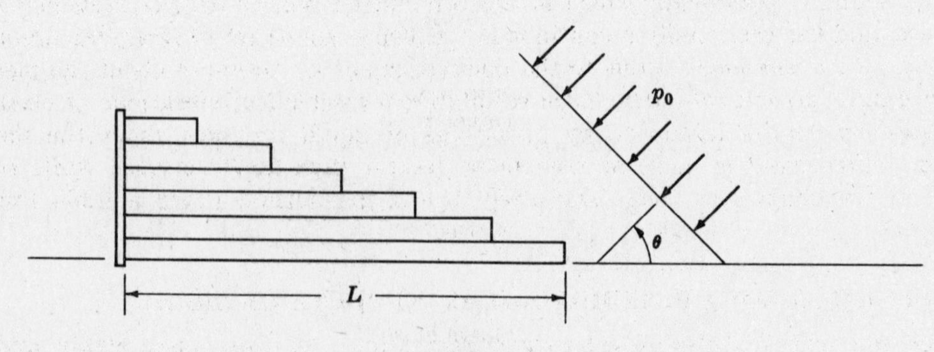

Fig. 5-20

5.45. A cardioid microphone has response of $2M$ at the axis, compute its responses at angles of 30°, 60°, 90°, 130° and 150°. *Ans.* $1.866M$, $1.5M$, M, $0.5M$, $0.134M$

Chapter 6

Sound and Hearing

NOMENCLATURE

f = frequency, cyc/sec
HL = hearing loss, db
I = sound intensity, watts/m^2
IL = intensity level, db
ISL = intensity spectrum level, db
LL = loudness level, phons
p = acoustic pressure, nt/m^2
PBL = pressure band level, db
PSL = pressure spectrum level, db
SIL = speech interference level, db
SL = sensation level, db
SPL = sound pressure level, db
ω = circular frequency, rad/sec
ρ = density, kg/m^3

INTRODUCTION

Noise, music and speech are the three basic categories of sound. The human voice as the natural sound source and the human ear as the natural sound receiver constitute the fundamental natural sound system. Basic understanding of sound and the human ear is therefore essential for acoustical studies and measurements.

NOISE

Noise is simply anything that we hear, and is subjectively defined as unpleasant or unwanted sound. Technically noise is the combined result of single-frequency sounds or pure tones, and has essentially a continuous frequency spectrum of irregular amplitude and waveform. *Airborne noise* is due to the fluctuations of air pressure about the mean atmospheric pressure, *structural-borne noise* results from mechanical vibrations of elastic bodies, and *liquid-borne noise* is caused by pulsations of liquid pressure about the mean static pressure. *Ultrasound* is noise of frequency greater than 20,000 cyc/sec while *infrasound* is noise of frequency less than 20 cyc/sec (below the normal lower audible limit of the human ear).

PHYSIOLOGICAL AND PSYCHOLOGICAL EFFECTS OF NOISE

Noise interferes with work, sleep and recreation. It also causes strain and fatigue, loss of appetite and indigestion, irritation and headache. High intensity noise has adverse cumulative effect on the human hearing mechanism, producing temporary or permanent deafness. Psychologically, noise adversely affects the output of workers, decreases their efficiency, and increases their liability to error because of distraction from work. Noise from machines causes wear and damage to the machines.

LOUDNESS

Loudness of a sound is the magnitude of the auditory sensation produced by the amplitude of the disturbances reaching the ear. Vibrational energy of sound is a physical property while loudness is a mental interpretation. Loudness of a sound is therefore a subjective quantity and cannot be measured exactly with any instrument. No absolute scale has been established for the measurement of loudness of a sound. A relative scale, based on the logarithm of the ratio of two intensities, is used.

The *sone* is an acoustic unit used to measure loudness of a sound. It is used to rank and compare loudness of sounds on a common basis as the ear hears them. A pure tone of frequency 1000 cyc/sec at a sound intensity level of 40 db is defined as having a loudness of one sone. A loudness of 0.001 sone or 1 millisone corresponds to the threshold of hearing. Unlike the phon, a loudness of 2 sones is twice as loud as a loudness of 1 sone.

The *phon* is an acoustic unit used to measure the overall loudness level of a noise. A pure tone of frequency 1000 cyc/sec at a sound intensity level of 1 db is defined as having a loudness level of 1 phon. All other tones will have a loudness level of n phons if they are judged by the ear to sound as loud as a pure tone of frequency 1000 cyc/sec at a sound intensity level of n db.

Like the decibel, a tone with a loudness level of 30 phons does not sound half as loud as a tone with a loudness level of 60 phons. A tone of frequency 500 cyc/sec at a loudness level of 40 phons, however, sounds exactly as loud to the ear as any other 40 phons tone at any other frequency.

Loudness level of a sound is defined as

$$\text{LL} = 10 \log \frac{I}{10^{-12}} \quad \text{phons}$$

where I is sound intensity in watts/m^2.

Figure 6-1 shows contours of equal loudness level in phons over the entire band of audible frequencies against intensity level in db or intensity in watts/m^2. The upper contour of 120 phons represents the *threshold of feeling* while the lower contour of zero phons represents the *threshold of hearing*. At low intensity levels the human ear is most sensitive to frequencies between 1000 and 5000 cyc/sec, and at very high intensity levels the response is more uniform.

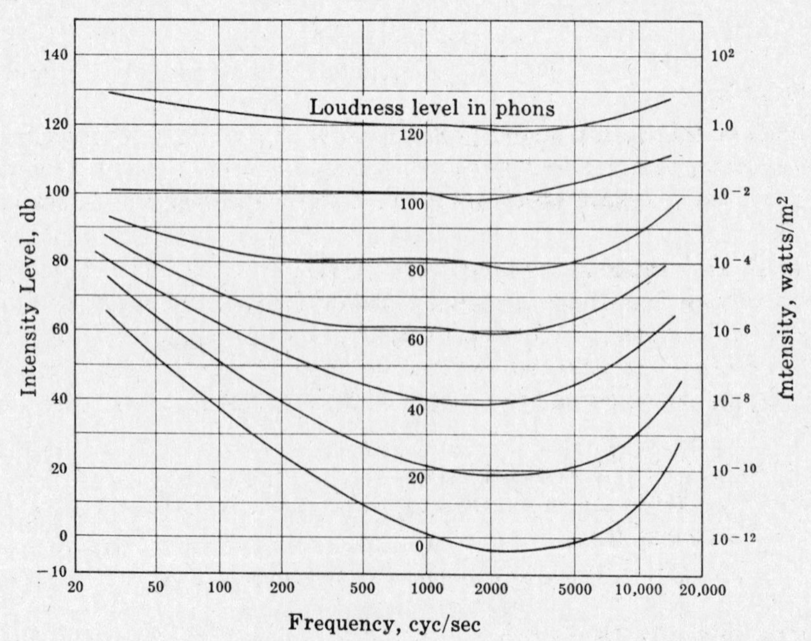

Fig. 6-1

Figure 6-2 is a plot of loudness level versus loudness in phons and sones respectively. (See Problems 6.1-6.6.)

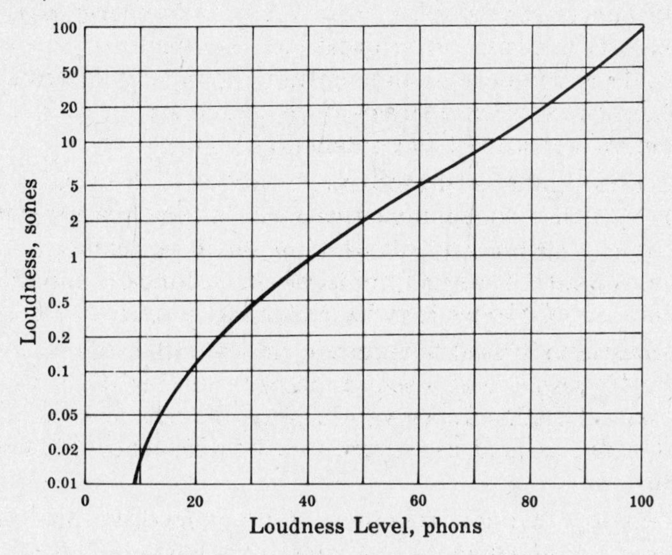

Fig. 6-2

The *noy* has been suggested as a possible acoustic unit to rank and compare the noisiness or annoyance of noises as the ear hears them. A noise, for example, may be judged subjectively by the ear to be louder but not necessarily more annoying than another noise half as loud.

Perceived-noise-level PNdb is a subjective scale developed to measure the unwantedness or noisiness of a noise, especially noises from jet aircraft. It not only represents the intensity of the noise but also its frequency spectrum. The noise spectrum is mathematically divided into a number of frequency bands and the sound pressure levels in these bands are determined. These are combined in some fashion after suitable weighting factors have been applied to each frequency band measurement. The result is perceived-noise-level in decibels.

NOISE ANALYSIS

In noise analysis the overall sound pressure level of a noise can be accurately measured by a *sound level meter* and a *sound analyzer,* while an *audio frequency spectrometer* and a *level recorder* plot the pressure spectrum level of the noise over the entire band of audible frequencies.

An *octave* is the interval between two frequencies having the ratio 2:1. The commonly used *octave bands* are 37.5-75, 75-150, 150-300, 300-600, 600-1200, 1200-2400, 2400-4800, 4800-9600 cyc/sec. A *one-third octave band* is a band of frequencies in which the ratio of the extreme frequencies is equal to the cube root of 2. A *narrow band* is a band whose width is less than one-third octave but not less than one percent of the center frequency.

Intensity spectrum level ISL at any particular frequency f of a noise is defined as the intensity level of the given noise contained within a band of frequencies 1 cyc/sec wide, centered on the frequency f.

$$\text{ISL} \quad = \quad 10 \log \frac{I}{I_0 \Delta f} \quad = \quad \text{IL} \; - \; 10 \log \Delta f \quad \text{db}$$

where I is sound intensity in watts/m^2, $I_0 = 10^{-12}$ watt/m^2 is the reference intensity, IL is intensity level in decibels, and Δf is the bandwidth in cyc/sec.

Pressure spectrum level PSL can be similarly defined as the sound pressure level contained within a band of frequencies 1 cyc/sec wide.

$$\text{PSL} = \text{SPL} - 10 \log \Delta f \quad \text{db}$$

where SPL is sound pressure level in decibels in the band of Δf cyc/sec width.

Pressure band level PBL is similarly given by

$$\text{PBL} = \text{PSL} + 10 \log \Delta f \quad \text{db}$$

(See Problems 6.7-6.12.)

White noise has a constant spectrum level over the entire band of audible frequencies, and need not be random or time-dependent. The amplitude of a *random noise* occurs, as a function of time, according to a Gaussian distribution curve. A random noise does not have a uniform frequency spectrum. *Pink noise* is characterized by equal energy per octave from 20 to 20,000 cyc/sec.

PITCH AND TIMBRE

Loudness, pitch and timbre are the three fundamental quantities which characterize a tone. From the physical point of view, *pitch* is the frequency of vibration of a pure tone. For a complex sound, pitch is characterized by its frequencies, and to some extent by the sound pressure and the wave form. To the human ear, pitch is that attribute of auditory sensation in terms of which sounds may be ranked and compared. In short, pitch is the mental counterpart of modes of vibration.

Sound intensity significantly affects pitch at very low and very high audio frequencies. When sound pressure is increased, the pitch of a low frequency tone will decrease whereas the pitch of a high frequency tone will increase. The *mel* is an acoustic unit used to describe the pitch of a sound. A pure tone of frequency 1000 cyc/sec and loudness level 40 phons is defined to have a pitch of 1000 mels.

Timbre or tone quality may be described as the instantaneous cross section of the tone, i.e. in terms of the number, intensity, distribution and phase of the harmonics. Intensity of overtones can produce changes in timbre whose subjective behavior is much more complex than that of loudness or pitch.

MUSIC

Music can be described as a highly subjective and complex mental sensation derived from listening to a succession or combination of different sounds produced by various vibrating bodies such as strings, membranes and air columns. Unlike noise, musical tones have simple harmonic structure with regular waveforms and shapes, and consist of a fundamental and harmonics of integral-related frequencies. *Musical acoustics* involves psychological and physical laws as well as aspects and phenomena of tone production.

SPEECH

Speech sounds are complex audible acoustic waves that provide the listeners with numerous clues. *Speech* concerns the structure of language and is characterized by the interpretive aspect, loudness, pitch, timbre and tempo. *Intelligibility of speech* is an indication of how well speech is recognized and understood. This depends on acoustic power delivered during the speech, speech characteristics, hearing acuity, and ambient noises.

Sound articulation is the percentage of the total number of speech sounds correctly recorded and identified. *Syllable articulation* is the number of syllables heard correctly from 100 speech syllables announced. Articulation generally increases rapidly with speech level until 70 db.

Speech interference level SIL in decibels is the arithmetic average of readings in the three octave frequency bands, i.e. 600-1200, 1200-2400, and 2400-4800 cyc/sec. A voice speech spectrogram shows a time series of frequency versus amplitude plots.

The *masking* of a sound can be described as the shift of the threshold of hearing of the host sound due to the presence of the masking sound. It is the reduction of the ability of a listener to hear one sound in the presence of other sounds. For a given frequency, the decibel difference between the background noises and the normal threshold of audibility is defined as the *degree of masking.*

In general, pure tones are used as the masked sounds. A tone of high pitch can easily be masked by a tone of low pitch. A continuous bland background noise tends to dull the edges of intermittent harsh sounds.

THE HUMAN VOICE

The mechanism of the human voice is a very low efficiency sound-producing system. It has four main parts: (1) a power generator that includes diaphragm, lungs, bronchi, trachea and associated muscles, (2) a vibrator called the larynx, (3) resonators (nose, mouth, throat and other voids) and sounding boards (chest, head and palate), and (4) articulators such as lips, tongue, teeth and palate.

The loudness of the human voice is dictated by the stream of air forced through the vocal cords from the lungs. The frequency of the human voice is controlled by the elasticity and vibration of the vocal cords, while the resonators govern the quality of the sound produced.

THE HUMAN EAR

The human hearing mechanism is essentially a very sensitive electroacoustic transducer responding to sound waves of a wide range of frequencies, intensities and waveforms. It translates acoustic pressure fluctuations into pulses in the auditory nerve. These pulses are carried into the brain which interprets and identifies them, and converts them into sensations – the perception of sound.

As the response of the human ear is a purely subjective quantity, it cannot be measured directly like other physical quantities. The response of the human ear varies with both frequency (20-20,000 cyc/sec) and sound intensity (10^{-12}-1 watt/m^2) at all values. However, the human ear is more sensitive to changes in frequency than to changes in sound intensity and more sensitive to sounds of low intensity than to those of high intensity. Because of its nonlinear responses to sound waves, the human ear actually creates sounds of various frequencies.

Hearing loss HL can be defined as the decibel difference between a patient's threshold of audibility and that for a person having normal hearing at a given frequency. It is actually a shift in sensation level.

$$\text{HL} \;=\; 10 \log \frac{I}{I_0} \quad \text{db}$$

where I is the threshold sound intensity for the patient's ear and I_0 is the threshold sound intensity for the normal ear.

Sensation level SL of a tone is the number in decibels by which it exceeds its threshold of hearing.

$$\text{SL} = 10 \log \frac{I}{I_t} \quad \text{db}$$

where I is the intensity of the tone and I_t is the intensity at the threshold of hearing.

The hearing mechanism is highly resilient to intensity changes and can be overloaded. *Deafness* is usually rated by the amount of hearing loss in decibels. *Conductive deafness* is hearing impairment due to abnormality or obstruction in the middle ear. *Nerve deafness* is the loss of hearing caused by nerve defect.

Hearing test employs an audiometer, an attenuator, an interrupter switch, and an earphone to determine the threshold of hearing, hearing defect and deterioration.

The ability of the human ear to identify and locate the direction of a source of sound with great accuracy is termed *binaural audition* or *auditory localization.* This is due to the difference in sound intensity at the two ears due to diffraction, and to the phase difference in sound arriving in different times at the two ears. (See Problems 6.13-6.16.)

Solved Problems

LOUDNESS

6.1. A pure tone of frequency 200 cyc/sec has an intensity level of 60 db. Determine its loudness level and loudness. To what intensity level must this pure tone be raised in order to increase its loudness to twice the original value?

The loudness level can be found from Fig. 6-1. The intersection of lines representing a frequency of 200 cyc/sec and an intensity level of 60 db yields a loudness level of 52 phons.

From Fig. 6-2, a loudness level of 52 phons corresponds to a loudness of 2.3 sones.

For a pure tone of twice the loudness, i.e. 4.6 sones, the corresponding loudness level is seen to be 60 phons. And from Fig. 6-1, a pure tone of frequency 200 cyc/sec and loudness level 60 phons corresponds to an intensity level 65 db.

6.2. The loudness level of a 1000 cyc/sec pure tone is 60 phons. How many such tones must be sounded together in order to produce a loudness level twice that produced by one tone?

The loudness level required is 120 phons. Then at a frequency of 1000 cyc/sec, the intensity level is 120 db. Using

$$\text{IL} = 10 \log (I/10^{-12}) \text{ db}$$

the intensity of one such tone is

$$60 = \log (I/10^{-12}), \qquad I = 10^{-12} \text{ antilog } 6 = 10^{-6} \text{ watt/m}^2$$

and the intensity of all the tones together would be

$$120 = \log (I/10^{-12}), \qquad I = 10^{-12} \text{ antilog } 12 = 1.0 \text{ watt/m}^2$$

Thus the number of tones required $= 1/10^{-6} = 10^6$.

6.3. A pure tone of intensity level 60 db and frequency 1000 cyc/sec is mixed with another pure tone of intensity level 50 db and frequency 1000 cyc/sec. Find the loudness level of this combination.

From Fig. 6-1, the first pure tone has loudness level 60 phons, and the second pure tone has loudness level 50 phons.

Since loudness level $\text{LL} = 10 \log (I/10^{-12})$ phons, where I is the sound intensity in watts/m^2,

$$(\text{LL})_1 = 10 \log (I/10^{-12}) = 60 \quad \text{or} \quad I_1 = 10^{-12} \text{ antilog } 6 = 10^{-6} \text{ watt/m}^2$$

$$(\text{LL})_2 = 10 \log (I/10^{-12}) = 50 \quad \text{or} \quad I_2 = 10^{-12} \text{ antilog } 5 = 10^{-5} \text{ watt/m}^2$$

Thus the sound intensity of the combination is $I = I_1 + I_2 = 1.1 \times 10^{-6}$ watt/m^2, and the loudness level of the combination is $\text{LL} = 10 \log (1.1 \times 10^{-6}/10^{-12}) = 60.44$ phons.

6.4. A pure tone of frequency 1000 cyc/sec has intensity level 60 db. Find the loudness level produced by two such tones operating simultaneously.

From Fig. 6-1, the loudness level of the tone is 60 phons. Then

$$10 \log (I/10^{-12}) = 60 \text{ phons} \quad \text{or} \quad I = 10^{-6} \text{ watt/m}^2$$

The intensity of two such tones is 2×10^{-6} watt/m^2, and the loudness level of two such tones is $10 \log (2 \times 10^{-6}/10^{-12}) = 63$ phons.

6.5. Given three pure tones with the following frequencies and intensity levels: 100 cyc/sec at 60 db, 500 cyc/sec at 70 db, and 1000 cyc/sec at 80 db. (*a*) Compute the total loudness in sones of these three pure tones. (*b*) What is the combined intensity level of these three pure tones? (*c*) Find the intensity level of a single 2000 cyc/sec pure tone which has the same loudness as all the three pure tones combined.

(*a*) The loudness level and loudness of a pure tone with known frequency in cyc/sec and intensity level in db can be found from Fig. 6-1 and Fig. 6-2. For the given pure tones, we have

Frequency cyc/sec	Intensity Level db	Loudness Level phons	Loudness sones
100	60	37	0.8
500	70	71	9.5
1000	80	80	18.5

The total loudness of these three pure tones is $0.8 + 9.5 + 18.5 = 28.8$ sones.

(*b*) The intensity level is defined as $\text{IL} = 10 \log (I/10^{-12})$ db, where I is the intensity in watts/m^2. The intensities of the three pure tones are found to be respectively 10^{-6}, 10^{-5} and 10^{-4} watts/m^2. Then the total intensity is 111×10^{-6} watt/m^2, and the combined level of these three pure tones is $10 \log (111 \times 10^{-6}/10^{-12}) = 80.47$ db.

(*c*) The total loudness of the combined tones (28.8 sones) corresponds to a loudness level of 87 phons. A pure tone of frequency 2000 cyc/sec and loudness level 87 phons has an intensity level of 86 db.

6.6. The frequencies and sound pressure levels of three pure tones are 200 cyc/sec at 64 db, 500 cyc/sec at 70 db, and 1000 cyc/sec at 74 db. (*a*) Which tone is the loudest? (*b*) What is their total loudness level in phons?

Sound pressure level in decibels relative to 0.0002 microbar can be expressed as

$$\text{SPL} = 20 \log p + 94 \text{ db}$$

where p is the acoustic pressure in nt/m^2. Then the acoustic pressures of the three pure tones are found to be $p_1 = 3.15 \times 10^{-2}$, $p_2 = 6.3 \times 10^{-2}$, $p_3 = 0.1$ nt/m^2.

Now intensity $I = p^2/\rho c$, where $\rho c = 415$ rayls is the characteristic impedance of air. Thus the intensities of the three pure tones are $I_1 = 2.38 \times 10^{-6}$, $I_2 = 9.56 \times 10^{-6}$, $I_3 = 24.1 \times 10^{-6}$ watts/m². The corresponding intensity levels in decibels are $(\text{IL})_1 = 63.8$ db, $(\text{IL})_2 = 69.8$ db, $(\text{IL})_3 = 73.8$ db.

(*a*) From Fig. 6-1 and Fig. 6-2, the loudness levels of the three pure tones and the corresponding loudness in sones are: 200 cyc/sec at 59 phons and 3.8 sones, 500 cyc/sec at 69 phons and 8.0 sones, and 1000 cyc/sec at 74 phons and 10 sones. Thus the loudest tone has a loudness of 10 sones, i.e. the pure tone of 1000 cyc/sec and intensity level 74 db.

(*b*) The total loudness is $3.8 + 8.0 + 10 = 21.8$ sones. The total loudness level in phons is therefore 83.

NOISE ANALYSIS

6.7. The sound intensity I_1 of each one-cycle band of a noise is $10^{-5}/f$ watts/m², where f is the center frequency of the band in cyc/sec. Determine the intensity spectrum level of the noise at 2000 cyc/sec and the intensity level of the noise between 1500 and 2500 cyc/sec.

The intensity spectrum level

$$\text{ISL} = 10 \log \frac{I}{I_0 \Delta f} = 10 \log \frac{10^{-5}/2000}{10^{-12}} = 37 \text{ db}$$

where I is the intensity in watts/m² and $\Delta f = 1$ cyc/sec is the bandwidth of the filter.

$$\text{IL} = \text{ISL} + 10 \log \Delta f = 37 + 10 \log 1000 = 67 \text{ db}$$

where $f = 2000$ cyc/sec and $\Delta f = 2500 - 1500 = 1000$ cyc/sec.

6.8. The acoustic pressure in each one-cycle band of a noise is expressed as $10/f$ nt/m², where f is the center frequency of the band in cyc/sec. Compute the pressure spectrum level of the noise at 1000 cyc/sec and the sound pressure level of a 50 cyc/sec bandwidth centered on a frequency of 2000 cyc/sec.

The pressure spectrum level of a noise is defined by

$$\text{PSL} = 20 \log \frac{p}{p_0 \Delta f} \text{ db}$$

where p is the pressure in nt/m², $p_0 = 0.0002$ microbar is the reference pressure, and $\Delta f = 1$ cyc/sec is the bandwidth of the filter. Thus at 1000 cyc/sec, $\text{PSL} = 20 \log \frac{10/1000}{2 \times 10^{-5}} = 54$ db; and at 2000 cyc/sec, $\text{PSL} = 48$ db.

The sound pressure level $\text{SPL} = \text{PSL} + 10 \log \Delta f = 48 + 10 \log 50 = 65$ db.

6.9. Figure 6-3 below shows the pressure spectrum levels of an office noise. Determine the overall pressure level of the office noise.

The mean pressure spectrum level in the frequency band 20-50 cyc/sec is approximately 63 db, so the corresponding pressure band level is

$$\text{PBL} = \text{PSL} + 10 \log \Delta f = 63 + 10 \log 30 = 77.79 \text{ db}$$

where sound pressure level $\text{SPL} = 20 \log p + 94$ db re 0.0002 microbar. Thus the sound pressure for this frequency band 20-50 cyc/sec is given by

$$77.79 = 20 \log p + 94 \quad \text{or} \quad p = 0.154 \text{ nt/m}^2$$

and the corresponding intensity is $I = p^2/\rho c = (0.154)^2/415 = 5.68 \times 10^{-5}$ watt/m².

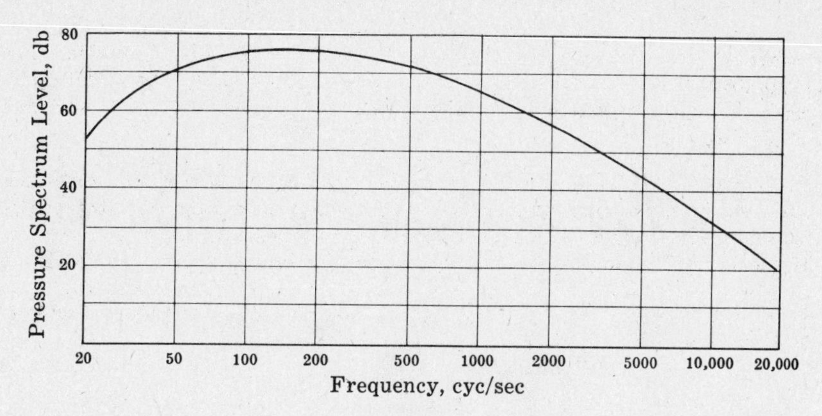

Fig. 6-3

The procedure is repeated for the other frequency bands, and the results are:

Frequency Band cyc/sec	Spectrum Level db	Band Level db	Pressure nt/m^2	Intensity watts/m^2
20-50	63	77.79	0.154	5.68×10^{-5}
50-100	72	89.00	0.560	0.75×10^{-3}
100-200	74	94.00	1.000	2.41×10^{-3}
200-500	72	96.90	1.390	0.0046
500-1000	66	93.10	0.9000	0.0019
1000-2000	60	90.000	0.630	9.5×10^{-4}
2000-5000	50	84.80	0.350	2.9×10^{-4}
5000-10,000	37	74.00	0.100	2.4×10^{-5}
10,000-20,000	26	66.00	0.040	3.7×10^{-6}

The intensity of the noise is the sum of the intensities of all bands of frequency and is found to be 0.0195 watt/m^2. The acoustic pressure of the noise is therefore given by

$$p^2 = 415(0.0195) = 8.15 \quad \text{or} \quad p = 2.85 \text{ nt/m}^2$$

Finally, the overall pressure level of the noise is

$$\text{SPL} = 20 \log 2.85 + 94 = 103.14 \text{ db}$$

This overall pressure level of a noise for the entire band of frequency, usually measured directly by means of sound level meters, is conveniently used for the rating of noise.

6.10. Figure 6-4 represents the frequency spectrum of white noise generated by an aircraft. Each line spectra has the same intensity level of 90 db. What is the intensity level of the white noise?

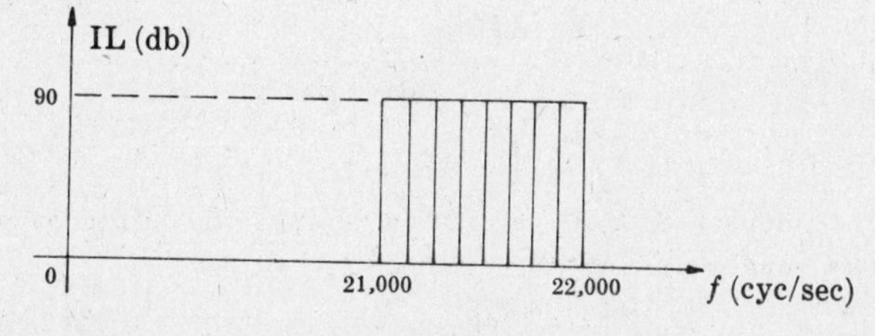

Fig. 6-4

(*a*) Assume each line spectra represents a single discrete frequency component; then the intensity level of the white noise is

$$(\text{IL})_t = 10 \log n + \text{IL} = 10 \log 1000 + 90 = 120 \text{ db}$$

where n is the number of tones having the same intensity level and IL is the intensity level of the tone in db.

(*b*) The intensity level of the white noise is also equal to the area under the intensity-frequency curve shown in Fig. 6-4, i.e.

$$(\text{IL})_t = 10 \log \int_{f_1}^{f_2} 10^{L/10}\, df = 10 \log 10^{L/10}(f_2 - f_1) = L + 10 \log (f_2 - f_1)$$

$$= 90 + 10 \log 1000 = 120 \text{ db}$$

where L is the intensity level of each line spectra.

6.11. A microphone with sensitivity -40 db relative to 1 volt per microbar is used to measure the spectrum level of a noise. If the open-circuit voltage is 0.01 volt and the bandwidth of the filter used with the microphone is 100 cyc/sec, find the pressure spectrum level PSL of the noise.

The sensitivity of the microphone is $20 \log (E/p) = -40$ db re 1 volt/microbar or $E/p = 0.01$ volt/microbar, where p is the acoustic pressure exerted on the microphone in microbars and E is the open-circuit voltage of the microphone in volts. Then $p = E/0.01 = 0.01/0.01 = 1$ microbar or 0.1 nt/m^2.

The sound pressure level of the noise is therefore

$$\text{SPL} = 20 \log (0.1/0.00002) = 74 \text{ db}$$

Thus $\text{PSL} = \text{SPL} - 10 \log \Delta f = 74 - 10 \log 100 = 54$ db, where Δf is the bandwidth of the filter used with the microphone in cyc/sec.

6.12. The noise spectrum of a certain machine is shown in Fig. 6-5. Compute the total sound intensity and the sound pressure level in the 150-300 cyc/sec band.

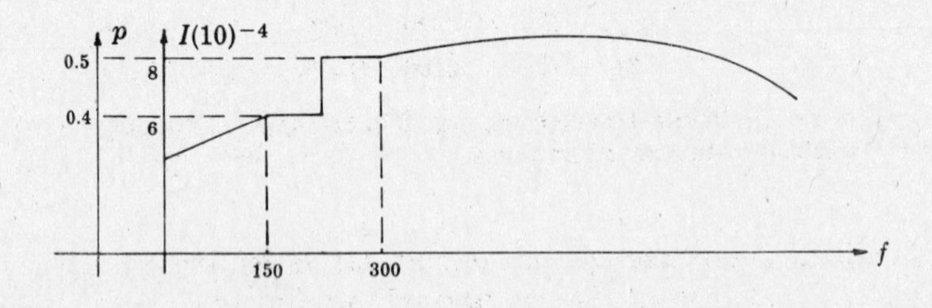

Fig. 6-5

In the 150-300 cyc/sec band, the average sound intensity for each 1 cyc/sec is

$$I_1 = 10^{-4}[6(75) + 8(75)]/150 = 9 \times 10^{-4} \text{ watt/m}^2$$

Hence intensity $I = I_1 \Delta f = (9 \times 10^{-4})(150) = 0.135$ watt/m^2.

Rewrite sound intensity as $(p^2/\rho c)_{\text{total}} = (p^2/\rho c)_1 \Delta f$. Since the characteristic impedance ρc must be the same, $p_{\text{total}} = p_1 \sqrt{\Delta f} = 0.45\sqrt{150} = 5.51$ nt/m^2 and $\text{SPL} = 20 \log 5.51/(2 \times 10^{-5}) = 108.88$ db, where $p_1 = [0.4(75) + 0.5(75)]/150 = 0.45$ nt/m^2 is the average sound pressure for each 1 cyc/sec band.

THE HUMAN EAR

6.13. Two pure tones of frequencies $f_1 = 300$ and $f_2 = 305$ cyc/sec are introduced simultaneously into the human ear. Determine the beats observed.

We hear a beat between the two fundamentals ($305 - 300 = 5$ cyc/sec) which varies from loud to soft and back to loud, five times a second.

Due to the nonlinear response of the ear, sounds of $2f_1 = 600$ cyc/sec (second harmonics) and $2f_2 = 610$ cyc/sec (second harmonics) are also produced. In addition to this 5 cyc/sec beat, we are aware of a 10 cyc/sec beat that arises from the beating of the second harmonics. Moreover, we hear beats of 15 cyc/sec, 20 cyc/sec, 25 cyc/sec, ... which come from the beating of the pairs of third, fourth, fifth, ... harmonics.

The higher harmonics, as a rule, have very little energy. Also the beats of the higher harmonics are too high to recognize.

6.14. Two pure tones of frequencies $f_1 = 1400$ and $f_2 = 800$ cyc/sec are introduced simultaneously into the human ear. Find the first, second and third order aural harmonics.

When two pure tones of different frequencies f_1 and f_2 are introduced simultaneously into the human ear, aural combination tones or aural harmonics will be produced in the ear and will be detected as the combination of the sums or differences of the two tones.

First order:

Summation tone: $f_1 + f_2 = 1400 + 800 = 2200$ cyc/sec

Difference tone: $f_1 - f_2 = 1400 - 800 = 600$ cyc/sec

Second order:

Summation tones: $2f_1 + f_2 = 2800 + 800 = 3600$ cyc/sec

$2f_2 + f_1 = 1600 + 1400 = 3000$ cyc/sec

Difference tones: $2f_1 - f_2 = 2800 - 800 = 2000$ cyc/sec

$2f_2 - f_1 = 1600 - 1400 = 200$ cyc/sec

Third order:

Summation tones: $3f_1 + f_2 = 4200 + 800 = 5000$ cyc/sec

$2f_1 + 2f_2 = 2800 + 1600 = 4400$ cyc/sec

$3f_2 + f_1 = 2400 + 1400 = 3800$ cyc/sec

Difference tones: $3f_1 - f_2 = 4200 - 800 = 3400$ cyc/sec

$2f_1 - 2f_2 = 2800 - 1600 = 1200$ cyc/sec

$3f_2 - f_1 = 2400 - 1400 = 1000$ cyc/sec

Other tones of multiple frequencies, e.g. $2f_1, 3f_1, 4f_1, \ldots, 2f_2, 3f_2, 4f_2, \ldots$ are possible but are weak in comparison with the other tones.

6.15. If the nonlinear response of the human ear is expressed as $r = a_1 p + a_2 p^2$ where $p = P_1 \cos \omega_1 t + P_2 \cos \omega_2 t$ is the sum of two harmonic sound waves, determine the amplitudes and frequencies of the response.

The nonlinear response is

$$\begin{aligned} r &= a_1(P_1 \cos \omega_1 t + P_2 \cos \omega_2 t) + a_2(P_1 \cos \omega_1 t + P_2 \cos \omega_2 t)^2 \\ &= a_1 P_1 \cos \omega_1 t + a_1 P_2 \cos \omega_2 t + a_2(P_1^2 \cos^2 \omega_1 t + P_2^2 \cos^2 \omega_2 t + 2P_1 P_2 \cos \omega_1 t \cos \omega_2 t) \end{aligned}$$

Now employing trigonometric identities

$$\cos^2 \omega_1 t = \tfrac{1}{2} + \tfrac{1}{2} \cos 2\omega_1 t, \qquad 2 \cos \omega_1 t \cos \omega_2 t = \cos(\omega_1 + \omega_2)t + \cos(\omega_1 - \omega_2)t$$

the response can be expressed as

$$\begin{aligned} r = {} & \tfrac{1}{2}(P_1^2 + P_2^2)a_2 + a_1 P_1 \cos \omega_1 t + a_1 P_2 \cos \omega_2 t + \tfrac{1}{2} a_2 P_1^2 \cos 2\omega_1 t \\ & + \tfrac{1}{2} a_2 P_2^2 \cos 2\omega_2 t + a_2 P_1 P_2 \cos(\omega_1 - \omega_2)t + a_2 P_1 P_2 \cos(\omega_1 + \omega_2)t \end{aligned}$$

6.16. Find the sensation levels of a pure tone of intensity level 40 db at 10,000, 5000, 2000, 1000, 500, 200 and 100 cyc/sec.

The sensation level of a tone is defined as the number of decibels by which it exceeds its threshold of hearing. From Fig. 6-1 and for a pure tone of intensity level 40 db at 10,000 cyc/sec, its intensity level is seen to exceed the threshold of hearing by 27 db. Thus the sensation level at 10,000 cyc/sec is 27 db.

The sensation levels at other frequencies are similarly found to be

Frequency, cyc/sec	5000	2000	1000	500	200	100
Sensation level, db	37	42	40	34	20	2

The sensation level at 10,000 cyc/sec can also be determined by

$$SL = 10 \log (I/I_t) = 10 \log (10^{-8}/2 \times 10^{-11}) = 27 \text{ db}$$

where I is the intensity in watts/m^2 of the tone at a particular frequency, and I_t is the threshold intensity also in watts/m^2 at the same frequency.

Supplementary Problems

LOUDNESS

6.17. A pure tone of frequency 1000 cyc/sec has intensity level 50 db. What loudness level will be produced by two such tones together? *Ans.* 53 phons

6.18. A pure tone of frequency 1000 cyc/sec has intensity level 50 db; another pure tone of frequency 1000 cyc/sec has intensity level 40 db. What loudness level will be produced by two such tones? *Ans.* 50.4 phons

6.19. A pure tone of frequency 1000 cyc/sec has intensity level 60 db. How many such tones, if all sound simultaneously, will produce a loudness level twice as great as that produced by one tone? *Ans.* 10^6

6.20. Find the difference in intensity of two pure tones at 1000 cyc/sec if one is twice as loud as the other. *Ans.* 3 db

6.21. The loudness of one pure tone is twice that of another. What is the difference in energy? *Ans.* 100 times

6.22. If the energy of a pure tone is increased 1000 times, how much is the loudness increased? *Ans.* 3 times

6.23. If the intensity of a pure tone at 1000 cyc/sec is increased 10 times, find the change in loudness. The initial loudness level of the tone is 40 phons. *Ans.* 1 sone

6.24. Show that a reduction of loudness level from 72 to 40 phons gives a noise one-tenth as loud.

6.25. If one sone corresponds to 40 phons, 2 sones to 50 phons, 4 sones to 60 phons, etc., show that $10 \log s = (p - 40) \log 2$, where s is the number of sones and p the number of phons.

NOISE ANALYSIS

6.26. Find the limiting sound pressure level in air. *Ans.* 194 db

6.27. A noise is generated by combining 100 identical pure tones. Each pure tone has intensity level 60 db. Determine the intensity level of the noise. *Ans.* 80 db

6.28. Show that a tone of sound pressure 1 nt/m^2 has 10^6 times more energy than a tone of the same frequency but having sound pressure 0.001 nt/m^2.

6.29. Show that the total intensity level of n identical pure tones, each at an intensity level of IL db, is $(\text{IL})_t = 10 \log n + \text{IL}$ db.

THE HUMAN EAR

6.30. Find the sensation level of a tone of intensity 10^{-6} watt/m^2 and frequency 50 cyc/sec.
Ans. 9 db

6.31. What is the minimum variation in sound pressure detected by the human ear?
Ans. 10^{-6} atmosphere (0.01 nt/m^2)

6.32. The nonlinear response of the human ear can be expressed as $r = a_1p + a_2p^2 + a_3p^3$, where $p = P_0 \cos \omega t$ is the harmonic acoustic pressure exerted on the ear. Determine the amplitudes and frequencies of the response.
Ans. $r = \frac{1}{2}a_2P_0^2 + (a_1P_0 + \frac{3}{4}a_3P_0^3) \cos \omega t + \frac{1}{2}a_2P_0^2 \cos 2\omega t + \frac{1}{4}a_3P_0^3 \cos 3\omega t$

Chapter 7

Architectural Acoustics

NOMENCLATURE

a = sound absorption, sabins or metric sabins
c = speed of sound in air, m/sec
E_0 = sound energy density, joules/m^3
I = sound intensity, watts/m^2
IL = intensity level, db
L = mean free path, m
L_s = space average sound pressure level, db
m = 2α, absorption coefficient for air, nepers/m
p = acoustic pressure, nt/m^2
R = room acoustics, ft^2 or m^2
RF = noise reduction factor, db
S = area, m^2
SPL = sound pressure level, db
T = reverberation time, sec
TL = transmission loss, db
V = volume, m^3
W = sound power, watts
ω = circular frequency, rad/sec
ρ = density, kg/m^3
α = sound absorption coefficient
$\bar{\alpha}$ = average sound absorption coefficient
α_e = effective sound absorption coefficient
τ = sound transmission coefficient

INTRODUCTION

Architectural acoustics deals basically with reverberation control, noise insulation and reduction, and sound distribution and absorption. It strives for the intelligibility of speech, the freedom from external unwanted noises, and the richness of music.

REVERBERATION

Reverberation is the persistence of sound in an enclosure as the result of continuous reflections of sound at the walls after the sound source has been turned off. As resonant free vibration with damping, reverberation depends on the size and shape of the enclosure as well as the frequency of the sound.

Reverberation time T at a specific frequency is the time in seconds for the sound pressure to decrease to 10^{-6} of its original value (or a 60 db drop) after the source is turned off.

$$T = 0.161V/a \text{ seconds (metric units)}$$

$$T = 0.049V/a \text{ seconds (English units)}$$

where V is the volume of the enclosure in m³ or ft³ and a is the total sound absorption of the enclosure in metric sabins or sabins. If reverberation time is too short, the sound may not be sufficiently loud in all portions of the enclosure. If it is too long, echoes will be present. Though the best intelligibility would be obtained with the shortest possible reverberation time, shorter reverberation time decreases sound intensity in the enclosure which in turn decreases intelligibility. Reverberation time is therefore an important measure of good room acoustics.

Reverberation chamber (or live room) is a specially constructed room with paddle-like turning vanes to cause uniform sound diffusion and with room surfaces having practically no sound absorption. The walls are highly reflective of sound waves, and consequently sound waves suffer very little loss at each reflection. These reflections will produce uniform sound energy distribution so that at any point in the room (not too close to the wall or the source) the sound appears to come equally from all directions. A reverberation chamber is used to measure the total sound power output of equipment, to establish the noise reduction coefficient, to test the sound control efficiency of materials and structures, and to calibrate microphones.

The *growth* of sound intensity in a reverberation chamber is given by

$$I(t) = \frac{W}{a}(1 - e^{-(ac/4V)t}) \text{ watts/m}^2$$

and the *decay* of sound intensity is similarly given by

$$I(t) = \tfrac{1}{4}E_0 c e^{-(ac/4V)t} \text{ watts/m}^2$$

where W is the sound power output in watts, a is the total sound absorption in metric sabins, c is the speed of sound in m/sec, V is the volume of the room in m³, and E_0 is the sound energy density in joules/m³ when the source is shut off. (See Problems 7.1-7.6.)

NOISE INSULATION AND REDUCTION

When noise at the source cannot be economically reduced below the objectionable range, noise insulation or soundproofing is required. This can be accomplished either by absorption or by reduction of the transmission of sound.

In buildings, airborne noise leaks through holes and cracks, weak or poorly-fitting doors and windows, air intakes and exhausts. It also sets panels and walls into vibration. Airborne noise can be reduced by breaking its transmission path, by using absorptive materials and directly surrounding the source with effective sound-absorbing devices or enclosures (e.g. sound barriers and silencers).

Transmission loss TL is airborne noise reduction. It is defined as the difference in decibels between the sound energy striking the surface separating two spaces and the sound energy transmitted. It cannot be measured directly, but is computed from sound pressure levels on both sides of the surfaces.

$$\text{TL} = 10 \log \frac{\sum S}{\sum S\tau} = (\text{SPL})_1 - (\text{SPL})_2 \quad \text{db}$$

where S is the area of the surface in m² and τ is the sound transmission coefficient.

Structural-borne noise is vibration of elastic bodies. It travels through walls, floors, columns, beams, pipes, ducts, and other solid structures. Since the amount of energy it carries is much greater than that of airborne noise, structural-borne noise should be suppressed at its source. Its transmission paths should be interrupted by resilient mounting insertions and sound plenums or traps. Walls should have discontinuities which are filled with air or absorptive materials.

Machine noise generally indicates poor balance, excessive clearance, turbulent flow or other improper working of some components of the machine. Most machine noises can be reduced and attenuated by proper redesign or using soundproofing enclosures lined with absorptive materials. Acoustical filters such as mufflers, plenum chambers, resonators, hydraulic filters, and sound traps should be employed wherever necessary. Sources should be properly isolated and vibration-mounted to reduce sound and vibration transmission.

Impact noise can be reduced by using carpets to cushion the impact areas of floors which are isolated from supporting structures by resilient mountings.

Space average sound pressure level L_s is defined as

$$L_s \quad = \quad 10 \log \frac{p_1^2 + p_2^2 + \cdots + p_n^2}{np_0^2} \quad \text{db}$$

where p_n are sound pressures in nt/m² and $p_0 = 0.00002$ nt/m² is the reference sound pressure.

Background noise requires similar acoustical treatments described for airborne noise. (See Problems 7.7-7.15.)

SOUND ABSORPTION

Sound absorption is a process in which sound energy is converted partly into heat (by frictional and viscous resistance of the pores and fibers of acoustical materials) and partly into mechanical vibration of the materials.

Unwanted sounds can be absorbed by draperies, carpets, suspended space absorbers, and interchangeable absorptive panels in rooms and buildings. Thin panels with air trapped behind them are employed to absorb sounds at low frequencies. Helmholtz resonators and resonator-panel absorbers are most efficient for sound absorption at their resonant frequencies. Mufflers impede the transmission of sound but permit the free flow of air.

The *sound absorption coefficient* α of a material is defined as the decimal fraction of perfect absorption that it has; e.g. $\alpha = 0.6$ means 60% absorption. It is the efficiency of a material in absorbing sound energy at a specified frequency, and varies with the angle of incidence and the thickness of the material. An open space is sometimes taken as a standard of unity absorption coefficient.

α is obtained by statistically averaging the ratio of absorbed to incident energy over all possible angles of incidence. The *average sound absorption coefficient* $\bar{\alpha}$ is determined by averaging the absorption coefficients over all the absorbing areas of the room.

Sound absorption a in *sabins* is the total area in square feet of perfectly absorbing material. Similarly, 1 *metric sabin* is one square meter of material having perfect sound absorption.

Noise reduction factor RF is given by

$$\text{RF} \quad = \quad \text{TL} + 10 \log (a/S) \quad \text{db}$$

where TL is transmission loss in decibels, a is the total sound absorption in sabins, and S is the area of the partition in ft².

The difference in noise level can be expressed as

$$(\text{db})_{\text{before}} - (\text{db})_{\text{after}} \quad = \quad 10 \log \frac{a_{\text{after}}}{a_{\text{before}}} \quad \text{db}$$

where the a's are sound absorption in sabins.

Acoustical materials used for sound absorption are characterized by reduction efficiency, porosity, flow resistance, propagation constant, and structure factor. Other factors such as flame resistance, light reflection, paintability, weather exposure, non-hygroscopicity, heat insulation, weight, ease of installation, and appearance should also be taken into consideration.

In general, sound intensity in an enclosure is inversely proportional to the amount of sound absorption present. If the enclosure is very large while the total sound absorption is small, the absorption of sound in air must be considered.

Anechoic chamber (or dead room) is characterized by highly absorptive wedges or long pyramids mounted to the walls of the room to absorb all incident sound energy. It simulates a *free field* or unbounded space. Complete soundproofing can be achieved by construction of an anechoic chamber with a floating floor vibration-mounted to another room. Accurate and consistent measurements of acoustic characteristics of equipment, absolute calibration of microphones, and sound radiation patterns of loudspeakers can be made inside the anechoic chamber.

The decay of sound intensity in an anechoic chamber is given by

$$I(t) \;=\; I_0 e^{(Sc/4V)\ln(1-\bar{\alpha})t} \quad \text{watts/m}^2$$

where I_0 is the sound intensity in watts/m^2 when the source is shut off, S is the total wall area in m^2, c is the speed of sound in air in m/sec, V is the volume of the room in m^3, and $\bar{\alpha}$ is the average sound absorption coefficient of the room. (See Problems 7.16-7.23.)

SOUND DISTRIBUTION

Sound distribution describes how the sound pressure level varies with position in an enclosure. To insure smooth growth and decay of sound, rooms and buildings are designed to have sound as evenly as possible distributed or diffused over the entire area by acoustical treatments such as the scattering effects of objects, irregularities of wall surface, random mounting of absorptive material, and reflecting surfaces and diffusers.

Model analysis with light rays, ultrasonic waves, or ordinary audio frequency sound is used to study sound distribution. Graphical construction of first reflections of the sound waves at various cross sections can also be used as in Fig. 7-1. (See Problems 7.24-7.26.)

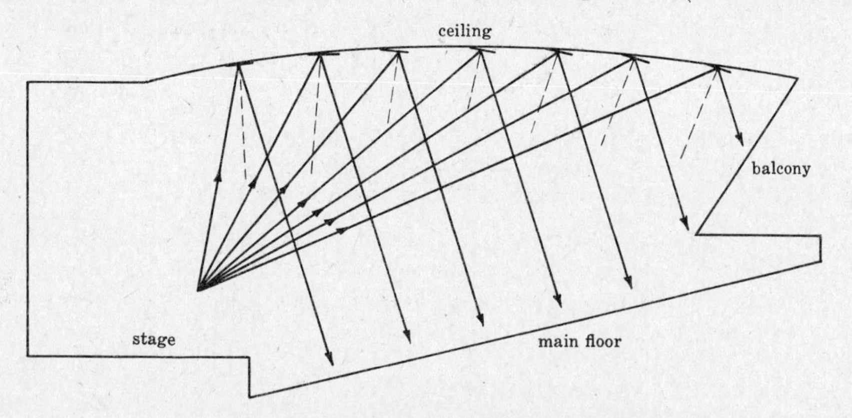

Fig. 7-1. Reflection of sound waves

ROOM ACOUSTICS

An acoustically well-designed room has good intelligibility of sounds of sufficient intensity (optimum reverberation time), freedom from extraneous and unwanted noises (soundproofing and reduction), and good sound distribution.

Sound that reaches a listener via two paths differing greatly in length produces an unpleasant fluttering effect called *echoing*. *Room flutter* occurs between a pair of parallel opposite walls that are smooth and highly reflective. The sound is reflected back and forth between the pair to produce multiple echoes. *Sound focusing* is concentration of sound at a point in an enclosure due to reflection of sound from curved or circular surfaces. The result is unequal distribution of sound. *Dead spot* is a region of deficiency of sound, i.e. practically nothing can be heard from there, and is due to destructive interference of two or more sound waves. Because of diffraction of sound, i.e. sound waves bending around an obstacle, the obstacle may prove to be an effective barrier if its size is comparable with the wavelength of sound. An *acoustic shadow* is formed on the other side of the obstacle.

Acoustical design of rooms should also encourage oblique waves because they decay most rapidly, but should discourage axial waves because they are most persistent.

Percentage articulation, which is sometimes used as an intelligibility rating of rooms, is determined from the shape and noise of the room, reverberation and loudness. *Room constant* R is another way to indicate and compare the acoustics of a room:

$$R = \frac{S}{1-\bar{\alpha}} \text{ ft}^2$$

where S is the total wall area of the room in ft², and $\bar{\alpha}$ is the average sound absorption coefficient. (See Problems 7.27-7.30.)

Solved Problems

REVERBERATION

7.1. Derive an expression for the rate of absorption of sound energy by the walls of an enclosure.

Consider the radiation of sound energy from an elementary volume dV within the enclosure toward an elementary surface area dS of the wall as shown in Fig. 7-2. dV is at a distance r from the elementary surface area dS, where r makes an angle θ with the normal to dS.

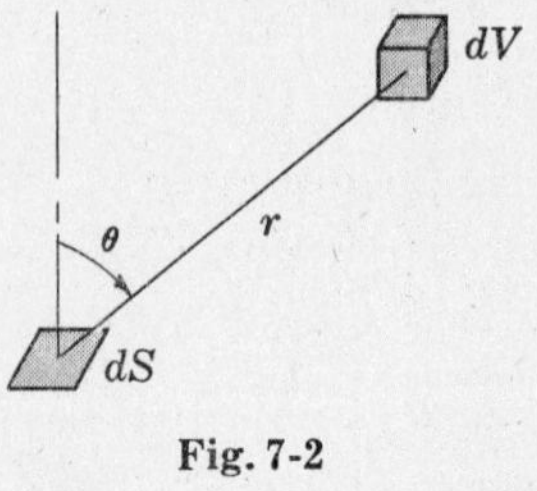

Now dV is radiating sound energy equally in all directions with velocity c, and the differential amount of energy striking dS is

$$dE_d = (dV\,E_d\,dS\cos\theta)/4\pi r^2$$

dV
θ
r
dS

Fig. 7-2

where E_d is the sound energy density in the enclosure, $E_d\,dV$ is the amount of energy in dV, $4\pi r^2$ is the surface area of a sphere of radius r surrounding dV, and $dS\cos\theta$ is the projected area of dS on any portion of the sphere.

Using spherical coordinates (r, θ, ϕ), $dV = r^2 \sin\theta\, dr\, d\theta\, d\phi$ and the expression for the differential amount of energy striking dS can be rewritten as

$$dE_d = (E_d\, dS \cos\theta \sin\theta\, dr\, d\theta\, d\phi)/4\pi$$

so the total differential amount of sound energy contribution to dS of a hemispherical shell of radius r and thickness dr is given by

$$dE_d = \frac{E_d\, dS}{4\pi}\int_0^{2\pi}\int_0^{\pi/2} \sin\theta \cos\theta\, d\theta\, d\phi\, dr = \tfrac{1}{4}E_d\, dS\, dr$$

But this total energy travels toward dS with velocity $c = dr/dt$. Hence the rate at which sound energy arrives at dS is

$$dE_d/dt = \tfrac{1}{4}E_d\, dS\, dr(c/dr) = \tfrac{1}{4}E_d c\, dS$$

or $\frac{1}{4}E_d c$ per unit area. The intensity I of such diffuse sound energy at the walls is therefore $I = \frac{1}{4}cE_d$.

If the enclosure has areas $S_1, S_2, S_3, \ldots$ having absorption coefficient $\alpha_1, \alpha_2, \alpha_3, \ldots$, then the rate at which sound energy is being absorbed by all these surfaces is $\frac{1}{4}cE_d(\alpha_1 S_1 + \alpha_2 S_2 + \cdots)$ or $\frac{1}{4}acE_d$, where a is the total sound absorption of the enclosure.

7.2. Derive expressions for the growth and decay of sound in a reverberation chamber.

In general, the rate of sound energy radiated from the source inside a reverberation chamber or live room must equal the rate of increase of sound energy in the medium throughout the interior of the room plus the rate of sound energy absorbed by the walls of the room. This condition can be expressed by the fundamental differential equation of growth of sound energy,

$$V\, dE_d/dt + \tfrac{1}{4}acE_d = W \qquad (1)$$

where V is the volume of the room, E_d is the sound energy density, a is the total absorption of the room, c is the speed of sound in air, and W is the rate of sound energy being produced. The first term represents the rate sound energy increases in the medium, and the second term is the rate of sound absorption obtained by the classical ray theory. (See Problem 7.1.)

Solution of (1) can be written as

$$E_d(t) = (4W/ac)e^{(ac/4V)t} + Ce^{-(ac/4V)t} \qquad (2)$$

For growth of sound, the initial sound energy is zero, i.e. $E_d(0) = 0$. Then from (2),

$$E_d(0) = 4W/ac + C = 0 \quad \text{or} \quad C = -4W/ac$$

and the expression for the growth of sound energy in a live room is

$$E_d(t) = (4W/ac)(1 - e^{-(ac/4V)t}) \qquad (3)$$

Since $I = E_d c/4$ and $E_d = p^2/\rho c^2$, we can express the growth of sound intensity and of acoustic pressure in a live room as

$$I(t) = \frac{W}{a}(1 - e^{-(ac/4V)t}), \qquad p^2(t) = \frac{4W\rho c}{a}(1 - e^{-(ac/4V)t})$$

As time t increases, the expressions for the growth of sound energy, sound intensity, and sound pressure approach their ultimate values of the steady state condition. These are

$$E_d = 4W/ac, \quad I = W/a, \quad p^2 = 4W\rho c/a$$

For decay of sound the source is shut off at time $t = 0$, and assume energy density at $t = 0$ equals E_0. From (2) with $W = 0$,

$$E_d(0) = E_0 = C$$

so

$$E_d(t) = Ce^{-(ac/4V)t} = E_0 e^{-(ac/4V)t} \qquad (4)$$

The corresponding expressions for the decay of sound intensity and sound pressure in a live room are similarly given by

$$I(t) = \tfrac{1}{4}E_0 ce^{-(ac/4V)t}, \qquad p^2(t) = 2\rho cE_0 e^{-(ac/4V)t}$$

7.3. Derive an expression for the reverberation time in a live room.

In Problem 7.2 we showed how sound grows and decays in a live room based on the assumptions of homogeneous sound energy density and continuous sound absorption by the walls. The expression for decay of sound is $E_d(t) = E_0 e^{-(ac/4V)t}$.

Now reverberation time is defined as the time interval during which the sound energy density falls from its steady state value to $1/10^6$ of this value, or a 60 db drop. We then have

$$E_d/E_0 = e^{-(ac/4V)T} = 10^{-6}, \qquad \ln e^{-(ac/4V)T} = \ln 10^{-6}$$

or $-(ac/4V)T = 2.3 \log 10^{-6}$, and so the reverberation time T is $T = -4V(2.3 \log 10^{-6})/ac$ sec.

In metric units ($c = 343$ m/sec at 20°C) $T = 0.161V/a$ sec, and in English units ($c = 1130$ ft/sec at 20°C) $T = 0.049V/a$ sec, where V is the volume of the enclosure either in m^3 or ft^3, and a is the total sound absorption of the enclosure either in metric sabins or sabins.

7.4. A room of volume 86 m^3 has a total sound absorption of 10 metric sabins. A sound source having 10 microwatts sound power output is turned on. (*a*) What is the sound intensity level inside the room at the end of 0.2 sec? (*b*) Determine the maximum sound intensity level attainable. (*c*) Find the decay rate of the sound intensity level when the source is turned off.

(*a*) For growth of sound in a live room,

$$I = \frac{W}{a}(1 - e^{-(ac/4V)t}) = 86.5 \times 10^{-8} \text{ watt/m}^2$$

where $W = 10 \times 10^{-6}$ watt is the rate of sound energy produced in the room, $a = 10$ metric sabins is the total sound absorption of the room, $c = 343$ m/sec is the speed of sound in air, and $V = 86$ m^3 is the volume of the room. Then the sound intensity level

$$\text{IL} = 10 \log \frac{86.5 \times 10^{-8}}{10^{-12}} = 59.4 \text{ db re } 10^{-12} \text{ watt}$$

(*b*) $I_{max} = W/a = 10^{-6}$ watt/m^2. Then $(\text{IL})_{max} = 10 \log 10^{-6}/10^{-12} = 60$ db re 10^{-12} watt.

(*c*) For decay of sound in a live room, the sound intensity at any time t is

$$I(t) = \tfrac{1}{4}E_0 c e^{-(ac/4V)t}$$

When the source is shut off, $I(0) = \frac{1}{4}E_0 c$ and $I(t)/I(0) = e^{-(ac/4V)t}$. The change in intensity level is thus

$$10 \log e^{-(ac/4V)t} = (10/2.3) \ln e^{-(ac/4V)t} = -1.09act/V = 43 \text{ db/sec}$$

Hence it takes 1.4 sec for the sound to die out completely after the source is turned off.

7.5. The internal dimensions of a reverberation chamber are $5 \times 6 \times 8$ ft and its average sound absorption coefficient is 0.04. (*a*) A sound source of 1.0 microwatt output is tested inside the chamber. Find the maximum sound pressure level produced. (*b*) A man goes into the chamber to make measurements. What will be the new sound pressure level if the equivalent sound absorption of the man is 9.41 sabins?

(*a*) Maximum sound pressure level will be obtained when steady state condition is reached inside the chamber. This condition is represented by a sound pressure of

$$p = \sqrt{4W\rho c/a} \text{ nt/m}^2$$

where W is the acoustic power output in watts, $\rho c = 415$ rayls is the characteristic impedance of air, and a is the total sound absorption in metric sabins.

Now total sound absorption is $a = \alpha \Sigma S = 0.04[2(30) + 2(40) + 2(48)] = 9.41$ sabins or $9.41/10.76 = 0.88$ metric sabins. Thus

$$p_{max} = \sqrt{4(10^{-6})415/0.88} = 0.0435 \text{ nt/m}^2$$

$$(\text{SPL})_{max} = 20 \log 0.0435/2(10^{-5}) = 66.8 \text{ db}$$

(*b*) When the man is inside the chamber, the total sound absorption becomes $9.41 + 9.41 = 18.82$ sabins or 1.76 metric sabins. This will change the sound pressure to

$$p_{max} = \sqrt{4(10^{-6})(415)/1.76} = 0.0308 \text{ nt/m}^2$$

and so

$$(\text{SPL})_{max} = 20 \log 0.0308/2(10^{-5}) = 63.8 \text{ db}$$

A 3 db drop in sound pressure level is observed because of the additional sound absorption. Since the sound pressure level inside the chamber can be accurately measured by a sound level meter, this procedure can be reversed to determine the amount of sound absorption of the man or sound absorption materials. In fact, reverberation chambers are often used to determine the sound absorption coefficients of different types of building materials.

7.6. A classroom is $4 \times 6 \times 10$ m and has a reverberation time of 1.5 sec. (*a*) What is the total sound absorption a of the classroom? (*b*) Forty students are in the classroom, and each is equivalent to 0.5 metric sabin sound absorption. Find the new reverberation time of the classroom. (*c*) If a speaker lectures with an acoustic power output of 10 microwatts, determine the sound pressure level in the classroom with and without the students.

(*a*)
$$a = 0.161V/T = 0.161(240)/1.5 = 25.8 \text{ metric sabins}$$

(*b*)
$$T = 0.161V/a = 0.161(240)/(25.8 + 20) = 0.85 \text{ sec}$$

(*c*) Using $p = \sqrt{4W\rho c/a}$ where $W = 10^{-5}$ watts and $\rho c = 415$ rayls, the sound pressures produced by the speaker with and without the students are respectively 0.0191 and 0.0254 nt/m². The corresponding sound pressure levels are

$$\text{SPL} = 20 \log 0.0191/(2 \times 10^{-5}) = 59.64 \text{ db}$$

$$\text{SPL} = 20 \log 0.0254/(2 \times 10^{-5}) = 62.12 \text{ db}$$

NOISE INSULATION AND REDUCTION

7.7. Sound transmission loss through solid panels can be evaluated in a specially constructed room as shown in Fig. 7-3. The sound energy produced in the source room travels through the test sample into the receiver room lined with absorptive materials. Derive an expression for the transmission loss.

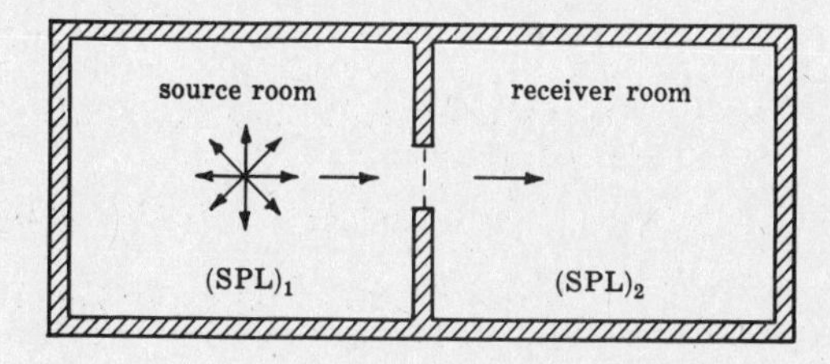

Fig. 7-3

Assume sound energy density is constant in the source room having sound pressure level $(\text{SPL})_1$. Now sound energy transmitted through the test piece (here we assume this is the only possible path for the transmission of sound from the source room to the receiver room) must therefore be equal to that absorbed by the wall surface of the receiver room at sound pressure level $(\text{SPL})_2$.

Since transmission loss can be defined as the ratio between the sound power striking the panel on one side and the sound power being transmitted from the other side of the panel,

$$\text{TL} = 10 \log (W_1/W_2) \text{ db}$$

where $W_1 = I_1S_1$ watts, $W_2 = I_2S_2$ watts, S_1 is the area of the test sample in m², S_2 is the area of the wall surface of the receiver room in m², I_1 is the intensity in the source room in watts/m², and I_2 is the intensity in the receiver room in watts/m². Then

$$\text{TL} = 10 \log \frac{I_1S_1}{I_2S_2} = 10 \log \frac{(p_1^2/\rho c)S_1}{(p_2^2/\rho c)S_2} = 10 \log \frac{p_1^2S_1}{p_2^2S_2} = 10 \log (p_1/p_2)^2 + 10 \log (S_1/S_2)$$

where

$$10 \log (p_1/p_2)^2 = 20 \log (p_1/p_2) = 20 \log \frac{p_1/p_0}{p_2/p_0} = 20 \log \frac{p_1}{p_0} - 20 \log \frac{p_2}{p_0} = (\text{SPL})_1 - (\text{SPL})_2$$

Thus the expression for transmission loss becomes

$$\text{TL} = (\text{SPL})_1 - (\text{SPL})_2 + 10 \log (S_1/S_2) \text{ db}$$

If $S_1 \ll S_2$, then $10 \log (S_1/S_2) \doteq 0$ and $\text{TL} = (\text{SPL})_1 - (\text{SPL})_2$ db. Hence by measuring the sound pressure levels with a sound level meter, the transmission loss of a given panel can be determined.

7.8. A 1×2.5 m door is located in a 4×7 m wall. The door has a transmission loss of 20 db while the wall has a transmission loss of 30 db. What is the transmission loss of the combination?

Using $\text{TL} = 10 \log (1/\tau)$, we have for the door $10 \log (1/\tau_d) = 20$ or $\tau_d = 0.01$, and for the wall $10 \log (1/\tau_w) = 30$ or $\tau_w = 0.001$. Hence the transmission loss of the combination is

$$\text{TL} = 10 \log (\Sigma S/\Sigma S\tau) = 10 \log \frac{28}{2.5(0.01) + 25.5(0.001)} = 27.47 \text{ db}$$

7.9. The space under a solid door is 1/100 of the total area of the door. If the noise level outside the room is 90 db, find the noise level inside the room with the door closed.

Assume the solid door does not transmit sound and that the space under the door is the only open space for sound transmission. The transmission loss through the space under the door is

$$\text{TL} = 10 \log \frac{\Sigma S}{\Sigma S\tau} = 10 \log \frac{S + 0.01S}{0(S) + 1.0(0.01S)} = 20 \text{ db}$$

where S is the area, and $\tau_w = 0$ and $\tau = 1.0$ are the transmissivities of the door and open space respectively. The noise level inside the room with the door closed is therefore $90 - 20 = 70$ db, i.e. only a 20 db drop in noise level.

If there is no space under the door, the theoretical noise level drop will be 90 db as there is no sound transmission at all. On the other hand, if the space under the door is reduced, say, to 1/1000 of the total area of the door, the transmission loss will be 30 db and the noise level inside the room will be 60 db.

7.10. An office is separated by a partition of area 100 m² having a transmission loss of 40 db. A door of area 2.5 m² having a transmission loss of 30 db is added to the partition. If the room adjoining the office has a noise level of 75 db, what will be the noise level in the office when the door is closed and when the door is open?

Transmission loss $\text{TL} = 10 \log (1/\tau)$ db, where τ is the transmissivity of the material. For the partition alone, we find $40 = 10 \log (1/\tau_w)$ or $\tau_w = 0.0001$; and for the door alone, $30 = 10 \log (1/\tau_d)$ or $\tau_d = 0.001$. Hence for the partition with the door built-in,

$$\text{TL} = 10 \log \frac{\Sigma S}{\Sigma S\tau} = 10 \log \frac{100}{97.5(0.0001) + 2.5(0.001)} = 39.1 \text{ db}$$

where $\Sigma S\tau$ is sometimes called the *transmittance* of the material. Thus the noise level in the office with the door closed is $75 - 39.1 = 35.9$ db.

With the door open, the transmissivity for the open space is 1.0 and the transmission loss becomes

$$\text{TL} = 10 \log \frac{100}{97.5(0.0001) + 2.5(1.0)} = 16.5 \text{ db}$$

Thus the noise level in the office with the door open is $75 - 16.5 = 58.5$ db.

7.11. A small fan radiates 20 microwatts of sound energy into a soundproof room having 10 metric sabins sound absorption. Assuming sound energy absorbed equals sound energy generated, calculate the sound intensity level in the room.

$$I = W/a = 20(10^{-6})/10 = 2\times 10^{-6} \text{ watt/m}^2$$

$$\text{IL} = 10 \log (2\times 10^{-6})/10^{-12} = 63 \text{ db re } 10^{-12} \text{ watt/m}^2$$

7.12. When the air conditioner is operating, the noise level in a room is observed to be 70 db. Additional acoustical materials of 50 metric sabins sound absorption are mounted to the ceiling of the room. What is the new noise level if the initial sound absorption of the room is 15 metric sabins?

Let the sound power output from the air conditioner be W watts. Then the sound intensities in the room before and after the addition of acoustical materials are $I_1 = W/a_1$ and $I_2 = W/a_2$ watts/m², where $a_1 = 15$ and $a_2 = 15 + 50 = 65$ metric sabins.

Since the initial noise level is 70 db, we have

$$(\text{IL})_1 = 10 \log (I_1/I_0) = 70 \quad \text{or} \quad I_1 = I_0 \text{ antilog } 7 \text{ watts/m}^2$$

where I_0 is the reference intensity in watts/m².

Now $W = a_1 I_1 = a_1 I_0 \text{ antilog } 7$, and so $I_2 = W/a_2 = (a_1/a_2) I_0 \text{ antilog } 7$. Then $(\text{IL})_2 = 10 \log (I_2/I_0) = 70 + 10 \log (a_1/a_2) = 63.62$ db.

The same result is obtained if we assume noise reduction is proportional to sound absorption, i.e.

$$\Delta(\text{IL}) = 10 \log (a_2/a_1) = 10 \log (65/15) = 6.38 \text{ db} \quad \text{and} \quad (\text{IL})_2 = 70 - 6.38 = 63.62 \text{ db}$$

7.13. Two adjoining rooms have sound intensity levels of 73 and 64 db respectively. What is the attenuation through the wall?

$$\text{Attenuation} = (\text{IL})_1 - (\text{IL})_2 = 73 - 64 = 9 \text{ db}$$

or $10 \log (I_1/I_2) = 9$ db, where I_1 and I_2 are the respective sound intensities.

7.14. A room has 100 metric sabins sound absorption and a total wall area of 200 m². If the average sound transmissivity is 0.05, find the noise-insulation factor.

The noise-insulation factor is $10 \log (a/\Sigma S\tau) = 10 \log \dfrac{100}{200(0.05)} = 10$ db.

7.15. A room of dimensions $3\times 5\times 7$ m has a reverberation time of 0.85 sec and 15 metric sabins sound absorption. A standard tapping machine is used at four different positions to excite the floor. The sound pressure level readings in octave bands are 82.3, 85.1, 79.8, and 80.4 db re 0.0002 microbar. Find the space average sound pressure level and the normalized impact sound level.

The space average sound pressure level is

$$L_s = 10 \log \frac{p_1^2 + p_2^2 + p_3^2 + p_4^2}{4p_0^2} \text{ db}$$

Now $(\text{SPL})_1 = 82.3 = 20 \log p_1/(2\times 10^{-5})$ or $p_1 = 0.264$ nt/m². Similarly, $p_2 = 0.362$, $p_3 = 0.19$, $p_4 = 0.2$ nt/m². Then $L_s = 82.16$ db, where $p_0 = 0.00002$ nt/m² is the reference pressure.

This differs from the average sound pressure level $\frac{1}{4}(82.3 + 85.1 + 79.8 + 80.4) = 81.9$ db by 0.26 db. This is a small difference. However, for large rooms the difference will be significant.

The normalized impact sound level $= L + 10 \log (a/a_0) = 82.16 + 10 \log (15/10) = 83.94$ db.

SOUND ABSORPTION

7.16. Derive expressions for the decay of sound and reverberation time in dead rooms.

By geometrical analysis involving classical ray theory, the average distance traversed by a sound wave between two successive reflections in an enclosure is found to be

$$L = 4V/S$$

where L is the mean free path, V is the volume of the enclosure, and S is the total wall area of the enclosure.

Traveling at speed c, the number of reflections the sound wave makes with the walls in any time t will be

$$n = ct/L = Sct/4V$$

Assuming an average sound absorption coefficient $\bar{\alpha}$ of the enclosure, the sound wave loses a fraction $\bar{\alpha}$ of its intensity at each reflection. The intensity after n reflections is therefore

$$I_n = I_0(1-\bar{\alpha})^n = I_0(1-\bar{\alpha})^{(Sc/4V)t}$$

or the decay of sound in dead rooms is

$$I(t) = I_0 e^{\ln(1-\bar{\alpha})^{(Sc/4V)t}} = I_0 e^{[(Sc/4V)\ln(1-\bar{\alpha})]t}$$

where the decaying factor is $-(Sc/4V)[-\ln(1-\bar{\alpha})]t$.

Comparing with the decay of sound in live rooms (see Problem 7.2), we have

$$ac/4V = (Sc/4V)\alpha_e \quad \text{or} \quad \alpha_e = -\ln(1-\bar{\alpha})$$

where α_e is the effective absorption coefficient.

The reverberation time in dead rooms can be obtained from that for live rooms by putting $\alpha_e = -\ln(1-\bar{\alpha})$, i.e.

$$T = \frac{0.049V}{S[-\ln(1-\bar{\alpha})]} \text{ in metric units} = \frac{0.161V}{S[-\ln(1-\bar{\alpha})]} \text{ in English units}$$

7.17. A small reverberation chamber $8 \times 9 \times 10$ ft is employed to measure the effective sound absorption coefficient of certain acoustical tile. The observed reverberation time is 5 sec or 1.0 sec when 40 ft² of acoustical tile is used to cover part of one wall of the chamber. Find the effective sound absorption coefficient of the tile.

The volume of the chamber is $V = 720$ ft³, and the total area of the wall surfaces is $S = 2(8)9 + 2(8)10 + 2(9)10 = 484$ ft².

Since reverberation time in a reverberation chamber is $T = 0.049V/S\alpha$ sec, the sound absorption coefficient of the chamber wall is

$$\alpha_1 = 0.049V/ST_1 = 0.049(720)/[484(5)] = 0.015$$

When acoustical tile of total sound absorption $S_2\alpha_2$ is added to part of one wall of the chamber (where S_2 is the area in ft² of the tile and α_2 is the effective sound absorption coefficient of the tile), the new reverberation time of the chamber becomes

$$T_2 = 0.049V/(S_1\alpha_1 + S_2\alpha_2) \text{ sec}$$

where $S_1 = S - S_2 = 484 - 40 = 444$ ft² is the new area of the wall surfaces of the chamber. Thus

$$\alpha_2 = \frac{0.049V - S_1\alpha_1T_2}{S_2T_2} = \frac{0.049(720) - 444(0.015)(1.0)}{40(1.0)} = 0.71$$

7.18. Find the reverberation time of an office which has a volume of 1600 m³ and a total sound absorption of 80 metric sabins. What is the sound absorption required for an optimum reverberation time of 1.2 sec?

Reverberation time $T = 0.161V/a = 0.161(1600)/80 = 3.22$ sec.

For an optimum reverberation time of 1.2 sec,

$$a = 0.161V/T = 0.161(1600)/1.2 = 216 \text{ metric sabins}$$

i.e. additional sound absorption required $= 216 - 80 = 136$ metric sabins.

7.19. Ten persons are talking in a room with total sound absorption of .975 metric sabins. If each person produces an acoustic power output of 10 microwatts, compare the background sound pressure level of the reverberant sound with the direct sound pressure level at a distance 0.3 m from the closest speaker.

The sound pressure in the live room is

$$p_r = \sqrt{4\rho cW/a} = \sqrt{4(415)(10^{-4})/0.975} = 0.41 \text{ nt/m}^2$$

where p_r is the background reverberant sound pressure, $a = 0.975$ metric sabin is the total sound absorption of the room, $\rho c = 415$ rayls is the characteristic impedance of air, and $W = 10^{-4}$ watt is the total acoustic power output. Then the reverberant sound pressure level becomes

$$(\text{SPL})_r = 20 \log(0.41/0.00002) = 86.4 \text{ db}$$

For the direct sound pressure, we have $I = W/4\pi r^2 = p_d^2/\rho c$ or

$$p_d = \sqrt{W\rho c/4\pi r^2} = \sqrt{10^{-5}(415)/[4\pi(0.3)^2]} = 0.061 \text{ nt/m}^2$$

and so

$$(\text{SPL})_d = 20 \log(0.061/0.00002) = 69.6 \text{ db}$$

It is apparent that the background reverberant sound presents an unpleasant high level noise which, for all practical purposes, completely masks the intelligibility of conversation. The situation can be remedied by reducing the acoustic power output of each person (i.e. speak softly), thereby lowering the background reverberant sound.

7.20. The observed reverberation time at 5000 cyc/sec in a reverberation chamber filled with dry air is 16 sec. With moist air, the reverberation time is 6 sec. If $\alpha/f^2 = 1.4(10^{-11})$ for dry air, determine the absorption coefficient (or attenuation constant) for the moist air.

We have shown that the intensity of a plane acoustic wave decreases according to

$$I(t) = I_0 e^{-2\alpha x} = I_0 e^{-mx}$$

where $m = 2\alpha$ in nepers/m is the absorption coefficient for air. But the decay of sound in a live room is $I(t) = I_0 e^{-act/4V}$, and when the effect of air absorption is incorporated we have

$$I(t) = I_0 e^{-(a/4V+m)ct}$$

and

$$\Delta\text{IL} = 10 \log(I/I_0) = (10/2.3) \ln e^{-(a/4V+m)ct} = -4.34(a/4V+m)ct$$

which represents the change in intensity level in decibels. The decay rate is therefore

$$D = 4.34(a/4V+m)c \text{ db/sec}$$

Now reverberation time is the period required for the level of the sound in the room to decay by 60 db, or

$$T = 60/D = \frac{60}{4.34(a/4V+m)c} = 0.161V/(a+4Vm) \text{ sec}$$

where $c = 343$ m/sec is the speed of sound in air, V is the volume of the room in m^3 and a is the total sound absorption in metric sabins. Also

$$T = 0.049V/(a+4Vm) \text{ sec}$$

where V is the volume of the room in ft^3 and a is the total sound absorption in sabins.

Since the volume and total sound absorption due to the wall surface of the room are constant, we can write expressions for the reverberation time for dry and moist air,

$$T = \frac{0.049V}{a+4Vm} \quad \text{or} \quad m = \frac{0.049V - Ta}{4VT} \quad \text{for dry air}$$

$$T' = \frac{0.049V}{a+4Vm'} \quad \text{or} \quad m' = \frac{0.049V - T'a}{4VT'} \quad \text{for moist air}$$

Combining,

$$m' - m = \frac{0.01225(T - T')}{TT'}$$

Now $\alpha/f^2 = 1.4 \times 10^{-11}$ or $\alpha = 1.4 \times 10^{-11} \times (5000)^2 = 3.5 \times 10^{-4}$ neper/ft. Thus

$$m' = \frac{0.01225(16-6)}{6(16)} + 7 \times 10^{-4} = 19.7 \times 10^{-4} \text{ neper/ft}$$

7.21. A room has an average sound absorption coefficient 0.5 and mean free path 10 m. Calculate the reverberation time of the room.

With an average sound absorption coefficient $\bar{\alpha} = 0.5$, the sound waves lose a fraction $\bar{\alpha}$ of their intensity at each reflection. The number of reflections required for the intensity to decrease to 10^{-6} of its original value is therefore $0.5^n = 10^{-6}$, from which $n = 20$.

Since we know the average free path is 10 m, the number of reflections made by a sound wave per sec is $n = ct/L = 343(1.0)/10 = 34.3$. Thus the reverberation time is $T = 34.3/20 = 1.72$ sec.

Conversely, we can measure reverberation time directly and use the information to calculate the number of reflections and the free mean path.

7.22. The volume of a room is 324 m³. The wall has area 122 m² and average sound absorption coefficient 0.03. The ceiling has area 98 m² and average sound absorption coefficient 0.8. The floor has area 98 m² and average sound absorption coefficient 0.06. Compute the reverberation time for this room.

The average sound absorption coefficient of the room is

$$\bar{\alpha} = \frac{\alpha_1 S_1 + \alpha_2 S_2 + \alpha_3 S_3}{S_1 + S_2 + S_3} = \frac{0.03(122) + 0.8(98) + 0.06(98)}{122 + 98 + 98} = 0.27$$

Then the total sound absorption of the room $a = 0.27(318) = 86$ metric sabins.

Reverberation time $T = 0.161V/a = 0.161(324)/86 = 0.6$ sec.

7.23. An office with a noise level of 72.5 db has originally 100 metric sabins sound absorption. Sound absorption material with a coefficient of 0.85 is applied to the ceiling of dimensions 20×40 m. What will be the resultant noise level?

Since the original sound absorption is 100 metric sabins, and $20(40)(0.85) = 680$ metric sabins are added, the total sound absorption is 680 metric sabins. (Here we assume the original sound absorption of 100 metric sabins is entirely due to the ceiling.) Sound reduction is therefore

$$(\text{db})_1 - (\text{db})_2 = 10 \log (680/100) = 8.34 \text{ db}$$

and the resultant noise level is $72.5 - 8.34 = 64.16$ db. A reduction of 5-10 db is considered satisfactory for most offices.

SOUND DISTRIBUTION

7.24. An electric motor is tested on a large hard surface inside an anechoic chamber. At a radius of 1 m from the motor, five readings of the noise level are taken near the centers of five equal areas on a hemispherical surface. These readings are 73, 72, 69, 70 and 68 db. What is the sound power output of the motor?

The noise level is $10 \log (I/I_0)$ db where $I_0 = 10^{-12}$ watt/m² is the reference intensity. Then $I_1 = I_0 \text{ antilog } 7.3 = 1.99 \times 10^{-5}$, $I_2 = 1.58 \times 10^{-5}$, $I_3 = 7.94 \times 10^{-6}$, $I_4 = 10^{-5}$, $I_5 = 6.28 \times 10^{-6}$ watt/m².

The area of a hemispherical surface is $2\pi r^2 = 6.28\ \text{m}^2$, and the area of each of the five segments is $1.26\ \text{m}^2$. The total acoustic power through all five segments is the acoustic output of the motor.

Now $W_1 = 1.26(1.99 \times 10^{-5}) = 25.2 \times 10^{-6}$, $W_2 = 20.1 \times 10^{-6}$, $W_3 = 10.1 \times 10^{-6}$, $W_4 = 12.7 \times 10^{-6}$, $W_5 = 8.0 \times 10^{-6}$ watt, and thus

$$W = W_1 + W_2 + W_3 + W_4 + W_5 = 76.1 \times 10^{-6} \text{ watt}$$

7.25. The sound pressure level of a machine in a reverberation chamber $3 \times 4 \times 5$ m is 70 db re 0.00002 nt/m^2. The reverberation time is 4 sec. Find the acoustic power output of the machine.

The maximum sound pressure level in a reverberation chamber is obtained when steady state condition is reached, i.e. $p_{\text{max}} = \sqrt{4W\rho c/a}$ nt/m^2 where W is the acoustic power output in watts, $\rho = 1.21$ kg/m^3 is the density of air, $c = 343$ m/sec is the speed of sound in air, and a is the total sound absorption in metric sabins.

Reverberation time $T = 55.2V/ac$ sec, where $V = 60$ m^3 is the volume of the chamber.

Upon eliminating the constant a from these two expressions,

$$W = \frac{13.8p^2V}{c^2T} = \frac{13.8(0.063)^2(60)}{1.21(343)^2(4)} = 5.84 \times 10^{-6} \text{ watt or } 5.84 \text{ microwatts}$$

since $\text{SPL} = 20 \log p/(2 \times 10^{-5})$ or $p = 0.063$ nt/m^2.

7.26. A room has dimensions $4 \times 5 \times 8$ m. Determine (*a*) the mean free path of a sound wave in this room, (*b*) the number of reflections per sec made by sound waves with the walls of this room, and (*c*) the decay rate of sound in this room.

(*a*) The mean free path L is the average distance a sound wave travels through the air between two successive encounters with the walls of the room.

$$L = 4V/S = 4(160)/184 = 3.48 \text{ m}$$

where $V = 4(5)8 = 160$ m^3 is the volume of the room, and $S = 2(4)5 + 2(4)8 + 2(5)8 = 184$ m^2 is the total wall surface area of the room.

(*b*) $n = c/L = 343/3.48 = 98.5$, where $c = 343$ m/sec is the speed of sound in air.

(*c*) The decay rate of sound depends on the total sound absorption of the room. If we assume a fairly dead room with an average sound absorption coefficient $\bar{a} = 0.6$, then

$$D = -\frac{1.08Sc \ln(1 - \bar{a})}{V} = -\frac{1.08(184)343 \ln 0.4}{160} = 179 \text{ db/sec}$$

ROOM ACOUSTICS

7.27. Compute the lowest characteristic frequencies associated with the axial sound waves, the tangential sound waves, and the oblique sound waves in a rectangular room of dimensions $3 \times 5 \times 7$ m.

The frequency equation for harmonic acoustic wave motion in a rectangular room is

$$f_{xyz} = \tfrac{1}{2}c\sqrt{(n_x/L_x)^2 + (n_y/L_y)^2 + (n_z/L_z)^2} \text{ cyc/sec}$$

where $c = 343$ m/sec is the speed of sound in air, the n's are the modes of vibration, and the L's are the lengths of the sides of the room.

The axial sound waves are those moving parallel to either one of the three rectangular axes, i.e. two of the n's are zero. The lowest characteristic frequency associated with the axial waves in the z direction is

$$f_{001} = (343/2)(1/7) = 24.4 \text{ cyc/sec}$$

The tangential waves are those moving parallel to the surfaces of either one of the walls, i.e. one of the n's is zero. The lowest characteristic frequency associated with the tangential wave in the yz plane is

$$f_{011} = (343/2)\sqrt{(1/5)^2 + (1/7)^2} = 42 \text{ cyc/sec}$$

The oblique waves are those striking all six walls of the room, i.e. none of the n's is zero. The lowest characteristic frequency associated with the oblique waves is

$$f_{111} = (343/2)\sqrt{(1/3)^2 + (1/5)^2 + (1/7)^2} = 66.1 \text{ cyc/sec}$$

It is apparent that axial waves are the most persistent while oblique waves decay most rapidly. Wall irregularities as well as irregular room shapes are therefore preferred because they discourage axial sound waves and encourage oblique sound waves.

7.28. Compute the characteristic frequencies associated with the first six principal modes of vibration in a rectangular room of dimensions $3 \times 5 \times 7$ m.

The frequency equation for harmonic wave motion in a rectangular enclosure is

$$f_{xyz} = \tfrac{1}{2}c\sqrt{(n_x/L_x)^2 + (n_y/L_y)^2 + (n_z/L_z)^2} \text{ cyc/sec}$$

For the first principal mode in the x direction, $n_y = n_z = 0$; in the y direction, $n_x = n_z = 0$; in the z direction, $n_x = n_y = 0$. Hence

$$f_{100} = (343/2)(1/3) = 57.2, \quad f_{010} = (343/2)(1/5) = 34.3, \quad f_{001} = (343/2)(1/7) = 24.2 \text{ cyc/sec}$$

Similarly, for the second principal mode in the x direction, $n_x = 2$, $n_y = n_z = 0$; in the y direction, $n_x = 0$, $n_y = 2$, $n_z = 0$; and in the z direction, $n_x = n_y = 0$, $n_z = 2$. Hence

$$f_{200} = (343/2)(2/3) = 114.4, \quad f_{020} = (343/2)(2/5) = 68.6, \quad f_{002} = (343/2)(2/7) = 49 \text{ cyc/sec}$$

7.29. For the fundamental mode of vibration, calculate the directional angles for the axial, tangential, and oblique waves in a rectangular enclosure of dimensions $3 \times 5 \times 7$ m.

Let the rectangular enclosure be the xyz coordinates with sides $L_x = 5$, $L_y = 3$, $L_z = 7$ m. The directional angles $\theta_x, \theta_y, \theta_z$ are the angles formed by the wave vector and the coordinate axes.

(1) Axial waves: x-axial, for the $(1,0,0)$ mode, $n_x = 1$, $n_y = n_z = 0$; y-axial, for the $(0,1,0)$ mode, $n_x = 0$, $n_y = 1$, $n_z = 0$; for the $(0,0,1)$ mode, $n_x = n_y = 0$, $n_z = 1$; the directional angles are respectively

$$\theta_x = 0,\ \theta_y = \theta_z = 90°; \quad \theta_x = \theta_z = 90°,\ \theta_y = 0; \quad \theta_x = \theta_y = 90°,\ \theta_z = 0$$

(2) Tangential waves: xy-tangential, $(1,1,0)$ mode, $n_x = n_y = 1$, $n_z = 0$;

$$\theta_x = \tan^{-1}(L_x/L_y) = 59°, \quad \theta_y = \tan^{-1}(L_y/L_x) = 31°, \quad \theta_z = 90°$$

yz-tangential, $(0,1,1)$ mode, $n_x = 0$, $n_y = n_z = 1$;

$$\theta_x = 90°, \quad \theta_y = \tan^{-1}(L_y/L_z) = 25.3°, \quad \theta_z = \tan^{-1}(L_z/L_y) = 64.7°$$

zx-tangential, $(1,0,1)$ mode, $n_x = n_z = 1$, $n_y = 0$;

$$\theta_x = \tan^{-1}(L_x/L_z) = 35.5°, \quad \theta_y = 90°, \quad \theta_z = \tan^{-1}(L_z/L_x) = 54.5°$$

(3) xyz-oblique waves: $(1,1,1)$ mode, $n_x = n_y = n_z = 1$;

$$\theta_x = \tan^{-1}(L_x/\sqrt{L_x^2 + L_y^2 + L_z^2}) = \tan^{-1}(5/9.1) = 28.9°$$

$$\theta_y = \tan^{-1}(3/9.1) = 18.3°, \quad \theta_z = \tan^{-1}(7/9.1) = 37.6°$$

For higher modes of vibration, e.g. the $(3,2,0)$ mode, the procedure for obtaining the directional angles is the same:

$$\theta_x = \tan^{-1}(3L_x/2L_y) = 48.9°, \quad \theta_y = \tan^{-1}(2L_y/3L_x) = 41.1°, \quad \theta_z = 90°$$

7.30. What is the room constant of an enclosure having a total surface area of 400 ft^2 and an average sound absorption coefficient of (*a*) $\bar{\alpha} = 0.2$, (*b*) $\bar{\alpha} = 0.8$?

Room constant $R = S\bar{\alpha}/(1-\bar{\alpha})$ ft^2 where S is the total wall area of the room in ft^2 and α is the average sound absorption coefficient. Substituting values, we find (*a*) $R = 100$ ft^2, (*b*) $R = 1600$ ft^2. The greater the room constant the better the room acoustics.

Supplementary Problems

REVERBERATION

7.31. A room $20 \times 40 \times 60$ ft has an average sound absorption coefficient 0.24. What is the reverberation time? *Ans.* $T = 1.0$ sec

7.32. What is the theoretical reverberation time if the sound absorption coefficient is (*a*) $\alpha = 1.0$, (*b*) $\alpha = 0$? *Ans.* (*a*) $T = 0$, (*b*) $T = \infty$

7.33. Show that Sabine's equation for the reverberation time $T = 0.049V/a$ will not be applicable for sound absorption coefficient $\alpha > 0.2$.

7.34. Show that Eyring's expression for reverberation time in a dead room, $T = \dfrac{0.049V}{S[-\ln(1-\bar{\alpha})]}$, yields identical value as given by Sabine's equation, $T = 0.049V/a$, for $\alpha = 0$.

7.35. The volume of a room is 1000 m^3 and its total wall area is 400 m^2. Calculate the reverberation time if 5% of incident sound energy is being absorbed at each reflection at the wall.
Ans. $T = 6.5$ sec

7.36. Derive an expression for the decay rate in db/sec of sound energy in a live room.
Ans. $372a/V$ (metric units), $1230a/V$ (English units)

7.37. A room of volume 400 ft^3 has 20 sabins absorption. Determine the reverberation time for both dry and humid air having a relative humidity of 40% at 75°F. The attenuation constant at 1500 cyc/sec is given as $m = 0.002$. *Ans.* 0.98, 0.85 sec

NOISE INSULATION AND REDUCTION

7.38. An office is planned in a building having an average noise level of 70 db. If the noise level in the office should be 45 db, what is the noise reduction required? *Ans.* 25 db

7.39. If the noise level outside a room is 65 db and its noise insulation is 30 db, find the noise level inside the room. *Ans.* 35 db

7.40. A wall 10×20 m with an initial transmission loss of 50 db has four windows built into it. The area of each window is 5 m^2 and its sound transmission coefficient is 0.01. What will be the new transmission loss of the wall with windows? *Ans.* 25.5 db

7.41. Sound waves of power level 70 db are incident on a concrete wall. Assuming 1/10,000 of the incident energy is transmitted through the wall, find the transmission loss of the wall and the reduced sound power level. *Ans.* TL = 40 db, SPL = 30 db

7.42. The noise level reduction of a noisy machine employing a partial enclosure is approximately given by $10 \log (A_t/A_0)$ db, where A_t is the total area of the enclosure and A_0 is the open area of the enclosure. Find the noise level reduction by an enclosure 20% opened. *Ans.* 7 db

SOUND ABSORPTION

7.43. Prove that if the total sound absorption in an enclosure is doubled, the average noise level will be decreased by 3 db.

7.44. An office has a noise level of 70 db with 10 metric sabins sound absorption. How much more absorption is needed to reduce the noise level to 60 db? *Ans.* 90 metric sabins

7.45. Derive an expression for the noise reduction level in dead rooms.

Ans. $10 \log \dfrac{A_2 \log (1-\alpha_2)}{A_1 \log (1-\alpha_1)}$ db

SOUND DISTRIBUTION

7.46. Compute the sound pressure level drop for a tenfold increase of distance from the source. Assume spherical acoustic wave propagation. *Ans.* 20 db

7.47. Show that the sound pressure level drop at each reflection by a sound wave in an enclosure is given by $10 \log 1/(1-\alpha)$ db.

7.48. If a sound system has efficiency 6%, find the power required to produce a sound pressure level of 100 db in an enclosure of volume 10,000 ft^3 and having a reverberation time 1.1 sec.
Ans. 1.85 watts

ROOM ACOUSTICS

7.49. Derive the expression for mean free path, $L = 4V/A$, by energy considerations.

7.50. Show that the sound pressure level in an enclosure can be obtained from the expression SPL $=$ PWL $+ 6.5 - 10 \log a$.

7.51. The dimensions of a rectangular room are $10 \times 15 \times 20$ m. Determine the characteristic frequency associated with the lowest degenerate normal mode of vibration of sound waves.
Ans. 55 cyc/sec

Chapter 8

Underwater Acoustics

NOMENCLATURE

a = absorption coefficient, db/m
A = transmission anomaly, db
c = speed of sound in air, m/sec
d = depth, m; directivity
D = diameter, m
E = voltage, volts
f = frequency, cyc/sec
g = velocity gradient, m/sec/m
H = transmission loss, db
I = sound intensity, watts/m^2
I_s = intensity level, db
k = wave number
L = length, m
p = acoustic pressure, nt/m^2
r = distance, m
S = area, m^2
SPL = sound pressure level, db
v = velocity, m/sec
W = sound power, watts
Y = Young's modulus, nt/m^2
ω = circular frequency, rad/sec
ρ = density, kg/m^3
λ = wavelength, m
α = absorption coefficient, nepers/m
σ = cavitation number

INTRODUCTION

Underwater acoustics deals with transmission of sound waves through water, taking into consideration the transmission losses, sound generation and reception, divergence and absorption, reflection and refraction, noises and reverberation.

UNDERWATER SOUND

As a medium for communication, water transmits sound waves far better than optical, radio or magnetic waves. The transmission of sound waves in water depends on variables such as temperature and pressure gradients, marine organisms, air bubbles, salt content and other nonhomogeneities.

Sound transmission losses in sea water are chiefly due to the following. (1) Divergence: outgoing spherical acoustic waves decrease in intensity according to the inverse square law. (2) Absorption: the dissipation of acoustic energy into the medium or boundaries because of viscous losses, heat conduction losses and molecular action. (3) Irreversible attenuation: losses caused by refraction, scattering, diffraction, interference, etc., commonly known as *transmission anomaly* A in decibels. The *total transmission loss* H in decibels is therefore given by

$$H = 20 \log r + ar + A \quad \text{db}$$

where r is the distance in meters between source and receiver, a is the absorption coefficient in db/m, and A is the transmission anomaly in decibels. (See Problems 8.13-8.15.)

REFRACTION

Refraction is the bending of sound waves because of velocity changes accompanying temperature and pressure changes. Since the velocity of sound is a function of temperature which varies linearly with depth, sound waves will be refracted downward in a circular arc. Because of the downward bending of sound waves, no sound waves will reach the surface of the sea, forming a *shadow zone* as shown in Fig. 8-1.

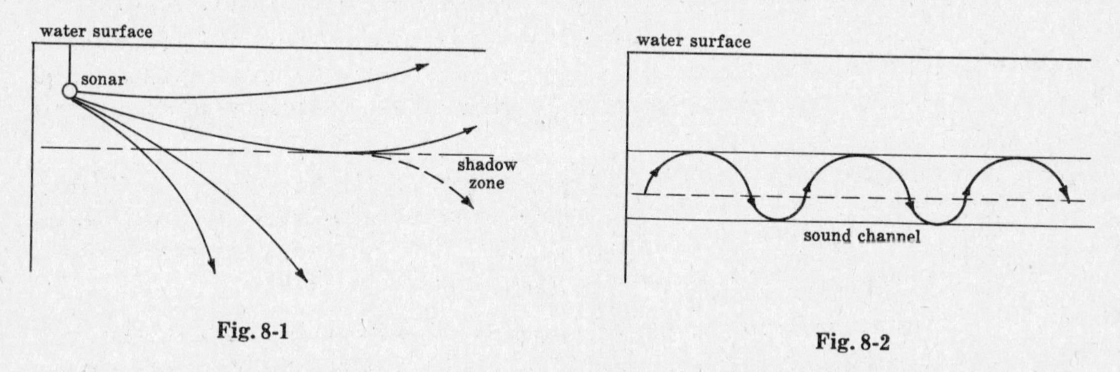

Fig. 8-1

Fig. 8-2

At great depths where the temperature is constant, sound velocity increases linearly with depth because of pressure. Here sound waves will be refracted upward and follow an arc of a circle. (See Problems 8.3-8.7.)

Sound channels are formed at great depths in the sea where the temperature is constant. Sound waves emitted at this constant temperature level will be refracted upward or downward along a narrow channel as shown in Fig. 8-2. This is due to temperature and pressure gradients. As a result, sound waves within sound channels spread out in a circle rather than a sphere, and propagate to much greater distances. (See Problems 8.8-8.11.)

REVERBERATION

Transmitted acoustic energy that returns to the listening hydrophone without intercepting an object or target is *reverberation*. Unlike ambient noise, reverberation or background scattering is directly related to the acoustic energy projected into water by the sound source. It is in general an unwanted signal and tends to interfere with the returned echo.

Volume reverberation is caused by the scattering of sound in the bounded and nonhomogeneous volume of the sea, while *surface* and *bottom reverberation* are due to reflections at the sea surface and bottom respectively.

Echo-sounding is based on the reflection of sound and the production of an echo. It locates submerged objects by sending out a sound wave and receiving the returned echo. *Passive listening* is used to detect sounds from an unknown direction by collecting sound waves while maintaining complete silence. This has greater detection range than echo-ranging, without the risk of revealing one's own position. (See Problems 8.12-8.17.)

AMBIENT NOISE

Ambient or *background noise* in the sea is a function of the state of agitation of the sea by natural agents such as wind and rain, and is often very unpredictable. Moreover, biological noises, man-made noises, noises from ships, and self-noise from sound systems all tend to mask the wanted signal.

UNDERWATER TRANSDUCERS

Hydrophone is an electroacoustic transducer that responds to sound waves in water and produces equivalent electric waves. Like microphones and other electroacoustic transducers, hydrophones should have good stability, high sensitivity and linear responses. They must be rugged to withstand high hydrostatic pressures and be independent of temperature. To meet high power and small displacement requirements, their faces should be large.

Hydrophone sensitivity in volts/microbar is the voltage generated at its terminals by unit sound pressure. It is a function of the angle measured from the acoustic axis of the hydrophone (or the axis of maximum sensitivity) and of the frequency of the signal generated.

Hydrophone directivity is an indication of the fraction of the total signal the hydrophone is permitted by its sensitivity pattern to convert into electrical energy. A hydrophone equally sensitive in all directions has a directivity factor of one and a directivity index zero. (See Microphone sensitivity and directivity of Chapter 5.)

Underwater sound projector, or simply *projector,* is an electroacoustic transducer used to generate sound in water. A projector converts electrical energy into acoustical energy in water through either magnetostrictive or piezoelectric effects. (For magnetostrictive or piezoelectric transducers, see Chapter 9.)

Sonars and passive sonars are underwater sound systems usually consisting of hydrophones, power amplifiers and readout devices. They are used to detect sounds in water. The sonar, for example, scans the water until its sound beam hits a target and produces an echo, whose reception at the sonar can be made to give information about the target. (See Problems 8.18-8.23.)

CAVITATION

If the existing pressure is reduced to less than the vapor pressure of the water, bubbles filled with water vapor are formed. These bubbles collapse when they are forced to move into a region of higher pressure. Their collapse or local boiling produces noises with accompanying vibration which is detrimental to the transmission of sound. This phenomenon is called *cavitation.*

A *cavitation number* σ is defined as

$$\sigma = \frac{2(p_0 - p_v)}{\rho v^2}$$

where p_0 is the ambient pressure in nt/m^2, p_v is the vapor pressure of water in nt/m^2, ρ is the density of water in kg/m^3, and v is the speed of the vehicle in m/sec. (See Problems 8.24-8.25.)

Solved Problems

UNDERWATER SOUND

8.1. What is the ratio of particle velocity in air to that in water if (*a*) acoustic pressure in air and in water are the same, (*b*) acoustic intensity in air and in water are the same?

(*a*) Particle velocity $v = p/\rho c$ m/sec where p is acoustic pressure in nt/m^2, and ρc is the characteristic impedance of the medium in rayls. Then

$$\frac{v_{\text{air}}}{v_{\text{water}}} = \frac{p/(\rho c)_{\text{air}}}{p/(\rho c)_{\text{water}}} = \frac{(\rho c)_{\text{water}}}{(\rho c)_{\text{air}}} = \frac{1{,}480{,}000}{415} = 3570$$

(*b*) Acoustic intensity $I = p^2/2\rho c$ watts/m^2. Then $p = \sqrt{2I\rho c}$, and

$$\frac{v_{\text{air}}}{v_{\text{water}}} = \frac{[\sqrt{2I\rho c}/\rho c]_{\text{air}}}{[\sqrt{2I\rho c}/\rho c]_{\text{water}}} = \frac{\sqrt{(\rho c)_{\text{water}}}}{\sqrt{(\rho c)_{\text{air}}}} = \frac{1{,}480{,}000}{415} = 59.8$$

8.2. Prove that the path of a sound wave through water having a constant positive velocity gradient g m/sec per meter is an arc of a circle of constant radius $R = c_0/g$ meters.

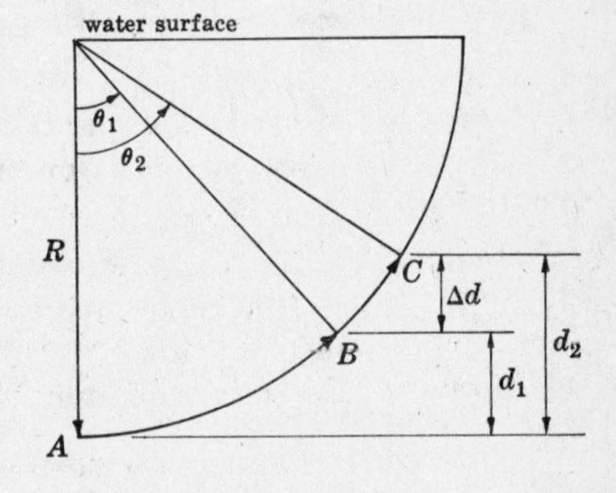

Fig. 8-3

Let R be the radius of an arc ABC of a circle as shown in Fig. 8-3. Then

$$d_1 = R(1 - \cos\theta_1), \qquad d_2 = R(1 - \cos\theta_2)$$

$$\Delta d = d_2 - d_1 = R(\cos\theta_1 - \cos\theta_2)$$

Since the water has a constant positive velocity gradient, the velocity of sound increases with depth.

$$c_2 = c_1 - g\,\Delta d \quad \text{or} \quad \Delta d = -(c_2 - c_1)/g$$

where c_2 is the speed of sound at point C, and c_1 is the speed of sound at point B.

Now Snell's law for a sound wave in a medium in which the velocity changes with depth is given by

$$c_0/(\cos\theta_0) = c_1/(\cos\theta_1) = c_2/(\cos\theta_2)$$

where c_0 is the speed of sound at A.

From the expressions for Δd and Snell's law, $R = c_0/g = c_n/(g\cos\theta_n)$.

8.3. Determine the path of a sound ray in a layer of water where the velocity of sound increases with depth.

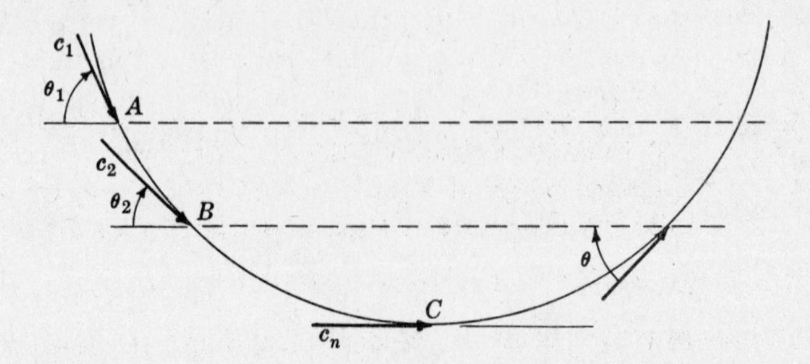

Fig. 8-4

Assume a sound ray at A has initial velocity c_1 and its path makes an angle θ_1 with the horizontal. At point B, assume the velocity of sound becomes c_2 and its path makes an angle θ_2 with the horizontal. Using Snell's law,

$$\frac{c_1}{\cos\theta_1} = \frac{c_2}{\cos\theta_2} \quad \text{or} \quad \cos\theta_2 = \frac{c_2\cos\theta_1}{c_1}$$

and so in general

$$\cos\theta_{n+1} = \left(\frac{c_{n+1}}{c_n}\right)\cos\theta_n$$

Now c_{n+1} is greater than c_n because the velocity of sound increases with depth. Thus

$$\cos\theta_{n+1} > \cos\theta_n \quad \text{or} \quad \theta_{n+1} < \theta_n$$

In words, the path of a sound ray traveling in water with constant positive velocity gradient is bending upward as shown in Fig. 8-4.

Finally at C the sound path becomes horizontal, and beyond this point c_{n+1} is smaller than c_n. So we have

$$\cos\theta_{n+1} < \cos\theta_n \quad \text{or} \quad \theta_{n+1} > \theta_n$$

i.e. the sound ray will continue to be refracted upward.

As long as the water has a constant positive velocity gradient, sound waves traveling in it will be refracted upward. This is true for any initial position of the sound ray.

8.4. A narrow beam of sound is produced horizontally in water having a constant velocity gradient of $-g$ m/sec per meter. Derive an expression for the horizontal distance traveled by the sound beam after it has reached a depth d meters.

In view of the negative constant velocity gradient, the water will refract sound. The narrow beam of sound will therefore follow the path of an arc of a circle whose radius is $R = c_0/g$, where c_0 is the velocity of sound and g is the velocity gradient. (See Problem 8.2.)

The sound beam at a depth d is tangent to the circle at point B and makes an angle θ_1 with the horizontal. From Fig. 8-5,

$$x^2 + (R-d)^2 = R^2 \quad \text{or} \quad x^2 = 2dR - d^2$$

where x is the horizontal distance traveled by the sound beam in reaching the depth d. Replacing R by c_0/g, this becomes

$$x^2 = 2dc_0/g - d^2$$

In general, the horizontal distance traveled by the sound beam is very much greater than the depth it reached, i.e. $x \gg d$, so the term d^2 can be neglected. Thus

$$x = \sqrt{2c_0 d/g}$$

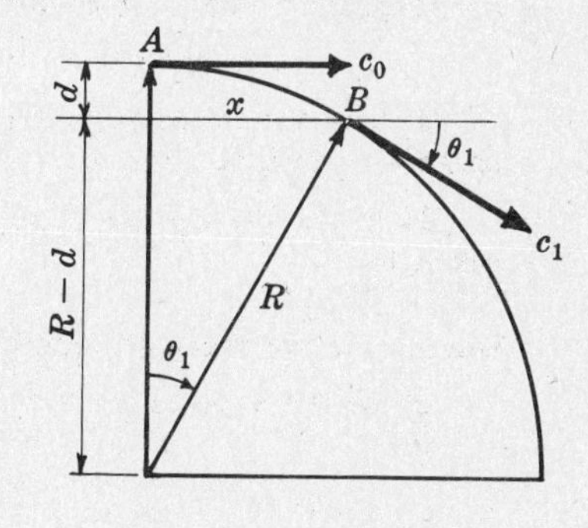

Fig. 8-5

8.5. The velocity of sound in sea water decreases uniformly from a value of 1500 m/sec at the surface to 1450 m/sec at a depth of 100 m. Determine (a) the velocity gradient, (b) the horizontal distance required for a horizontal sound ray at the surface to reach a depth of 100 m, and (c) the angle of such a sound ray upon reaching this level.

(a) Velocity gradient $g = (c_2 - c_1)/d = (1450-1500)/100 = -0.5$ m/sec per meter

(b) Horizontal distance $x = \sqrt{2c_1 d/g} = \sqrt{2(1500)(-100)/(-0.5)} = 775$ m

(c) Since $\theta_1 = 0$, the downward angle $\theta_2 = \cos^{-1}(c_2/c_1) = \cos^{-1}(1450/1500) = 10°$

8.6. Given an isothermal layer of sea water at 20°C and thickness 50 m. (*a*) If a sound ray leaves a sonar transducer at a depth of 10 m in a horizontal direction, what is the horizontal distance traveled by this ray before it reaches the surface of the water? (*b*) Find the downward angle of a sound ray that will become horizontal at the bottom of the isothermal layer and the horizontal distance this ray has traveled in reaching this position.

(*a*) Assume the speed of sound in water at the given temperature to be 1500 m/sec. Since the temperature is constant and the pressure is not, the speed of sound will increase 0.017 m/sec per meter increase in depth. In other words, the velocity gradient is due to hydrostatic pressure and is approximately $g = 0.017$ m/sec per meter. Hence the horizontal distance traveled is (see Problem 8.4)

$$x = \sqrt{2c_0 d/g} = \sqrt{2(1500.17)(10)/0.017} = 1330 \text{ m}$$

(*b*) Since $\cos\theta_1 = 1$ and $c_1 = 1500 + 50(0.017)$, the required downward angle is

$$\theta_0 = \cos^{-1}(c_0/c_1) = \cos^{-1}(1500.17/1500.85) = 1.5°$$

and
$$x = \sqrt{2c_1 d/g} = \sqrt{2(1500.85)(40)/0.017} = 2660 \text{ m}$$

8.7. A destroyer is searching for an enemy submarine in water having a constant velocity gradient of -0.1 m/sec per meter. Its sonar transducer is at a depth of 10 m where the velocity of sound is 1500 m/sec. The sonar detects a submarine at a horizontal distance of 800 m and at a downward angle of 10°. What is the depth of the submarine?

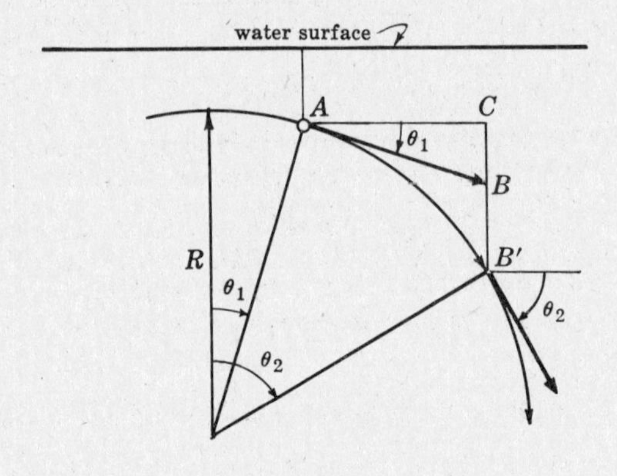

Fig. 8-6

From Fig. 8-6, the apparent depth CB of the submarine from the sonar is $800 \tan 10° = 141$ m, and so the apparent depth is $10 + 141 = 151$ m below the surface of the water.

Because of the constant negative velocity gradient of the water, the narrow sound beam AB from the sonar will actually bend downward in an arc of a circle of radius

$$R = c_0/(-g) = 1500/0.1 = 15{,}000 \text{ m}$$

The inclination of the sound beam at point B' is θ_2, which is greater than θ_1 because of refraction by water. Then

$$AC = R\sin\theta_2 - R\sin\theta_1 = 800$$

$$\sin\theta_2 = \frac{800 + R\sin\theta_1}{R} = \frac{800 + 15{,}000 \sin 10°}{15{,}000} \quad \text{or} \quad \theta_2 = 13.2°$$

The depth between the sonar transducer and the submarine is

$$CB' = R\cos\theta_1 - R\cos\theta_2 = 15{,}000(0.985 - 0.973) = 180 \text{ m}$$

Hence the true depth of the submarine below the surface of the water is $180 + 10 = 190$ m. This value is considerably different from that obtained earlier without taking refraction into consideration. On the other hand, if the constant velocity gradient is positive, the sound beam will bend upward. The true depth of the submarine is then less than its apparent depth.

SOUND CHANNELS

8.8. A surface sound channel is formed by a water layer of thickness 100 m and velocity gradients as shown in Fig. 8-7 below. Determine (*a*) the maximum angles with which a sound ray may cross the axis of the sound channel and remain within the channel and (*b*) the horizontal distance these sound rays cross the axis of the channel.

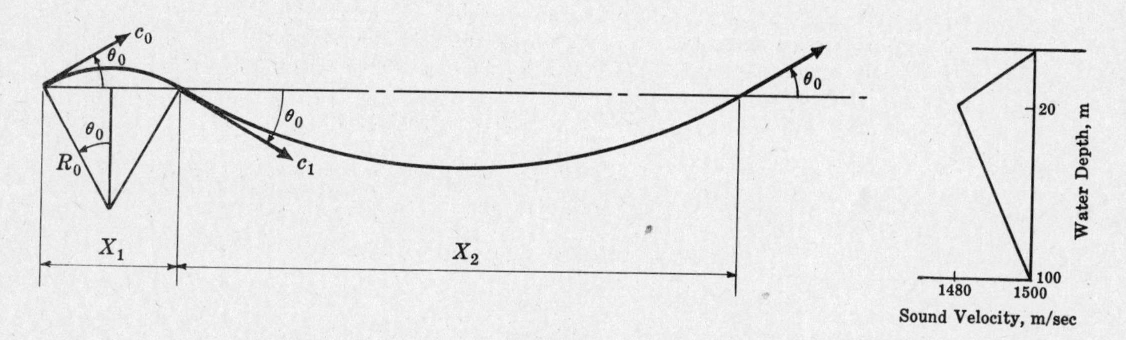

Fig. 8-7

(*a*) The constant velocity gradient in the upper channel is $g' = (1480 - 1500)/20 = -1.0$ m/sec per meter, and in the lower channel is $g'' = (1500 - 1480)/80 = 0.25$ m/sec per meter.

In water with constant velocity gradient, the horizontal distance traveled by a sound ray in reaching a depth d is $x = \sqrt{2c_0 d/g}$. For the upper and lower channels, we obtain respectively

$$x_0 = \sqrt{2c_0 d_0/g'} = \sqrt{2(1500)(20)/1.0} = 246 \text{ m}, \quad x_1 = \sqrt{2c_1 d_1/g''} = \sqrt{2(1500)(80)/0.25} = 980 \text{ m}$$

The radius R of the arc of a circle traveled by the sound ray is $R = c_0/g$. Thus in the upper and lower channels respectively, $R_0 = c_0/g' = 1500$ m, $R_1 = c_1/g'' = 6000$ m.

But $\sin\theta_0 = x_0/R_0 = 246/1500$ or $\theta_0 = 9.4°$, and $\sin\theta_1 = x_1/R_1 = 980/6000$ or $\theta_1 = 9.4°$. Therefore the maximum angle with which a sound ray may cross the axis of the channel in either direction and still remain within the channel is the same and is 9.4°. Also, a sound ray that once crosses the axis of the sound channel at $\theta_0 = 9.4°$ will continue to recross the axis at this same angle.

(*b*) The horizontal distances at the first and second crossings of the axis of the channel are

$$X_1 = 2R_0 \sin\theta_0 = 492 \text{ m}, \quad X_2 = 2R_1 \sin\theta_0 = 1960 \text{ m}$$

8.9. Referring to Problem 8.8, find (*a*) the time required for a sound ray to travel to the second crossing if it crosses the axis of the channel at the maximum angle, (*b*) the time required for a sound ray to travel the same distance as in part (*a*) along the axis of the channel, and (*c*) the difference in the time required.

(*a*) Along the axis of the channel the velocity of sound is at its minimum value of 1480 m/sec, so the time required to travel to the second crossing is $t_1 = x/c_m = (492 + 1960)/1480 = 1.66$ sec.

(*b*) The mean horizontal velocity of the sound ray crossing the axis of the channel at angle θ_0 is (see Problem 8.11)

$$c_x = c_m(1 + \theta_0^2/6) = 1480(1 + 0.164^2/6) = 1486.67 \text{ m/sec}$$

where $\theta_0 = 9.4° = 0.164$ rad. Hence $t_2 = (492 + 1960)/1486.67 = 1.65$ sec.

(*c*) The time difference for such a short distance is $t_1 - t_2 = 0.01$ sec. It is clear that for great distances the difference will be appreciable. Moreover, at 26.7 m below the axis of the channel and at 6.67 m above the axis of the channel, the speed of sound will equal the mean horizontal velocity of the sound ray crossing the axis at 9.4°.

8.10. Figure 8-8 shows the velocity profile of a portion of the sea. Determine the path of a sound wave traveling in it.

Assume the sound wave is initially horizontal at the surface of the sea. For the first layer of water from 0 to 200 m depth, the velocity of sound decreases linearly with depth. The velocity gradient is

$$g_1 = (1450 - 1500)/200 = -0.25 \text{ (m/sec)/m}$$

and the radius of the path is

$$R_1 = c_0/(-g_1) = 1500/0.25 = 6000 \text{ m}$$

Since $\theta_0 = 0$, the angle θ_1 this sound ray makes with the horizontal at the depth of 200 m is $\theta_1 = \cos^{-1} 1450/1500 = 15°$ and the horizontal distance it travels in reaching the second layer is

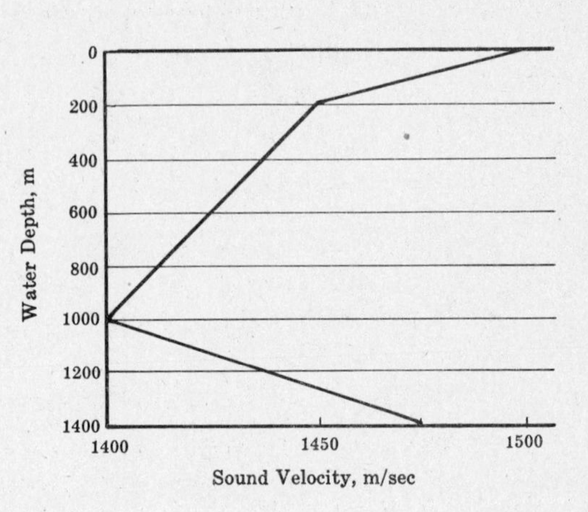

Fig. 8-8

$$x_1 = R_1 \sin\theta_1 = 6000 \sin 15° = 1550 \text{ m}$$

Similarly for the second layer of water,

$$g_2 = (1400 - 1450)/800 = -0.0625 \text{ (m/sec)/m}, \qquad R_2 = c_0/(-g_2) = 1500/0.0625 = 24{,}000 \text{ m}$$

$$\theta_2 = \cos^{-1} c_2/c_0 = \cos^{-1} 1400/1500 = 21°$$

and

$$x_2 = R_2(\sin 21° - \sin 15°) = 2400 \text{ m}$$

Below this depth of 1000 m, the temperature is constant. The velocity of sound, however, increases at a constant rate of 0.017 m/sec per meter increase in depth because of increasing hydrostatic pressure. The sound ray therefore bends along a radius

$$R_3 = c_0/(-g_3) = 1500/(-0.017) = -88{,}200 \text{ m}$$

Thus the sound ray will become horizontal at a depth of $(1500 - 1400)/0.017 + 1000 = 6890$ m, and

$$x_3 = R_3 \sin\theta_2 = 88{,}200 \sin 21° = 31{,}800 \text{ m}$$

Upon reaching this maximum depth of 6890 m and a velocity of 1500 m/sec, the sound ray begins an upward path similar to the downward path as shown in Fig. 8-9. The total horizontal distance traveled by this sound ray is

$$x = 2(x_1 + x_2 + x_3) = 2(1550 + 2400 + 31{,}800) = 72{,}000 \text{ m}$$

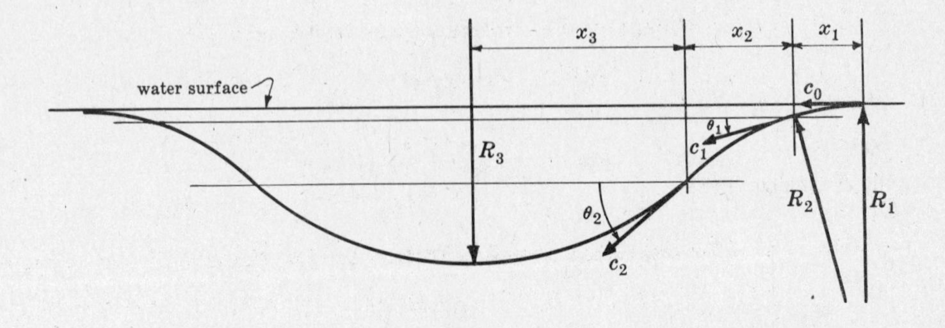

Fig. 8-9

8.11. Derive an expression for the mean horizontal velocity of sound rays crossing the axis of a sound channel at an angle θ_0.

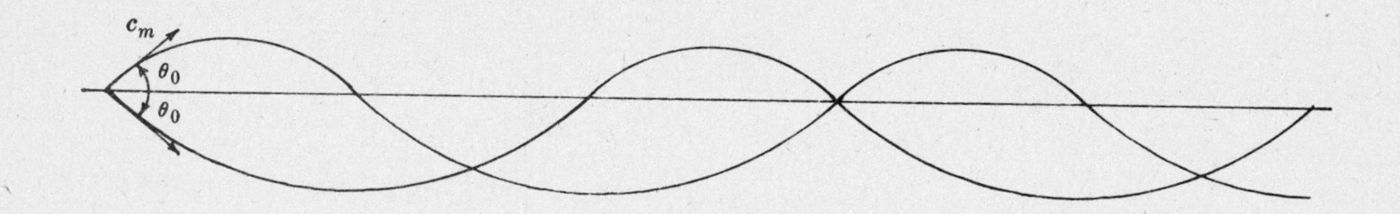

Fig. 8-10

At the axis of the sound channel, the velocity of sound is a minimum, i.e. $c = c_m$. At any other point in the sound channel, the velocity of sound is, by Snell's law, $c = (c_m \cos\theta)/(\cos\theta_0)$ and its horizontal component is $c_x = c\cos\theta = (c_m \cos^2\theta)/(\cos\theta_0)$. Hence the average value is

$$\bar{c}_x = \frac{c_m}{\cos\theta_0}\int_0^{\theta_0} \frac{1}{\theta_0}\cos^2\theta\,d\theta = \frac{c_m}{\theta_0\cos\theta_0}\left[\tfrac{1}{2}\theta + \tfrac{1}{2}\sin\theta\cos\theta\right]_0^{\theta_0} = \frac{c_m}{2}\left(\frac{1}{\cos\theta_0} + \frac{\sin\theta_0}{\theta_0}\right)$$

From their series expansions, $\sin\theta_0 = \theta_0 - \theta_0^3/6$, $\cos\theta_0 = 1 - \theta_0^2/2$, and so

$$\bar{c}_x = c_m(1 + \theta_0^2/6)$$

where θ_0 is in radians. Thus the mean horizontal velocity of sound rays crossing the axis of the sound channel is always greater than the minimum velocity of sound c_m at the axis of the channel.

SOUND TRANSMISSION LOSSES

8.12. For propagation of spherical acoustic waves through an unbounded and homogeneous body of sea water, derive an expression for the transmission loss in decibels due to divergence and absorption.

Because of divergence and absorption, the sound pressure amplitudes at distances r_1 and r_2 from the sound source can be written as

$$p_1 = \frac{P}{r_1}e^{-\alpha r_1}, \qquad p_2 = \frac{P}{r_2}e^{-\alpha r_2}$$

where P is the pressure amplitude at the sound source, and α is the absorption coefficient in nepers/m.

The sound pressure levels at these two points are

$$(\text{SPL})_1 = 20\log\frac{p_1}{p_0}\ \text{db}, \qquad (\text{SPL})_2 = 20\log\frac{p_2}{p_0}\ \text{db}$$

where p_0 is the reference sound pressure.

The difference in sound pressure level between these two points is

$$\begin{aligned}(\text{SPL})_1 - (\text{SPL})_2 &= 20\log\frac{P}{r_1 p_0}e^{-\alpha r_1} - 20\log\frac{P}{r_2 p_0}e^{-\alpha r_2} \\ &= 20\log\frac{r_2}{r_1} + 20\log e^{\alpha(r_2 - r_1)} \\ &= 20\log\frac{r_2}{r_1} + 8.7\alpha(r_2 - r_1)\ \ \text{db}\end{aligned}$$

If $r_1 = 1$ m, then the transmission loss from r_1 to r_2 or simply a distance r meters is

$$(\text{SPL})_1 - (\text{SPL})_2 = 20\log r_2 + 8.7\alpha(r_2 - 1)\ \ \text{db}$$

or

$$H = 20\log r + ar\ \ \text{db}$$

where a is the absorption constant in db/m for sound waves in sea water.

The spatial rate of transmission loss is

$$\frac{dH}{dr} = \frac{20}{2.3}\frac{d(\ln r)}{dr} + a = \frac{8.7}{r} + a$$

When the rate of transmission loss caused by divergence is equal to the rate of transmission loss caused by absorption, we have

$$\frac{dH}{dr} = 0 \quad \text{or} \quad r = r_c = \frac{8.7}{a}$$

where r_c is sometimes known as the *crossover range.*

ECHO-SOUNDING

8.13. Derive an expression for the intensity level of the returning wave in underwater echo ranging.

Underwater echo ranging is a process in which a sonar transducer scans the water until the emitted sound beam hits a submerged object. The object then produces an echo whose reception at the source can be made to give information about the object.

Let I'_s be the intensity at a distance of 1 m from the sound source; then the intensity at a distance of r meters from the same sound source is I'_s/r^2 watts/m².

If the underwater object is at a distance r meters from the sound source with a perfectly reflecting surface of cross-sectional area S m², the sound energy received by the object will be at the rate of I'_sS/r^2 watts. Assume the sound energy received by the object will be radiated back equally well in all directions, i.e. a sphere of area $4\pi r^2$. The sound intensity of the returning waves at the source is

$$I_e = \frac{I'_sS/r^2}{4\pi r^2} = \frac{I'_s}{r^4}(D/4)^2 \quad \text{watts/m}^2$$

where D is the diameter of the underwater object in meters, and $S = \pi r^2$ m². In decibel notation,

$$I_e = 10 \log I'_s/I_0 + 10 \log (D/4)^2 - \log r^4 = I_s + 20 \log D/4 - 40 \log r$$

where I_s is the intensity level of the transmitted signal at 1 m from the sound source, $20 \log D/4$ represents the transmission gain or target strength due to the reflection of the underwater object, and $40 \log r$ is the loss due to divergence.

The effects of directivity d, refraction $2A$ and absorption $2ar$ can be incorporated into the expression for the intensity level of the returning echo signal:

$$I_e = I_s + 20 \log D/4 + d - 40 \log r - 2A - 2ar \quad \text{db}$$

8.14. Determine the intensity level and sound pressure level of the returned echo from a submerged object of average diameter 40 m at a distance of 3000 m from a transmitting source. The sound source radiates 1500 watts of acoustic power at a frequency of 20 kc/sec in a beam of 20 db directivity index. The transmission anomaly is 10 db.

The intensity level of the returned echo from a submerged object in underwater echo ranging is given by (see Problem 8.13)

$$I_e = I_s + 20 \log D/4 + d - 40 \log r - 2A - 2ar = 24.8 \text{ db}$$

where $I_s = 1500/4\pi = 120$ watts/m² or 140 db re 10^{-12} watt/m², $20 \log D/4 = 20 \log 40/4 = 20$ db, $d = 20$ db, $40 \log r = 40 \log 3000 = 139.2$ db is the loss due to divergence, $2A = 10$ db, $2ar = 2(0.001)3000 = 6$ db is the absorption loss at 20 kc/sec frequency.

$$p = \sqrt{I\rho c} = \sqrt{(3.08 \times 10^{-10})(1{,}480{,}000)} = 2.14 \times 10^{-2} \text{ nt/m}^2$$

where $24.8 = 10 \log I/I_0$ or $I = 10^{-12}$ antilog $2.48 = 3.08 \times 10^{-10}$ watt/m². Hence

$$\text{SPL} = 20 \log (2.14 \times 10^{-2})/(2 \times 10^{-4}) = 82.1 \text{ db re 1 microbar}$$

8.15. A sonar transducer has a source level of 100 db re 1 microbar. Calculate the sound pressure level produced by the transducer at a distance of 4000 m.

At 1 m from the source, $I = W/4\pi = p^2/\rho c$ watts/m², where p is the effective sound pressure in nt/m², W is the total acoustic power output in watts, and $\rho c = 1{,}480{,}000$ rayls is the characteristic impedance of water. Thus $p = \sqrt{\rho c W/4\pi} = 344 W^{1/2}$ nt/m² and SPL $= 20 \log 10p = 20 \log 344 W^{1/2} = 71 + 10 \log W$ db or $W = 794$ watts.

At $r = 4000$ m, we have $I = W/4\pi r^2 = 3.95 \times 10^{-6}$ watt/m², $p = \sqrt{\rho c I} = 2.41$ nt/m², and SPL $= 20 \log 24.1 = 27.6$ db re 1 microbar.

If other losses are neglected, we have transmission loss due to divergence $H = 20 \log r = 20 \log 4000 = 72$ db, and SPL $= 100 - 72 = 28$ db at $r = 4000$ m.

8.16. A sonar transducer has an intensity level of 125 db re 1 microbar and generates output pulses of 0.05 second duration. It has a receiving directivity of 20 db while radiating acoustic energy at a frequency of 20 kc/sec. Compute the reverberation level produced by scatterers of a density and size $n\sigma = 10^{-5}$ per meter at a range of 2000 m from a submerged object.

The reverberation level produced by scatterers in sea water is

$$I_R \;=\; I_s + 10 \log n\sigma + 10 \log \tfrac{1}{2} c\Delta t - d - 20 \log r - 2ar \;=\; -11.3 \text{ db re 1 microbar}$$

where $I_s = 125$ db, $10 \log n\sigma = -50$ db, $10 \log \frac{1}{2} c\Delta t = 10 \log \frac{1}{2}(1480)(0.05) = 15.7$ db, $d = 20$ db, $20 \log r = 20 \log 2000 = 66$ db, and $2ar = 2(0.004)(2000) = 16$ db.

8.17. A sonar transducer produces an axial sound pressure level of 50 db re 1 microbar at a distance of 1000 m in sea water. If the absorption constant has a value of 0.01 db/m, find the axial sound pressure level at 1 m and at 2000 m. At what distance will the axial sound pressure level be reduced to 0 db? At what distance will the transmission loss caused by spherical divergence be equal to that caused by absorption?

Assume the transmission anomaly $A = 0$; then transmission loss in sea water due to spherical divergence and absorption is

$$H \;=\; 20 \log r + ar \text{ db}$$

and at a distance of 1000 m, $H = 20 \log 1000 + 0.01(1000) = 70$ db where $a = 0.01$ db/m is the absorption constant. Thus

$$(\text{SPL})_1 \;=\; 70 + 50 \;=\; 120 \text{ db re 1 microbar}$$

At 2000 m, $H = 20 \log 2000 + 0.01(2000) = 86$ db and so

$$(\text{SPL})_{2000} \;=\; 120 - 86 \;=\; 34 \text{ db re 1 microbar}$$

When the axial sound pressure level is zero, we have the total transmission loss of 120 db, i.e. $120 = 20 \log r + 0.01r$ or $r = 4700$ m.

Transmission loss caused by spherical divergence is $20 \log r$ while transmission loss caused by absorption is ar. If they are equal, we have $20 \log r = 0.01r$ or $r = 7800$ m.

When the rate of transmission loss due to spherical divergence equals the rate of transmission loss resulting from absorption, we have

$$\frac{dH}{dr} \;=\; \frac{20}{2.3}\frac{d(\ln r)}{dr} + a \;=\; 8.7/r + a \;=\; 0$$

Thus $r_c = 8.7/a = 8.7/0.01 = 870$ m.

UNDERWATER TRANSDUCERS

8.18. In order to collect more sound underwater, two microphones M_1 and M_2 are used with their tubes leading into the common tube C as shown in Fig. 8-11. If sound waves come from the left, find an expression for the length of tube A for maximum sound intensity received at C.

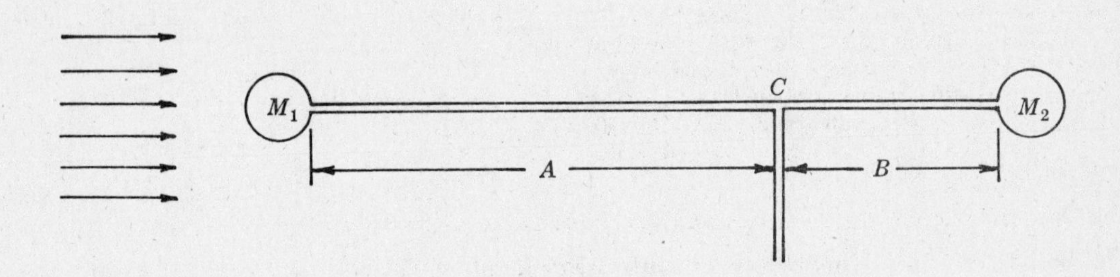

Fig. 8-11

Since sound waves come from the left, microphone M_1 will be excited first; sound propagates down tube A toward C with velocity c_a of sound in air. The remaining sound waves travel through water with velocity c_w of sound in water and excite microphone M_2. The resulting sound waves then travel down tube B toward C with velocity c_a.

For maximum sound intensity received at C, sound waves coming from microphones M_1 and M_2 should arrive at C in phase, i.e.

$$\frac{A}{\lambda_a} = \frac{A+B}{\lambda_w} + \frac{B}{\lambda_a}$$

where λ_a and λ_w are the wavelengths of sound in air and in water respectively. Thus

$$A = \frac{B(\lambda_a + \lambda_w)}{\lambda_w - \lambda_a} \quad \text{or} \quad A = \frac{B(c_a + c_w)}{c_w - c_a}$$

where $c_a = f\lambda_a$ and $c_w = f\lambda_w$.

8.19. A sonar transducer has a maximum detection range of 4000 m operating at 20 kc/sec frequency on a given submerged object. Determine its new maximum detection range if (*a*) the source level is increased by 30 db, (*b*) the operating frequency is reduced to 10 kc/sec.

(*a*) The general expression for returned echo signal level is

$$I_e = I_s + T - 2H \text{ db}$$

where I_s is the source strength, T is the target strength, and $H = 20 \log r + ar$ is the loss due to divergence and absorption. For the initial 4000 m range,

$$I_e = I_s + T - 2[20 \log 4000 + 0.00373(4000)] = I_s + T - 2(86.9) \text{ db}$$

where $a = 0.00373$ db/m is the absorption coefficient at 20 kc/sec.

Now the source strength I_s is increased by 30 db while the echo and target strengths remain the same. Then $2H = 2(86.9) + 30$ and

$$H = 20 \log r + 0.00373r = 86.9 + 15 = 101.9 \quad \text{or} \quad r = 6700 \text{ m}$$

(*b*) At 10 kc/sec, the absorption coefficient is found to be $a = 0.001$ db/m. Then

$$H = 20 \log r + 0.001r = 86.9 \quad \text{or} \quad r = 8100 \text{ m}$$

8.20. Sound waves are produced at a depth d below the surface of the sea. Derive an expression for the intensity at point P a distance L from the source S as shown in Fig. 8-12.

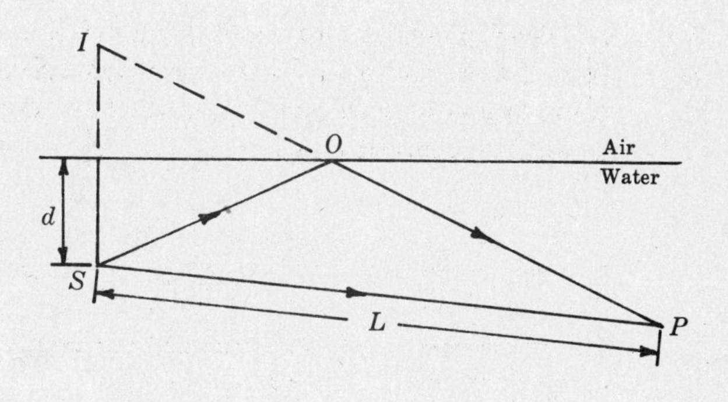

Fig. 8-12

For a homogeneous medium, sound waves reach P via two paths: SP directly from the source S, and SOP after reflection at O on the boundary surface. From the acoustic mirror phenomenon, the sound ray OP appears to come from the acoustic image I directly opposite the source S. The total effect at P is therefore the sum of the direct and reflected waves.

Let p_1 be the acoustic pressure at P due to the direct wave alone,

$$p_1 = P_1 \cos(\omega t - \theta_1) \text{ nt/m}^2$$

where P_1 is the pressure amplitude, and $\theta_1 = \omega L/c$ is the phase difference between the pressure at S and that at P. Similarly let p_2 be the acoustic pressure at P due to the reflected wave alone,

$$p_2 = P_2 \cos(\omega t - \theta_2 - 180°) \text{ nt/m}^2$$

where $\theta_2 = \omega(IP)/c$ is the phase difference between the pressure at the image I and the receiver P, and 180° is the phase change due to reflection at the interface (from water to air). The resultant pressure at P is therefore

$$p = p_1 + p_2 = P_1 \cos(\omega t - \theta_1) + P_2 \cos(\omega t - \theta_2 - 180°)$$

Now

$$P_1 \cos(\omega t - \theta_1) = P_1 \cos\omega t \cos\theta_1 + P_1 \sin\omega t \sin\theta_1$$

$$P_2 \cos(\omega t - \theta_2 - 180°) = P_2 \cos\omega t \cos(\theta_2 + 180°) + P_2 \sin\omega t \sin(\theta_2 + 180°)$$

$$\cos(\theta_2 + 180°) = -\cos\theta_2$$

$$\sin(\theta_2 + 180°) = -\sin\theta_2$$

We have

$$p = \cos\omega t(P_1 \cos\theta_1 - P_2 \cos\theta_2) + \sin\omega t(P_1 \sin\theta_1 - P_2 \sin\theta_2)$$

or

$$p = P\cos(\omega t - \phi)$$

where $P = \sqrt{A^2 + B^2}$, $\phi = \tan^{-1}(B/A)$, $A = P_1 \cos\theta_1 - P_2 \cos\theta_2$, and $B = P_1 \sin\theta_1 - P_2 \sin\theta_2$.

The intensity of the resultant radiation at P becomes

$$I = p^2/2\rho c = (A^2 + B^2)/2\rho c \text{ watts/m}^2$$

where ρc is the characteristic impedance of water in rayls.

But

$$A^2 = P_1^2 \cos^2\theta_1 + P_2^2 \cos^2\theta_2 - 2P_1P_2 \cos\theta_1 \cos\theta_2$$

$$B^2 = P_1^2 \sin^2\theta_1 + P_2^2 \sin^2\theta_2 - 2P_1P_2 \sin\theta_1 \sin\theta_2$$

$$A^2 + B^2 = P_1^2 + P_2^2 - 2P_1P_2 \cos(\theta_1 - \theta_2)$$

and finally

$$I = \frac{P_1^2 + P_2^2 - 2P_1P_2\cos(\theta_1 - \theta_2)}{2\rho c}$$

It is convenient to express this intensity in terms of the intensity $I_0 = p_1^2/2\rho c$ produced at P by the direct radiation from source S. Then

$$I = \frac{p_1^2}{2\rho c}\left[\frac{P_1^2}{P_1^2} + \frac{P_2^2}{P_1^2} - \frac{2P_1P_2\cos(\theta_1 - \theta_2)}{P_1^2}\right] = I_0[1 + R^2 - 2R\cos(\theta_1 - \theta_2)]$$

where $R = P_2/P_1$ is the ratio of the pressure amplitudes due to reflected and direct waves.

Depending on the values of the phase angles θ_1 and θ_2, $\cos(\theta_1 - \theta_2)$ will fluctuate between -1 and $+1$. The resultant intensity is seen to fluctuate between $I_0(1+R)^2$ and $I_0(1-R)^2$. Furthermore, if the source and the receiver are close together near the surface, the phase angles are essentially zero and R approximates unity; then the resultant intensity will fluctuate between zero and $4I_0$.

8.21. An underwater magnetostrictive sound transducer has a nickel rod of radius 0.01 m and length 0.1 m. What is the frequency of vibration of the transducer?

When magnetostrictive effect is employed for transducer action, the rod is driven to vibrate longitudinally at its fundamental natural frequency,

$$f_0 = 1/2L\sqrt{Y/\rho} = 24{,}500 \text{ cyc/sec}$$

where $L = 0.1$ m is the length of the rod, $Y = 2.1 \times 10^{11}$ nt/m² is Young's modulus for nickel, and $\rho = 8.78 \times 10^3$ kg/m³ is the density of nickel. (Transducers are usually driven at their fundamental resonant frequencies, i.e. those of the rod, in order to obtain maximum efficiency.)

8.22. Obtain an expression for the resultant voltage E generated by two small omnidirectional hydrophones whose electrical outputs are connected in series as shown in Fig. 8-13. Sound waves are incident on a perpendicular plane through the center of the line connecting the hydrophones.

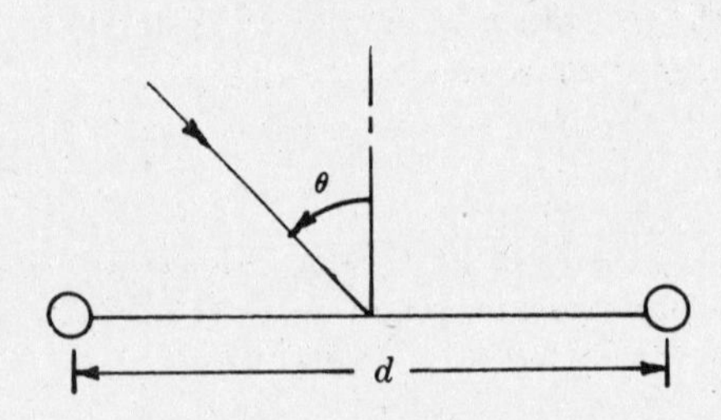

Fig. 8-13

For an angle of incidence θ, the phase difference of the hydrophones is given by (see Fig. 8-13)

$$\phi = (2\pi d/\lambda) \sin\theta = kd \sin\theta$$

where $k = \omega/c = 2\pi/\lambda$ is the wave number, and λ is the wavelength.

Now the voltage generated is $E_\theta = E_0 \cos\frac{1}{2}\phi$, where E_0 is the voltage generated when $\theta = 0$. Thus

$$E_\theta = E_0 \cos\tfrac{1}{2}kd\sin\theta$$

This result can also be obtained from the general expression for voltage E_θ generated by a line array consisting of n equally spaced small omnidirectional hydrophones whose electrical outputs are connected in series as (see Problem 5.28)

$$E = E_0 \frac{\sin(\frac{1}{2}knd\sin\theta)}{n\sin(\frac{1}{2}knd\sin\theta)}$$

Here $n = 2$ and so
$$E = E_0 \frac{\sin(kd\sin\theta)}{2\sin(\frac{1}{2}kd\sin\theta)} = E_0\cos(\tfrac{1}{2}kd\sin\theta)$$

since $\sin(kd\sin\theta) = \sin 2(\frac{1}{2}kd\sin\theta) = 2\sin(\frac{1}{2}kd\sin\theta)\cos(\frac{1}{2}kd\sin\theta)$.

8.23. The observed frequency of a returned echo signal from a submarine is 40,400 cyc/sec while the driving frequency supplied to the sonar transducer aboard a destroyer is 40,000 cyc/sec. If the destroyer is speeding at 40 knots, find the speed of the submarine.

Because of the Doppler effect, we can express the observed frequency as

$$f' = f(1 + 2v_{s/r}/c) \text{ cyc/sec}$$

where f is the actual frequency produced at the source in cyc/sec, $v_{s/r}$ is the relative speed between source and receiver in m/sec, and $c = 1500$ m/sec is the speed of sound in water. Then

$$40{,}400 = 40{,}000(1 + 2v_{s/r}/1500) \quad \text{or} \quad v_{s/r} = 7.5 \text{ m/sec}$$

and the speed of the submarine is $v_r = v_s - v_{s/r} = 20.4 - 7.5 = 12.9$ m/sec or 25.2 knots.

CAVITATION

8.24. Compute the maximum sound pressure level allowed in water without causing cavitation.

Cavitation results from negative instantaneous pressure in water. Because water supports very little tension, it will break away forming turbulences and eddies when pressure becomes negative. Now total pressure at a point in water is equal to atmospheric plus the hydrostatic pressure of the water. Atmospheric pressure is approximately $p = 10^5$ nt/m² or $p_{rms} = 10^5/\sqrt{2} = 0.7 \times 10^5$ nt/m², and so SPL $= 20 \log (0.7 \times 10^5/0.0002) = 116.9$ db re 1 microbar.

In order to prevent cavitation, a hydrophone should not produce sound pressure amplitude greater than the instantaneous pressure (atmospheric + hydrostatic) it is subjected to. The minimum value is 116.9 db re 1 microbar at the surface of the water and increases with depth because of increasing hydrostatic pressure. On the other hand, cavitation is sometimes deliberately induced for the destruction of liquid-borne organisms, in the dispersion of liquid-borne particles, the production of colloidal suspensions and emulsions, and the cleaning of metal parts.

8.25. A vehicle in water is at a depth of 30 m and moves at a speed of 30 knots. What is the required cavitation number such that cavitation will not take place?

$$\text{Cavitation number } \sigma = \frac{2(p_0 - p_v)}{\rho v^2} = 3.14$$

where $p_0 = 10^5 + 30(3.28)^3(62.4)/0.225 = 392{,}000$ nt/m² is the ambient pressure (atmospheric + hydrostatic pressure), $p_v = 2400$ nt/m² is the vapor pressure of water at 20°C, $\rho = 1061$ kg/m³ is the density of water at 20°C, and $v = 30$ knots or 15.3 m/sec is the speed of the vehicle in water.

Supplementary Problems

UNDERWATER TRANSMISSION

8.26. A simple underwater sound source radiates 10 watts of acoustic power at a frequency of 500 cyc/sec. Find the acoustic intensity and sound pressure at a distance of 5 m from the source.
Ans. $I = 0.032$ watt/m², $p = 22$ nt/m²

8.27. For plane acoustic waves in water, show that SPL = IL if the pressure fluctuations and particle displacements are in phase.

8.28. Show that low frequency sound waves are better than high frequency sound waves for underwater communications.

8.29. Sound pressure level for underwater acoustics is usually given 1 microbar as the reference pressure. Compared to the usual sound reference pressure of 0.00002 nt/m², what will be the sound pressure level? *Ans.* 74 db higher

8.30. Transmission anomaly produced by increased divergence at layer depth is often given by

$$A = 20 \log \frac{\sin \theta_1 + \sin \theta_2}{2 \sin \theta_1} \text{ db}$$

where θ_1 is the angle of incidence and θ_2 is the angle of transmittance. If $c_1 = 1500$ m/sec, $c_2 = 1450$ m/sec, and $\theta_1 = 5°$, find the value of transmission anomaly. *Ans.* 6 db

REFRACTION

8.31. Prove that in water having a constant negative velocity gradient, the path of sound waves will be refracted downward.

8.32. A submarine is at a depth of 180 m where the velocity of sound is 1500 m/sec. Its sonar transducer detects a surface vessel at an upward angle of 10° with the horizontal. What is the horizontal range of the vessel from the submarine? *Ans.* 800 m

8.33. A sonar transducer operating at a 50 kc/sec frequency has a source strength of 140 db. What will be the echo signal strength returned from a spherical object of 40 m radius at a distance of 1000 m? *Ans.* 16 db

UNDERWATER TRANSDUCERS

8.34. Three hydrophones A, B, C shown in Fig. 8-14 are used to detect the location of a submerged object in water. By observing the time of arrival of sound from the object O, show that the position of the object is given by the intersection of two hyperbolas with A and B, B and C as foci respectively.

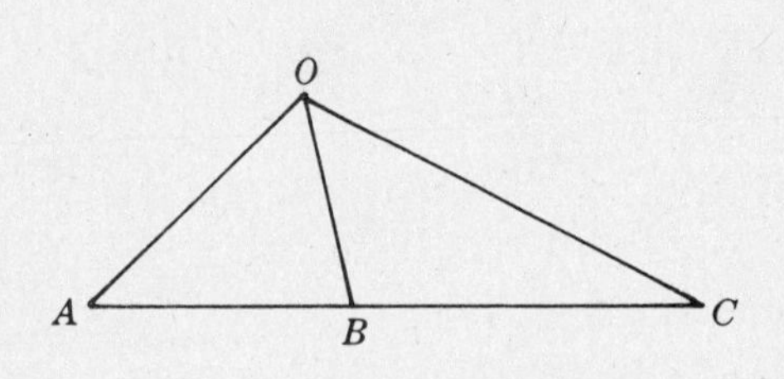

Fig. 8-14

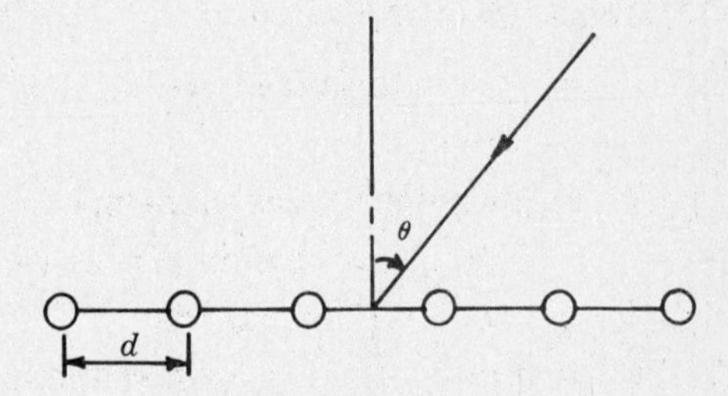

Fig. 8-15

8.35. Prove that the resultant voltage E_θ generated by a line array consisting of n equally spaced small omnidirectional hydrophones whose electrical outputs are connected in series is given by

$$E_\theta = E_0 \frac{\sin(\frac{1}{2}nkd \sin\theta)}{n \sin(\frac{1}{2}kd \sin\theta)}$$

where E_0 is the voltage generated when $\theta = 0$ and k is the wave number. (See Fig. 8-15.)

8.36. If sound waves of frequency 10 kc/sec are incident at an angle of 10° to the normal of a line array of six omnidirectional hydrophones spaced equally at 0.1 m apart, what will be the electrical phase difference between the signals produced in adjacent hydrophones? The electrical outputs of all hydrophones are connected in series. *Ans.* 40°

8.37. Show that the Doppler effect will give rise to the following expression for frequency received from a submerged object by the sonar transducer aboard a surface vessel:

$$f' = f(1 + 2V/c)$$

where f' is the observed frequency, f the actual frequency, V the speed between the object and the sonar transducer, and c the speed of sound in water.

8.38. Given the velocity V_d of a destroyer and the angle θ_d it makes with the line to a submarine as shown in Fig. 8-16. The speed of sound in water is c, and the frequencies of the source and the returning echo are f_d and f_s respectively. Find an expression for the velocity of the submarine.

Ans. $$V_s = \frac{2f_d V_d \cos\theta_d - c(f_s - f_d)}{2f_d \cos\theta_s}$$

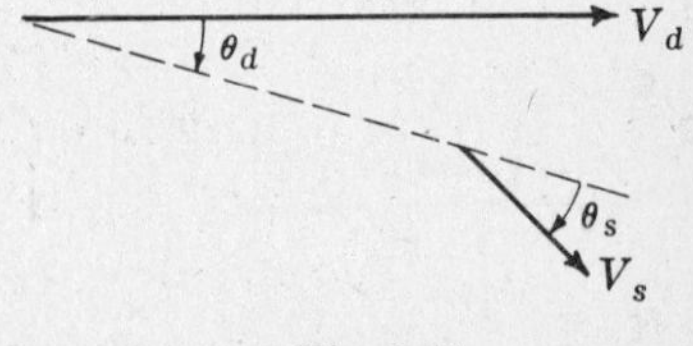

Fig. 8-16

8.39. The sonar transducer aboard a destroyer radiates sound waves of frequency 20 kc/sec. If the relative speed between the destroyer and a submarine is 7.5 m/sec, find the frequency observed by the sonar transducer. *Ans.* $f' = 20{,}200$ cyc/sec

CAVITATION

8.40. Compute the minimum power per unit area required to produce cavitation at (*a*) the surface of the sea, (*b*) a depth of 30 m. *Ans.* (*a*) 3000 watts/m^2, (*b*) 50,000 watts/m^2

8.41. In order to prevent cavitation, a hydrophone should not produce sound pressure amplitude greater than the hydrostatic pressure it is subjected to. For a hydrostatic pressure of 100,000 nt/m^2, compute the highest sound intensity allowed. *Ans.* $I = 1690$ watts/m^2

Chapter 9

Ultrasonics

NOMENCLATURE

a = acceleration, m/sec^2
A = area, m^2
B = flux density, webers/m^2
c = speed of sound in air, m/sec
C_0 = capacitance, farads
d_{12} = piezoelectric strain coefficient, m/volt
E = voltage, volts
e_{11} = piezoelectric stress coefficient, coulombs/m^2
f = frequency, cyc/sec
I = sound intensity, watts/m^2
k = coefficient of electromechanical coupling
K = proportional material constant, m^4/weber^2
L = length, m; inductance, henrys
p = acoustic pressure, nt/m^2
Q = quality factor
r = radius, m
R = resistance, ohms
s_{22} = compliance coefficient, m^2/nt
S = area, m^2
t = thickness, m; time, sec
Y = Young's modulus of elasticity, nt/m^2
W = sound power, watts
Z = impedance, rayls
ω = circular frequency, rad/sec
ρ = density, kg/m^3
λ = wavelength, m
α_t = sound power transmission coefficient
σ = stress, nt/m^2
ϕ = transformation factor, coulomb/m
ϵ_x = clamped dielectric constant
ϵ_0 = permittivity of free space, farads/m
μ = Poisson's ratio
μ_i = incremental permeability of the material, henrys/m
μ_0 = permeability of free space, henrys/m
Λ = magnetostriction constant

INTRODUCTION

Ultrasonics is the study of sound waves of frequencies higher than the upper hearing limit of the human ear (frequency region above 20 kc/sec) and has become synonymous with the applications and effects of ultrasonic vibration for other purposes than the excitation of the hearing mechanism. In fact, ultrasonic energy has been applied to gases, liquids and solids to produce desired changes and effects or to improve a product or a process.

The upper frequency limit for the propagation of ultrasonic waves is thermal lattice vibrations beyond which the material cannot follow the input sound. The smallest wavelength of sound is therefore twice the interatomic distance, and for metal this is approximately equal to 2×10^{-10} m. This occurs at a frequency of 1.25×10^{13} cyc/sec which corresponds to the twenty-first harmonic of a 10-megacycle quartz crystal. At such high frequencies, ultrasonic wave periods become comparable with relaxation time.

High-amplitude ultrasonic waves are sometimes called *sonic*, and *hypersound* refers to waves having frequencies greater than 10^{13} cyc/sec.

WAVE TYPES

Rayleigh surface waves propagate over a surface without influencing the bulk of the medium below the surface. They are produced from unbalanced forces at the surface of a solid and generate an elliptical motion of the medium whose amplitude decreases exponentially as the depth below the surface increases. Ultrasonic Rayleigh waves can be propagated along the surface of the test object to detect flaws or cracks on or near the surface of the test object.

Waves produced in a thin plate whose thickness is comparable to the wavelength are known as *Lamb waves*. They are very complex waves, moving in asymmetrical or symmetrical modes, and are employed to locate nonbonded areas in laminated structures, radial cracks in tubing, and for quality control of sheet and plate stock.

ULTRASONIC TRANSDUCERS

Basically there are three types of ultrasonic transducers: (1) *gas-driven* transducers, e.g. whistles, sirens; (2) *liquid-driven* transducers, e.g. hydrodynamic oscillators, vibrating blade transducer; and (3) *electromechanical* transducers, e.g. piezoelectric and magnetostrictive transducers. They are classified according to the form of energy used to excite them into mechanical vibration and the medium into which the wave is to be propagated.

The *quality factor* Q (or the quality of resonance) of a system determines the frequency responses of the system, i.e. for a low Q the frequency bandwidth is wide and for a high Q the frequency bandwidth is narrow. The magnification of ultrasonic transducer is approximately equal to the quality factor Q. (For water, quartz, and water, $Q = 7$.)

PIEZOELECTRIC TRANSDUCERS

If an alternating electric field is applied along the axis of a piezoelectric crystal, the latter will expand and contract along the axis (see Fig. 9-1). As the frequency of the applied electric field approaches the natural frequency of any longitudinal mode of vibration of the crystal, the amplitude of the resulting mechanical vibration of the crystal becomes significantly large. These types of mechanical vibration from piezoelectric crystals have been utilized to produce ultrasonic vibrations at frequencies ranging from 5 kc/sec to 200 kc/sec.

Piezoelectric transducers are usually made from quartz, tourmaline, Rochelle salt, ammonium dihydrogen phosphate, barium titanate, and ceramics having strong ferroelectric properties.

These transducers provide stable ultrasound of narrow bandwidth over a wide range of frequencies. Some piezoelectric materials, however, are hygroscopic and are incapable of sustaining high power densities without fracture, and some exhibit instabilities. (See Problems 9.1-9.8.)

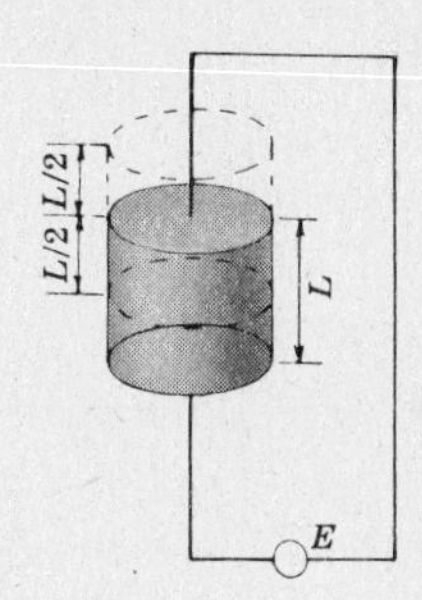

Fig. 9-1. Piezoelectric transducer

MAGNETOSTRICTIVE TRANSDUCERS

When a rod of magnetizable material is exposed to a magnetic field which varies in magnitude, the rod changes in length. An alternating current passing through a coil surrounding such a rod will cause it to vibrate longitudinally. These small forced vibration amplitudes will increase very greatly if the frequency of the applied current coincides with one of the normal longitudinal modes of vibration of the rod. Ultrasonic sound of this frequency is radiated.

Magnetostrictive transducers are usually made from alloys of iron, nickel, and cobalt. They are mechanically rugged and capable of producing large acoustical power with fairly good efficiency, e.g. 60%. Their deficiencies are low upper frequency limit because of extreme length required and conversion losses due to hysteresis and eddy-currents. (See Problems 9.9-9.12.)

ELECTROMAGNETIC TRANSDUCERS

Electromagnetic transducers, like most of the loudspeakers, generate ultrasound from the movement of a coil carrying a varying voltage in a magnetic field of constant intensity. (See Chapter 5.)

ABSORPTION

The absorption of ultrasonic energy by gases is due to viscosity effect and heat conduction. However, the delay in attainment of equilibrium between translational, rotational and vibrational energy of molecules also plays an important role in the absorption of ultrasonic energy.

In solids, the absorption of ultrasonic waves may be attributed to lattice imperfection, ferromagnetic and ferroelectric properties, electron-photon interactions, thermal effects, grain boundary losses, thermoelastic and structural relaxation, acoustoelectric effect, and nuclear magnetic resonance.

Ultrasound can be propagated to much greater distances in water and at much higher frequencies than in gases and solids. Attenuation and absorption of ultrasonic waves in water are comparatively low.

APPLICATIONS

As one of the important *nondestructive* testing methods, ultrasonics plays an essential part in flaw detection, process improvement, control and monitoring, and measurement of mechanical, physical, chemical and metallurgical properties of materials.

By means of a transducer, ultrasonic energy is converted into high frequency mechanical vibration of the medium through a proper coupling element such as a horn.

In industry, ultrasonics is widely used for metal processing such as solidification, precipitation, agglomeration, emulsification, dispersion. Many different types of ultrasonic devices, generators and detectors are currently being used.

In medicine, ultrasonics is employed for tumor detection, biological measurements and diagnostic work.

In underwater applications, ultrasonics is employed to measure water depth in the mapping of the ocean floor, and to detect submerged objects such as fish, submarines and mines.

Ultrasonics is also used for traffic control, fabric cleaning, aging of wines, packing of cement, counting and sorting, and dispersion of fog. (See Problems 9.14-9.17.)

Solved Problems

PIEZOELECTRIC TRANSDUCERS

9.1. An X-cut quartz crystal of thickness 0.001 m is vibrating at resonance. Find the fundamental frequency.

For longitudinal wave motion, $c = \sqrt{Y/\rho} = \sqrt{7.9 \times 10^{10}/2650} = 5460$ m/sec. Since $t = \frac{1}{2}\lambda = \frac{1}{2}(c/f)$,

$$f_1 = c/2t = 5460/0.002 = 2730 \text{ kc/sec}$$

9.2. An X-cut quartz piezoelectric transducer is to be operated in contact with water and with air. Determine the maximum intensity at resonance.

The maximum intensity at resonance is given by

$$I_{\text{max}} = \tfrac{1}{2}\rho c \left[\frac{\sigma_{\text{max}}}{(\rho c)_Q}\right]^2 \quad \text{watts/m}^2$$

where ρc is the characteristic impedance of the medium in contact with the transducer, $\sigma_{\text{max}} = 7600$ nt/m^2 is the maximum stress allowable for quartz, and $(\rho c)_Q = 14.5 \times 10^6$ rayls is the characteristic impedance of X-cut quartz.

When in contact with water, $\rho c = 1.48 \times 10^6$ and $I_{\text{max}} = 0.2$ watt/m^2.

When in contact with air, $\rho c = 415$ rayls and $I_{\text{max}} = 0.58$ watt/cm^2.

9.3. Determine the maximum acceleration and displacement of a quartz ultrasonic transducer radiating sound of 5 watts/cm^2 intensity and 20,000 cyc/sec frequency into water.

$$\text{Sound intensity } I = p^2/2\rho c \text{ watts/m}^2$$

or sound pressure $p = \sqrt{2\rho c I} = \sqrt{2(1.48 \times 10^6)(5 \times 10^4)} = 3.85 \times 10^5$ nt/m^2.

Since force $= pA = ma = t\rho A a$, the maximum acceleration is

$$a_{\text{max}} = p/\rho t = (3.85 \times 10^5)/2650(0.001) = 1.45 \times 10^5 \text{ m/sec}^2$$

where $t = 0.001\ m$ is the thickness of the quartz.

The corresponding maximum displacement is

$$d_{\text{max}} = a/\omega^2 = (1.45 \times 10^5)/[(6.28)(20,000)]^2 = 9.4 \times 10^{-6} \text{ m}$$

9.4. A plated X-cut quartz crystal of dimensions $L_x = 0.001$ m, $L_y = 0.02$ m, $L_z = 0.005$ m is used as a longitudinal ultrasonic transducer. Find the longitudinal strain in the unstrained crystal when 120 volts is applied between the plated surfaces. If the crystal is constrained so that it cannot move longitudinally, find the resulting stress.

The simplified equation for longitudinal piezoelectric vibrators is given by

$$\partial\eta/\partial y = -s_{22}F_y/S_y + d_{12}E_x/L_x$$

where $\partial\eta/\partial y$ = the longitudinal strain,

s_{22} = 1.27×10^{-11} m²/nt is the compliance coefficient,

F_y = the compressional force in newtons in the y direction,

S_y = L_xL_z = 0.000005 m² is the cross-sectional area,

d_{12} = 2.3×10^{-12} m/volt is the piezoelectric strain coefficient,

E_x = 120 volts is the applied voltage.

For the unconstrained crystal, $F_y = 0$; the longitudinal strain is

$$\partial\eta/\partial y = (2.3 \times 10^{-12})(120)/0.001 = 2.76 \times 10^{-7}$$

and when the crystal is constrained the resulting stress is

$$F_y/S_y = (2.76 \times 10^{-7})/(1.27 \times 10^{-11}) = 2.18 \times 10^{4} \text{ nt/m}^2$$

9.5. If the crystal of Problem 9.4 is fastened to a rigid backing plate at one end and radiating sound into water at the other end, find its fundamental frequency of longitudinal vibrations. Determine the acoustic power radiated when the crystal is driven at its fundamental frequency by an rms voltage of 120 volts.

The fundamental frequency $f_1 = c_y/4L_y = 68{,}100$ cyc/sec and the acoustic power radiated $W = \phi^2E^2/\rho cS_y = 1.6 \times 10^{-3}$ watts, where $c_y = 5450$ m/sec is the longitudinal wave velocity of sound in quartz, $L_y = 0.02$ m and $L_z = 0.005$ m are the dimensions of the crystal, $\phi = d_{12}L_z/22 = 9.1 \times 10^{-4}$ coulomb/m is the transformation factor for the crystal, $E = 120$ volts, $c = 1.48 \times 10^6$ rayls is the characteristic impedance of water, and $S_y = 5 \times 10^{-6}$ m² is the cross-sectional area of the crystal.

9.6. Compare the quality factor Q of a longitudinally vibrating quartz crystal radiating ultrasound into water and into air.

$$Q_{\text{water}} = \frac{\pi(\rho c)_{\text{quartz}}}{4(\rho c)_{\text{water}}} = \frac{3.14(1.45 \times 10^7)}{4(1.48 \times 10^6)} = 7.5$$

$$Q_{\text{air}} = \frac{\pi(\rho c)_{\text{quartz}}}{4(\rho c)_{\text{air}}} = \frac{3.14(1.45 \times 10^7)}{4(415)} = 27{,}500$$

The very large value of Q_{air} indicates that the resonance curve of a quartz crystal driven in air should be sharply peaked.

9.7. An X-cut quartz crystal has dimensions $L_x = 0.001$ m, $L_y = 0.02$ m, $L_z = 0.005$ m. The crystal is used in an air-back electroacoustic ultrasonic transducer. Find the elements of the equivalent circuit.

The circuit components near resonance for an air-back quartz electroacoustic transducer are determined as follows (see Fig. 9-2).

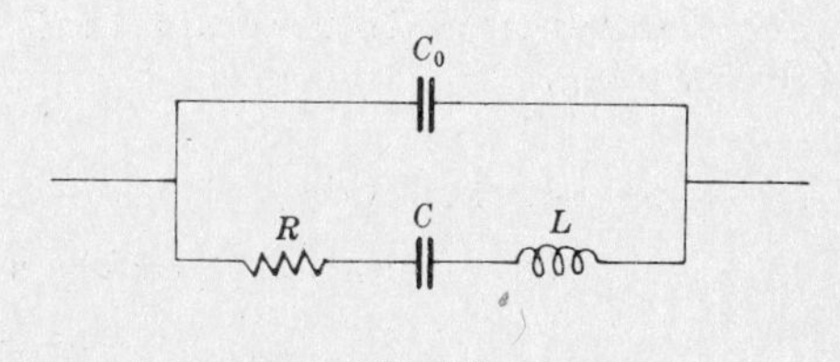

Fig. 9-2

$$C_0 = \epsilon_x'\epsilon_0L_yL_z/L_x = 4.45(8.85 \times 10^{-12})(0.02)(0.005)/0.001 = 3.94 \times 10^{-12} \text{ farads}$$

where $\epsilon_x' = 4.45$ is the clamped dielectric constant, $\epsilon_0 = 8.85 \times 10^{-12}$ farads/m is the permittivity of free space.

$$R = \rho_0 c_0 S_y/\phi^2 = 2500 \text{ ohms}$$

where $\rho_0 c_0 = 415$ rayls is the characteristic impedance of air, $\phi = d_{12}L_z/s_{22} = 9.1 \times 10^{-4}$ coulomb/m is the transformation factor, $d_{12} = 2.3 \times 10^{-12}$ m/volt is the piezoelectric strain coefficient, $s_{22} = 1.27 \times 10^{-11}$ m²/nt is the compliance coefficient, and $S_y = 0.001(0.005)$ m² is the cross-sectional area.

$$C = 8\phi^2 s_{22} L_y/\pi^2 S_y = 3.44 \times 10^{-14} \text{ farads} \quad \text{and} \quad L = L_y S_y = 2\phi^2 = 160 \text{ henrys}$$

where $\rho = 2650$ kg/m³ is the density of quartz.

9.8. The dimensions of an air-back barium titanate transducer are $L_x = 0.01$ m, $L_y = 0.02$ m, $L_z = 0.03$ m. Determine the fundamental frequency of this transducer and the acoustic power produced into water if 100 volts is applied.

The fundamental frequency is $f_1 = c_x/2L_x = 5200/[2(0.01)] = 260$ kc/sec and the acoustic power produced is $W = \phi^2 E_x^2/\rho c S_x = 41.4$ watts, where $\phi = 2e_{11}S_x/L_x = 1.92$ coulombs/m is the transformation factor, $\rho c = 1.48 \times 10^6$ rayls is the characteristic impedance of water, $e_{11} =$ 16 coulombs/m² is the piezoelectric stress coefficient for barium titanate, and $S_x = 0.02(0.03)$ m² is the area.

MAGNETOSTRICTIVE TRANSDUCERS

9.9. A magnetostrictive transducer is made of a duraluminum rod of length 0.13 m and diameter 0.015 m. It is supported at its center as shown in Fig. 9-3. Find its fundamental frequency of longitudinal vibration.

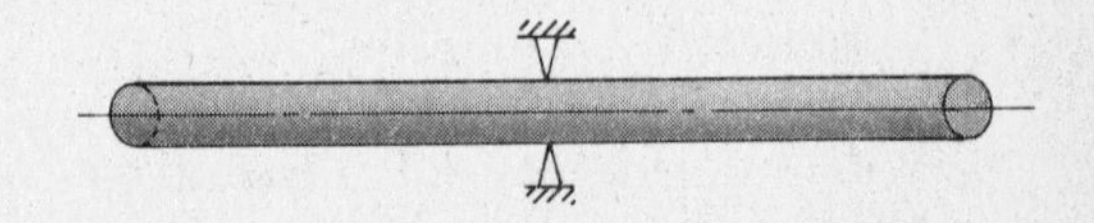

Fig. 9-3

The fundamental frequency $f_1 = \dfrac{1-(\mu\pi r)^2}{8L^3}\sqrt{\dfrac{Y}{\rho}} = 250$ kc/sec where $\mu = 0.31$ is Poisson's ratio, $Y = 21 \times 10^{10}$ nt/m² is Young's modulus, $r = 0.0075$ m is the radius of the rod, $L = 0.13$ m is the length of the rod, and $\rho = 8800$ kg/m³ is the density of the rod.

9.10. A magnetostrictive hydrophone is made of a nickel rod of length 0.2 m clamped at the center. Compute its fundamental frequency of longitudinal vibration.

Since the rod is clamped at the center and is vibrating at the fundamental mode, there must be a node at the center and antinodes at the free ends.

Wavelength $\lambda = 2L = 2(0.2) = 0.4$ m, and speed of wave propagation $c = f\lambda$. Then $f_1 = c/\lambda_1 = 4900/0.4 = 12{,}250$ cyc/sec.

9.11. A magnetostrictive steel vibrator is used as a drilling driver. The cross-sectional area is 0.0004 m² and the maximum allowable strain is 8×10^{-4}. What is the maximum driving force at the end of the driver?

$$\text{Maximum driving force} = AY(\partial\epsilon/\partial x) = 0.0004(19.5 \times 10^{10})(8 \times 10^{-4}) = 624 \text{ nt}$$

where $Y = 19.5 \times 10^{10}$ nt/m² is Young's modulus for the steel.

9.12. A longitudinal magnetostrictive ultrasonic transducer is constructed from a nickel tube of length 0.05 m, inner radius 0.005 m, and wall thickness 0.0002 m. If the proportional material constant is given as $K = -1.0 \times 10^{-4}$ m⁴/weber² and a polarizing flux density $B_0 = 0.3$ weber/m² is applied to the tube, find (*a*) the magnetostriction constant of the tube, (*b*) the permanent change in length of the tube, and (*c*) the coefficient of electromechanical coupling. (*d*) For an additional magnetic flux density of 0.03 weber/m^2, what will be the new length of the tube?

(*a*) Magnetostriction constant $\Lambda = 2YKB_0 = -12.6 \times 10^6$ where $Y = 21 \times 10^{10}$ nt/m^2 is Young's modulus of the nickel.

(*b*) Since static strain produced is proportional to the square of the polarizing flux density, i.e. $(\partial\epsilon/\partial x)_m = KB^2 = -9 \times 10^{-6}$, then the permanent change in length of the tube is $\Delta L = L(\partial\epsilon/\partial x)_m = -4.5 \times 10^{-7}$ m (contraction).

(*c*) The coefficient of electromechanical coupling is $k = \sqrt{\mu_i\mu_0\Lambda^2/Y} = 0.31$ where $\mu_i = 100$ henrys/m is the incremental permeability of the material, and $\mu_0 = 1.26 \times 10^{-6}$ henrys/m is the permeability of free space.

(*d*) For additional application of magnetic flux density $B_i = 0.03$ weber/m^2, the force equation becomes $F_{xi} = -SY(\partial\epsilon/\partial x)_i + \Lambda SB_i$; and with no restraining force, the strain is $(\partial\epsilon/\partial x)_i = \Lambda B_i/Y = -1.8 \times 10^{-6}$. Thus $L = L(\partial\epsilon/\partial x)_i = -9 \times 10^{-8}$ m (contraction). The new length of the nickel tube will be $[0.05 - (4.5 \times 10^{-7} + 9 \times 10^{-8})]$ m.

ELECTROMAGNETIC TRANSDUCERS

9.13. An electromagnetic transducer consisting of a steel rod of length 0.1 m and radius 0.05 m and carrying electric current in a magnetic field is employed to generate ultrasound. If the rod is elastically supported at its center to allow radial vibrations, determine its frequency at half-wave resonance.

$$\text{Frequency at half-wave resonance} = \frac{1 - \mu^2\pi^2 r^2}{8L^3}\sqrt{\frac{Y}{\rho}} = 630 \text{ kc/sec}$$

where $\mu = 0.28$ is the Poisson's ratio, $r = 0.05$ m is the radius of the rod, $L = 0.1$ m is the length of the rod, $Y = 19.5 \times 10^{10}$ nt/m^2 is Young's modulus for the steel, and $\rho = 7700$ kg/m^3 is the density of the steel.

APPLICATIONS

9.14. Delay time of $t = 10^{-6}$ second is designed for a computer for storing information to be extracted. If a copper wire of diameter 10^{-6} m is used as the ultrasonic delay line, find its length.

When an electrical signal is converted into an ultrasonic wave, the latter will be propagated through the copper wire at a speed of $c = 3700$ m/sec. At the end of the wire, the wave is reconverted back into its original forms. Thus the length required $= ct = 0.0037$ m.

9.15. Compute the transmitted pressure ratio and the sound power transmission coefficient for sound waves from water into lucite at normal incidence.

$$\text{Transmitted pressure ratio } p_t/p_i = 2\rho_2 c_2/(\rho_2 c_2 + \rho_1 c_1) = 1.4$$

where $\rho_1 = 998$ kg/m^3 is the density of water, $\rho_2 = 1200$ kg/m^3 is the density of lucite, $c_1 = 1480$ m/sec is the speed of sound in water, and $c_2 = 2650$ m/sec is the speed of sound in lucite.

$$\text{Sound power transmission coefficient } \alpha_t = \frac{4\rho_1 c_1 \rho_2 c_2}{(\rho_1 c_1 + \rho_2 c_2)^2} = 0.87$$

The acoustic pressure is seen to increase by 40% when it crosses the boundary while the intensity drops 13%. This is partly due to the crowding of energy into a smaller cross section of wavefront and partly due to change in density or velocity of sound.

9.16. The ultrasonic pulse-echo method is employed to detect possible defects in a steel bar of thickness 0.2 m. If the pulse arrival times are 30 and 80 microseconds, determine the defect.

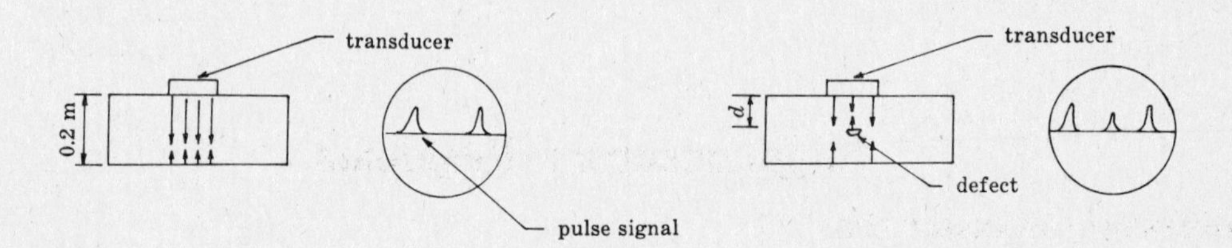

Fig. 9-4

In the pulse-echo method, a pulse of ultrasonic energy (commercial flaw-detector uses 1000 pulses/sec of 1.6 Mc/sec ultrasound) is sent out from the transducer into the test object as shown in Fig. 9-4. The sound wave is reflected back from the boundary of a defect as a reflected pulse, properly detected by the transducer and displayed by an oscilloscope.

Now the time taken by the reflected pulse from the boundary is $t_b = 80 \times 10^{-6}$ sec $= 2(0.2)/c$, from which $c = 5000$ m/sec. Hence the depth of the defect from the surface of the steel bar is $d = 5000(15 \times 10^{-6}) = 0.075$ m. The size of the defect can be mapped by moving the transducer around in the area where the initial indication of the defect is found.

In the similar transmission method for flaw-detection, a pulse of ultrasonic energy is sent into the test object through the source transducer and detected by the receiver transducer on the opposite side of the test object. If there is a defect in the test object, the receiver transducer will detect its presence from the reduced strength in the pulse.

Another way to detect flaws by ultrasonics is the *resonance method*. Here ultrasonic waves of various frequencies are sent into the test object by the transducer until a standing wave is set up in the test object as shown in Fig. 9-5. This indicates that the frequency of the oscillator driving the transducer coincides with a resonant frequency of the test object, resulting in a momentary increase in the energy drawn by the transducer. Large defects and unbonded areas in composite materials can set up standing wave patterns and thus be detected.

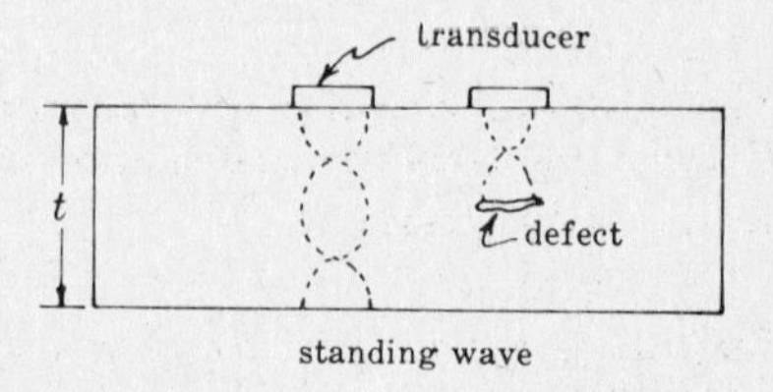

Fig. 9-5

9.17. An ultrasonic transducer is employed to measure the thickness of a steel plate. If the difference of two adjacent harmonics is found to be 56,000 cyc/sec, find the thickness of the plate.

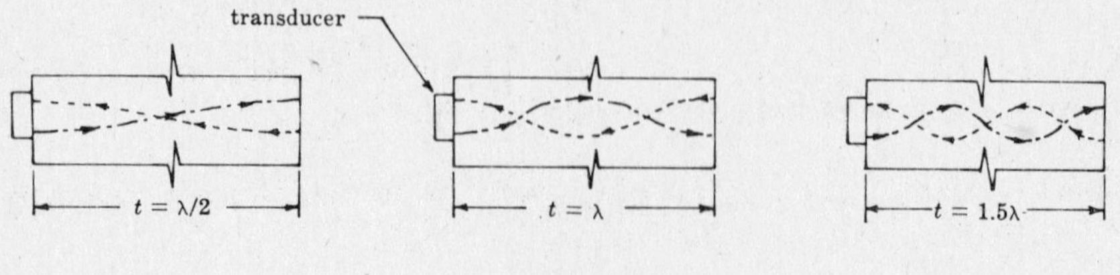

Fig. 9-6

As shown in Fig. 9-6, the fundamental frequency at which thickness resonance vibration will be produced is given by $f_1 = c/2t$ where f_1 is the fundamental frequency in cyc/sec, $c = 5050$ m/sec is the speed of sound in steel, and t is the thickness of the steel plate in meters.

Thickness resonance also occurs at all harmonics of the fundamental frequency, i.e. $f_2 = 2f_1$, $f_3 = 3f_1$, ..., $f_n = nf_1$. Since the frequency difference between two adjacent harmonics is numerically equal to the fundamental frequency,

$$t = \frac{c}{2(f_n - f_{n-1})} = \frac{5050}{2(56,000)} = 0.045 \text{ m}$$

Supplementary Problems

PIEZOELECTRIC TRANSDUCERS

9.18. For an X-cut quartz rod excited at the X faces and vibrating along Y with a node at the center, show that the fundamental frequency is given by $2720/y$ kc/sec, where y is length in mm.

9.19. For vibration in the direction of its thickness, what thickness must a free quartz plate have in order to obtain a fundamental frequency of 50 Mc/sec? *Ans.* 0.055 mm

9.20. An X-cut quartz crystal is radiating ultrasonic waves of frequency 1 Mc/sec into water on one side and into air on the other. If the radiating surface has a diameter of 0.05 m, determine the radiation resistance. *Ans.* $R_L = 50,000$ ohms

MAGNETOSTRICTIVE TRANSDUCERS

9.21. Show that when a polarizing flux is present in a rod of magnetostrictive material, its fundamental frequency of longitudinal vibration is reduced.

9.22. A rod free at both ends is vibrating vigorously at its fundamental longitudinal frequency by magnetostriction. Find the breaking point of the rod.
Ans. At the center (node or zero displacement)

9.23. The thickness vibration of a barium titanate generator is in a form of a circular bowl. If its thickness is 0.0064 m, find its approximate frequency of thickness vibration.
Ans. $f = 400$ kc/sec

9.24. One end of a magnetostrictive transducer is connected to a diaphragm while the other end is free. Show that the ratio of the particle velocity at the free end of the rod to that of the mass for the fundamental mode of vibration is $\operatorname{sech} k_1 L$.

APPLICATIONS

9.25. Only little energy can enter the human body when exposed to ultrasonic waves. Why?
Ans. Mismatching impedances

9.26. Show that the maximum rate of decrease in intensity of ultrasound is 6 db for each doubling of the distance from the source.

9.27. For an amplitude of 10^{-3} m, compute the values of acoustic intensity at a frequency of 1 Mc/sec in water and in air. *Ans.* 0.293, 8.45×10^{-5} watt/cm^2

9.28. For ultrasound waves traveling from steel to water at normal incidence, determine the pressure amplitude ratios. *Ans.* $p_r/p_i = 0.935$, $p_t/p_r = 0.061$

9.29. In ultrasonic cleaning, the transducer of impedance Z_1 is coupled to the volume liquid in the tank of impedance Z_2 so as to reduce loss of efficiency due to mismatching of impedances ($Z_1 > Z_2$ in general). For maximum efficiency, what should be the impedance of the coupling element?
Ans. $Z = \sqrt{Z_1 Z_2}$

INDEX